Nanoscale Sensors and their Applications in Biomedical Imaging

Mohd. Mohsin • Neha Soleja
Editors

Nanoscale Sensors and their Applications in Biomedical Imaging

Springer

Editors
Mohd. Mohsin
Department of Biosciences
Jamia Millia Islamia
New Delhi, Delhi, India

Neha Soleja
Department of Biosciences
Jamia Millia Islamia
New Delhi, Delhi, India

ISBN 978-981-97-3143-5 ISBN 978-981-97-3144-2 (eBook)
https://doi.org/10.1007/978-981-97-3144-2

This Springer imprint is published by the registered company Springer Nature Singapore Pte Ltd.
The registered company address is: 152 Beach Road, #21-01/04 Gateway East, Singapore 189721, Singapore

If disposing of this product, please recycle the paper.

About the Book

This book explores the field of nanobiology, which focuses on manipulating and studying materials on the nanoscale range for various applications, particularly in high-end manufacturing and biomedical contexts. This book emphasizes the development of nanoscale sensors as powerful tools for detecting and monitoring biological, chemical, and physical events, bridging the gap between biological systems and information systems, and proving valuable in areas such as disease diagnostics, drug discovery, drug delivery, and environmental sciences. By providing superior sensitivity and characterization of molecular interactions, nanoscale sensors enhance our understanding of essential biological processes and contribute to advancements in medicine, pharmacology, tissue engineering, and agriculture. The integration of nanotechnology and biosensors has paved the way for novel insights into self-control mechanisms in living systems through continuous real-time monitoring. Furthermore, this book highlights the significance of multimodality imaging in medical diagnostics and the role of nanobiotechnology in advancing fluorescence imaging techniques for real-time observation of biomolecules, emphasizing the potential of nanoscale sensors to revolutionize diagnostic and therapeutic applications. Overall, this book provides a comprehensive analysis of the diverse applications of nanoscale sensors in biomedical imaging.

Contents

Editors and Contributors

About the Editors

Mohd. Mohsin is an assistant professor at Jamia Millia Islamia, New Delhi, India, where he has been serving since 2014. He holds a PhD from Hamdard University, New Delhi, India. In 2009, he qualified for the National Eligibility Test of the Council of Scientific and Industrial Research (CSIR) and Junior Research Fellowships of the Indian Council of Medical Research. Specializing in nanobiotechnology, molecular biology, and metabolic engineering, his research focuses on genetically encoded fluorescence resonance energy transfer (FRET)-based nanosensors for healthcare and cellular environmental monitoring. He has presented his research work at various national and international conferences and published more than 30 research papers in reputable international journals. He also holds two approved patents that have been published. He received a high impact factor (>5) publication award in 2014. His contributions to the scientific community have been recognized with the Young Scientist Award in 2015 and an Innovative Research Award in 2017 from the Government of India. He also serves as an editorial member and reviewer for several journals. His career has been marked by various awards, scholarships, and fellowships, which highlight his dedication and expertise in nanobiotechnology research and education.

Neha Soleja is the research associate of the Indian Council of Medical Research at the Department of Biosciences, Jamia Millia Islamia, New Delhi, India. She earned her PhD in 2020 from the Department of Biosciences of Jamia Millia Islamia, Delhi, India, with a thesis titled "Construction of a Family of FRET-Based Nanosensors to Monitor the Flux of Metal Ions." Dr. Soleja's research primarily revolves around the design and implementation of genetically encoded nanosensors to enable real-time quantification of metabolites at the subcellular level. Her work has led to the publication of 14 scholarly articles in prestigious international journals. She has two patents to her credit. Dr. Soleja possesses a diverse skill set, encompassing various cutting-edge techniques in biochemical, biophysical, and molecular biology research. She has also participated in several workshops, seminars, and conferences, where she has presented her research findings and received awards and recognition for her work. With her multidisciplinary skill set and passion for advancing knowledge, she remains at the forefront of research in the field of nanobiotechnology.

Contributors

Altaf Ahmad Department of Botany, Aligarh Muslim University, Aligarh, Uttar Pradesh, India

Shahzad Ahmad Department of Medical Elementology and Toxicology, Jamia Hamdard, New Delhi, India

Zainab Ahmed Department of Biosciences, Jamia Millia Islamia, New Delhi, Delhi, India

Shahenvaz Alam Enzyme and Microbial Biochemistry Laboratory, Department of Chemistry, Indian Institute of Technology Delhi, New Delhi, India

Department of Biochemistry and Molecular Biology, University of Georgia, Athens, Georgia, USA

Aayaz Ali Department of Chemistry, Jamia Millia Islamia, Okhla, New Delhi, India

Asghar Ali Clinical Biochemistry Laboratory, Department of Biochemistry, School of Chemical and Life Sciences, Jamia Hamdard, New Delhi, India

Rashid Ali Department of Chemistry, Jamia Millia Islamia, Okhla, New Delhi, India

Shakeel Alvi Department of Chemistry, Jamia Millia Islamia, Okhla, New Delhi, India

Reshma Bano Department of Biosciences, Jamia Millia Islamia, New Delhi, Delhi, India

Kanchan Birat Jamia Hamdard, New Delhi, Delhi, India

Neha Chaurasiya Department of Botany, Aligarh Muslim University, Aligarh, Uttar Pradesh, India

Syeda Warisul Fatima Enzyme and Microbial Biochemistry Laboratory, Department of Chemistry, Indian Institute of Technology Delhi, New Delhi, India

Department of Radiation Oncology, Emory University School of Medicine, Atlanta, GA, USA

Tarab Fatima Centre for Nanoscience and Nanotechnology, Jamia Millia Islamia, New Delhi, India

M. Gorshkov Samara National Research University, Samara, Russia

A. Gorshkova Samara National Research University, Samara, Russia

Mohd. Rahil Hasan Department of Biotechnology, Jamia Hamdard, New Delhi, India

Samina Husain Centre for Nanoscience and Nanotechnology, Jamia Millia Islamia, New Delhi, India

Irfan Department of Biosciences, Jamia Millia Islamia, New Delhi, Delhi, India

Mohan Kamthan Clinical Biochemistry Laboratory, Department of Biochemistry, School of Chemical and Life Sciences, Jamia Hamdard, New Delhi, India

Sunil K. Khare Enzyme and Microbial Biochemistry Laboratory, Department of Chemistry, Indian Institute of Technology Delhi, New Delhi, India

Sana Masroor Department of Biosciences, Jamia Millia Islamia, New Delhi, Delhi, India

Prabhash Mishra Samara National Research University, Samara, Russia

Department of Applied Science and Humanities, Faculty of Engineering and Technology, Jamia Millia Islamia (Central University), New Delhi, India

Mohd. Mohsin Department of Biosciences, Jamia Millia Islamia, New Delhi, Delhi, India

Imra Mumtaz Department of Botany, Aligarh Muslim University, Aligarh, Uttar Pradesh, India

Romana Naaz Department of Applied Sciences and Humanities, Faculty of Engineering and Technology, Jamia Millia Islamia, New Delhi, India

Ubaid Mushtaq Naikoo Department of Biotechnology, Jamia Hamdard, New Delhi, India

Jagriti Narang Department of Biotechnology, Jamia Hamdard, New Delhi, India

Parul Narwal University Institute of Biotechnology, Chandigarh University, Mohali, Punjab, India

Rahila Nazir Department of Biosciences, Jamia Millia Islamia, New Delhi, Delhi, India

Johns Hopkins University, Baltimore, MD, USA

Neelam Prabha Negi University Institute of Biotechnology, Chandigarh University, Mohali, Punjab, India

V. Pavelyev Samara National Research University, Samara, Russia

IPSI RAS – Branch of the FSRC "Crystallography and Photonics" RAS, Samara, Russia

Saman Saim Rahman Clinical Biochemistry Laboratory, Department of Biochemistry, School of Chemical and Life Sciences, Jamia Hamdard, New Delhi, India

Prachi Sharma Samara National Research University, Samara, Russia

School of Electronics Engineering (SENSE), Vellore Institute of Technology (VIT) Vellore, India

Weqar Ahmed Siddiqi Department of Applied Sciences and Humanities, Faculty of Engineering and Technology, Jamia Millia Islamia, New Delhi, India

Yashda Singh Department of Biotechnology, Jamia Hamdard, New Delhi, India

Neha Soleja Department of Biosciences, Jamia Millia Islamia, New Delhi, Delhi, India

Nishant Tripathi Samara National Research University, Samara, Russia

Fluorescence-Based Nanosensors for Detection and Quantification of Nanovesicles at the Nanoscale

1

Reshma Bano, Neha Soleja, and Mohd. Mohsin

Abstract

Nanovesicles, including exosomes, microvesicles, and apoptotic bodies, serve as crucial mediators of intercellular communication by transporting biological material between cells. Despite their shared impact on host immunity, nanovesicles exhibit diverse biogenesis and distinct protein compositions. Exosomes play a crucial role in cellular cargo transport and are gaining attention as potential diagnostic markers and therapeutic targets for cancer. However, isolating highly pure subpopulations of nanovesicles poses a significant challenge due to their similarity in size to other contaminants. Researchers are continuously developing innovative approaches for nanovesicle isolation and analysis to overcome these challenges. Recent literature highlights a shift toward embracing novel nanosensors for the isolation, quantification, and characterization of nanovesicles. This chapter explores advancements in fluorescence sensors, revolutionizing nanovesicle analysis by reducing sample volume requirements and enhancing isolation precision. The combination of biosensor and fluorescent technologies demonstrates unparalleled selectivity and sensitivity in detecting and quantifying extracellular vesicles (EVs), offering boundless promise in biomedicine and clinical translation. These advancements represent a significant leap forward in understanding nanovesicle biology and harnessing their potential for diagnostic and therapeutic applications.

Keywords

Nanovesicles · Exosomes · Fluorescence sensors · Nanotechnology · Diagnostic markers

R. Bano · N. Soleja · M. Mohsin (✉)
Department of Biosciences, Jamia Millia Islamia, New Delhi, Delhi, India
e-mail: mmohsin1@jmi.ac.in

M. Mohsin, N. Soleja (eds.), *Nanoscale Sensors and their Applications in Biomedical Imaging*, https://doi.org/10.1007/978-981-97-3144-2_1

Abbreviations

ESCRT	Endosomal sorting complexes required for transport
EVs	Extracellular vesicles
ExoSen	Exosome sensor
FRET	Fluorescence resonance energy transfer
GO/PDA	Graphene oxide/polydopamine (GO/PDA)
GQDs	Graphene quantum dots
hMFEX	Homogeneous magneto-fluorescent exosome
ILV	Intraluminal vesicle
MIG-6	Mitogen-inducible gene-6
MVBs	Multivesicular bodies
PDMS	Polydimethylsiloxane
PLL	Polylysine
TR-FIA	Time-resolved fluorescence immunoassay
UCNPs	Upconversion nanoparticles

1.1 Introduction

Nanovesicles, also known as extracellular vesicles (EVs), are a heterogeneous collection of membrane-bound structures that transfer biological material between cells. Nanovesicles have been classified into three main categories: exosomes (with a diameter 50–120 nm), microvesicles (ranging between 100 and 1000 nm), and apoptotic bodies (50–5000 nm) (Valadi et al. 2007). These vesicles are present in biological fluids (e.g., blood, saliva, milk, urine, sperm, and amniotic fluid) and are considered to store significant information (Chiriacò et al. 2018). Nanovesicles are effective regulators of tumorigenesis by transporting cargo molecules among cancer cells and tumor stroma's cells (Pegtel et al. 2010). Despite their shared impact on host immunity, nanovesicles have diverse biogenesis and distinct protein compositions (Kosaka et al. 2010). Additionally, within living systems, exosomes act as the "intermediate fellow" involved in the transport of cellular cargo consisting of macromolecules such as proteins, lipids, and nucleic acids. Thus, among various types of nanovesicles, exosomes function as mediators of intracellular communication. Formerly released, these vesicles can interact with other cells in the instant locality or at distant vicinity by transfer via circulation. Exosomes have gathered attention as potential diagnostic markers and therapeutic targets for cancer. These nanovesicles are not only produced by cancer cells but also play crucial roles in various physiological functions in healthy cells (Raposo and Stoorvogel 2013).

Researchers employ various techniques, for isolating and characterizing nanovesicles, including ultracentrifugation, ultrafiltration, immunoaffinity, polymer sedimentation, and size exclusion chromatography (Kurian et al. 2021). However, a significant challenge lies in obtaining highly pure subpopulations of nanovesicles. This difficulty arises due to the similarity in size between nanovesicles and other contaminants, such as lipids, small organelles, microparticles, immune complex, and calcium-phosphate microprecipitates (Gurunathan et al. 2019; Van der Pol et al. 2016). Moreover, distinguishing specific classes of nanovesicles becomes even more complex because of overlapping size, density, and cargo (Reiner et al. 2017). However, isolating nanovesicles, including exosomes poses significant challenges due to the complex nature of the sample matrix and physiochemical properties involved. To overcome these challenges, researchers are continuously working on innovative approaches for exosome isolation and analysis.

Recent literature indicates a remarkable shift toward embracing novel nanosensors for the isolation, quantification, and characterization of nanovesicles. This chapter delves into the exciting advancements in fluorescence sensors, revolutionizing the analysis of nanovesicles by not only reducing sample volume requirements but also enhancing the precision of the isolation process. The combination of biosensor and fluorescent technologies has been a fascinating area to explore, and the studies discussed here constantly go above and beyond expectations by demonstrating the unparalleled selectivity and sensitivity of biosensors in detecting and quantifying EVs. Beyond simple expectations, the application of biosensors in the field of EV research holds boundless promise. The integration of cutting-edge technology with these small, but powerful, biological sensor opens up unprecedented opportunities in biomedicine and clinical translation.

1.2 Biogenesis of Nanovesicles

Nanovesicles released by all cell types (like normal cells, cancer, and immune cells) play a role in cellular communication through the transfer of dynamic biomolecules such as proteins, nucleic acids, lipids, and miRNAs (Raposo and Stoorvogel 2013). The major advances in nanovesicle research have primarily come from studying exosomes. Exosome biogenesis occurs via intraluminal vesicle (ILV) budding and multivesicular body (MVB) formation. MVBs can fuse with the plasma membrane after early to late endosome maturation and release the enclosed ILVs (then raised to as exosomes) in the extracellular medium (Fig. 1.1). Since cancer exosomes assist in tumor growth and progression, understanding the processes involved in their synthesis could lead to novel therapeutic options. The well-studied process of MVB formation is the endosomal sorting complexes required for transport (ESCRT) machinery, which is responsible for sorting ubiquitinated proteins into ILVs (Henne et al. 2011). Despite the substantial investigation into the role of ESCRT in MVB biogenesis, it is still unknown whether MVBs produced by ubiquitin-dependent pathways are destined for lysosomes for cargo decomposition or can merge with the plasma membrane to release exosomes (Christ et al. 2017). Another group of nanovesicles, known as microvesicles, falls into a larger size range, ranging from 100 to 1000 nm in diameter. There can be some overlap in size between microvesicles and exosomes. Microvesicles are generated when the plasma membrane undergoes a regulated outward budding, a process called ectocytosis. As a result, microvesicles are enveloped by lipids and proteins from the plasma membrane and contain various cytoplasmic proteins and cargo actively recruited during their

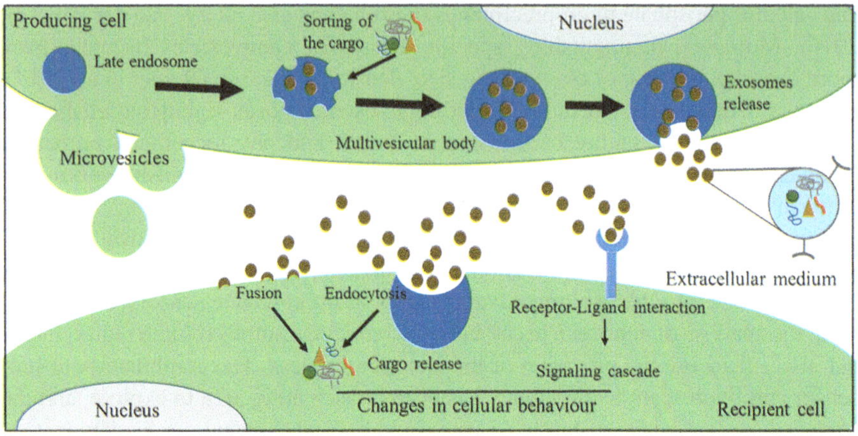

Fig. 1.1 Nanovesicles: biogenesis, release, and uptake by cells

formation. Apoptosis, the programmed cell death process, results in the fragmentation of cells into various membranous structures known as apoptotic bodies (Raiborg and Stenmark 2009). These apoptotic bodies can be significantly larger in size as compared with other vesicles. However, for the scope of this chapter, our primary focus will be on exosomes, and we will not explore the exploration of microvesicles and apoptotic bodies.

1.3 Challenges in Nanovesicles Analysis

Analyzing nanovesicles presents several common difficulties:

1. Their small size and biochemical complexity.
2. The accuracy of measurements greatly relies on the purity of the isolated sample, making pre-analytical methodologies critical.

Common isolation techniques involve gel permeation chromatography, ultracentrifugation, immunoaffinity capture, and polymer-facilitated precipitation, each used to isolate nanovesicles with its limitations. Due to contamination in complex biological samples like blood plasma, obtaining highly pure nanovesicles preparations is particularly challenging. The heterogeneity of nanovesicle populations and the existence of subpopulations with different phenotypes further add complexity. Additionally, determining unknown concentrations and correlating specific nanovesicles with their origins pose additional hurdles (Lane et al. 2018).

By selecting appropriate fluorophores and target molecules, researchers can study different EV populations and origins with greater precision and specificity. As the field of nanovesicle research progresses, standardizing protocols for nanovesicle isolation and enrichment will be crucial for broader clinical applications.

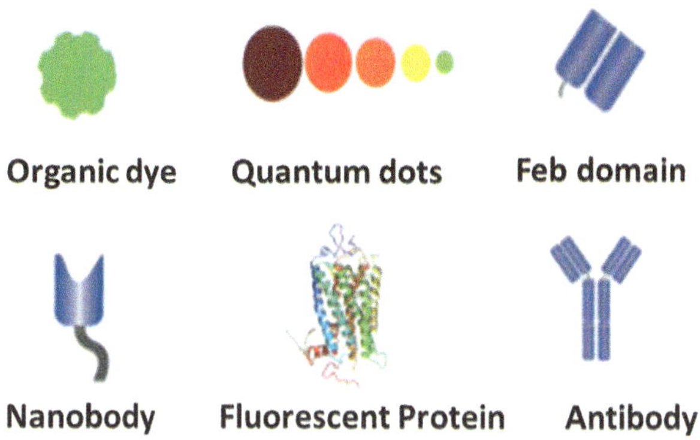

Fig. 1.2 Different types of fluorescent labels and molecular probes

Fluorescence-based methods, with their capacity for multiplexed analysis and precise detection, show immense promise in overcoming the challenges and unlocking the full potential of nanovesicles in various scientific and clinical endeavors (Ramirez et al. 2018).

1.4 Fluorescence-Based Nanosensors for Detection of Nanovesicles

Fluorescence-based methods have gained popularity in developing biosensors due to their high sensitivity. Several laboratory-based techniques for exosome detection, including flow cytometry, real-time polymerase chain reaction (rt-PCR), and fluorescence spectroscopy, have been significantly improved (Yang et al. 2019). However, in clinical settings, challenges such as the lack of trained professionals and space constraints hinder rapid and point-of-care detection (Sin et al. 2014). To address these issues, researchers have been focusing on developing more portable and smart fluorescence-based biosensors for exosome detection. The various fluorescent probes that may be used in different fluorescent biosensors are shown in Fig. 1.2.

Some important fluorescence-based nanosensors are described below:

1.4.1 Immunoaffinity-Based Fluorescence Detection

1.4.1.1 Time-Resolved Fluorescence Immunoassay

Biotinylated capture antibodies were bound to a streptavidin-coated 96-well plate, creating a time-resolved fluorescence immunoassay (TR-FIA) with high sensitivity and low background signals. The TR-FIA serves as a foundation for developing disease-specific detection assays by testing various antibodies against exosome membrane markers. The measured TR-FIA signal depends on the number of exosomes present, the quantity of antigens per vesicle, and the antibody's affinity to the epitope. For instance, the high signals of CD9 and CD63 in the medium could indicate either a substantial secretion of exosomes or a high concentration of antigens on each exosome. As long as the TR-FIA can consistently and accurately detect the signal associated with PCa-related exosomes, it holds immense potential as a valuable diagnostic tool (Duijvesz et al. 2015).

1.4.1.2 Magneto-Fluorescent Nanosensor

An innovative and integrated approach, called the homogeneous magneto-fluorescent exosome (hMFEX) nanosensor, has been developed for rapid and on-site analysis of tumor-derived exosomes. This pioneering technique combines exosome isolation and detection on a single platform, streamlining the process and eradicating the need for expensive reagents and equipment. Subsequently, an aptamer-triggered assembly of DNA three-way junctions is initiated in a homogenous solution, alongside aggregation-induced emission luminogens and graphene oxide. This unique combination results in an amplified fluorescence signal, enabling the precise quantification of the target subtype of tumor-derived exosomes in plasma. This novel homogenous approach not only ensures high specificity and desirable sensitivity but also offers a cost-effective solution for point-of-care cancer diagnosis (Li et al. 2020b).

1.4.1.3 Microfluidic ExoChip nanosensor

Microfluidic device called "ExoChip" has been engineered for the isolation, detection, and quantification of exosomes directly on the chip at nanoscale levels. To selectively isolate exosomes expressing CD63 on their surface, the chip was fabricated using polydimethylsiloxane (PDMS) and functionalized with CD63 antibodies as shown in Fig. 1.3. Detection was achieved using a fluorescent carbocyanine dye (DiO) that specifically labels exosomes, and their relative fluorescence intensities were quantified at excitation and emission wavelengths of 485 nm and 510 nm,

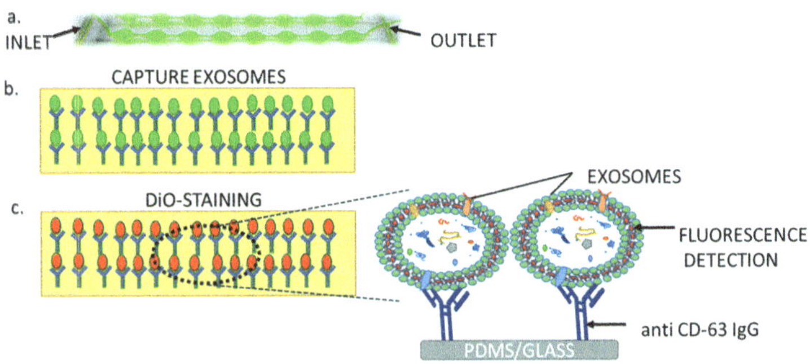

Fig. 1.3 Exosomes are efficiently captured by passing serum through an ExoChip coated with specific anti-CD-63 IgG antibodies. To visualize the captured exosomes, the ExoChip is treated with a membrane-specific dye called DiO (Bano et al. 2023)

respectively, using a BioTek Synergy Neo multi-purpose plate reader (Bano et al. 2023).

1.4.1.4 A Dual-Signal Fluorescent Biosensing Tool

The researchers have developed a highly sensitive fluorescent biosensing platform employing a dual-signal amplification strategy for the ultrasensitive detection of leukemia cell-derived exosomes (Fig. 1.4). This innovative protocol consists of three essential steps. First, they capture leukemia-derived exosomes, which contain CD63 and nucleolin, using anti-CD63 antibody-modified magnetic bead conjugates (MB-CD63). Subsequently, a DNA primer with a nucleolin-recognition aptamer (AS1411) is introduced to bind to the exosomes, initiating a rolling circle amplification (RCA) reaction. The RCA reaction generates multiple repeat sequences that can hybridize with gold nanoparticle (GNP)–DNA–fluorescent dye (FAM) conjugates (GNP–DNA–FAM). In the final step, the researchers introduce nicking endonuclease (Nb·BbvCI) to facilitate target recycling. As a result, FAM is released from the GNP–DNA–FAM conjugates, transforming it from a quenched to an emission state. This continuous accumulation of fluorescence signals enhances the sensitivity of exosome detection.

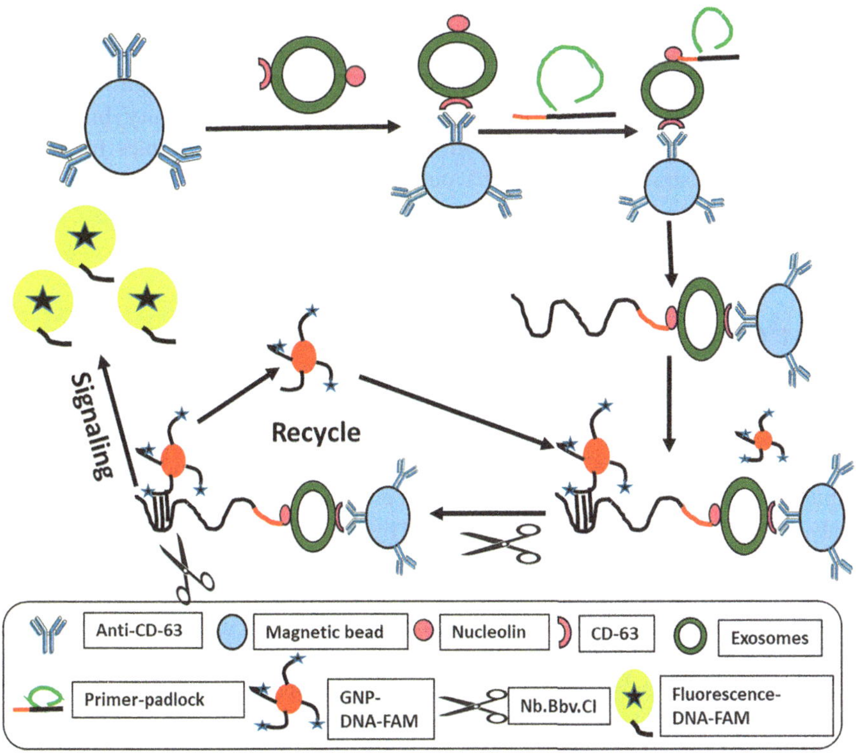

Fig. 1.4 Diagrammatic representation of the dual-signal fluorescent biosensing platform

The developed dual-signal amplification platform allows for the detection of exosomes at an impressively low concentration. Furthermore, the researchers successfully applied this method to detect exosomes in spiked serum samples, indicating its promising potential for clinical applications. The sensitive and efficient detection of leukemia cell-derived exosomes using this biosensing platform could have significant implications for early disease diagnosis and monitoring, offering a valuable tool for advancing liquid biopsy-based diagnostics in the field of leukemia research and clinical practice (Huang et al. 2018).

1.4.1.5 A Nanostructured Chip Coated with Graphene Oxide/Polydopamine

A novel microfluidic platform for exosome analysis has been developed, utilizing a unique graphene oxide/polydopamine (GO/PDA) nano-interface. Particularly, this study presents the first instance of a 3D nanoporous PDA surface coating formed through the microfluidic layer-by-layer deposition of GO and PDA, a prominent achievement. The nanostructured GO/PDA interface was found to significantly enhance the efficiency of exosome immuno-capture while also effectively mitigating non-specific exosome adsorption. Utilizing this nano-interface, the scientists introduced an exosome enzyme-linked immunosorbent assay (ELISA) with extraordinary sensitivity. Impressively, they achieved a surprisingly low detection limit coupled with an expansive dynamic range spanning four orders of magnitude. This achievement outperforms existing methods both in terms of sensitivity and the breadth of detection range. Exosome multiplexed analysis and high-throughput clinical sample screening are both feasible with the chip's versatility. In order to accelerate exosome studies along with the medical use of exosomes for illness detection and treatment, this platform would offer a valuable tool (Zhang et al. 2016).

1.4.2 Nanoparticle-Based Biosensors

1.4.2.1 CdSe Quantum Dot-Based Sensor

A group of researchers has introduced a stripping voltammetric immunoassay designed for the electrochemical detection of disease-specific exosomes. In this method, quantum dots are employed as signal amplifiers. The immunoassay encompasses a sequence of three consecutive steps. Initially, bulk exosome populations are captured magnetically using magnetic beads coated with a generic tetraspanin antibody, such as CD9 or CD63. This is followed by the identification of disease-specific exosomes using antibodies linked to cancer-related markers. In their study, the researchers utilized biotinylated HER-2 and FAM134B antibodies functionalized with CdSe quantum dots (CdSeQD) as markers for breast and colon cancer. The subsequent steps involve thorough magnetic washing and purification. The CdSe quantum dots are then subjected to acid dissolution, and the quantification of Cd^{2+} ions is carried out using anodic stripping voltammetry on a bare glassy carbon working electrode. This intricate process allowed for the sensitive detection of a remarkable 100 exosomes per μL in cancer cell lines. Additionally, the researchers extended their method's applicability to a small group of serum samples ($n = 9$) collected from patients diagnosed with colorectal adenocarcinoma. These findings propose a potentially robust bioassay capable of quantifying disease-specific exosomes in clinical samples. The novel approach of utilizing quantum dots as signal amplifiers and employing anodic stripping voltammetry demonstrates the research group's innovative progresses in the field of electrochemical exosome detection (Boriachek et al. 2017).

1.4.2.2 PMO Graphene Quantum Dots FET Biosensor

In contrast to conventional DNA probe immobilization on flat surfaces, the utilization of nanoparticle-based DNA probes offers enhanced anchoring of RNA molecules to the sensor surface, leading to improved detection sensitivity. This study presents a novel approach: phosphorodiamidate morpholino oligomer (PMO)-functionalized graphene quantum dots (GQDs) integrated with reduced graphene oxide (RGO) field-effect transistor (FET) biosensors for highly sensitive exosomal microRNA detection. The process involves constructing the RGO FET, depositing a polylysine (PLL) onto its surface, and attaching GQDs-PMO to the PLL-coated area. This hybrid facilitates exosomal microRNA detection. Fluorescence and zeta potential measurements confirm successful GQDs-PMO integration; reduced fluorescence at 365 nm indicates PMO-induced quenching. It detects exosomal miRNAs in plasma, distinguishing breast cancer from healthy samples. GQDs enhance FET sensor sensitivity. This advancement positions the FET-based biosensor as a promising tool for the early diagnosis of breast cancer, offering an innovative avenue for advancing cancer diagnostics (Li et al. 2022).

1.4.2.3 Facile Fluorescent Aptasensor

In this study, researchers have developed a simple yet effective fluorescent aptasensor for the evaluation of exosomal tumor-associated proteins. This innovative approach combines aptamers, aggregation-induced emission luminogens (AIEgens), and graphene oxide (GO) to serve as recognition elements, fluorescent dyes, and quenchers, respectively. The mechanism behind this aptasensor involves the interaction of numerous TPE-TAs (aggregation-induced emission luminogens) with one aptamer, leading to the rapid formation of aggregates. This process significantly enhances the fluorescent signal, resulting in amplified detection. When tumor-derived exosomes are absent, the GO component binds to the TPE-TAs/aptamer complex, inducing fluorescence quenching. However, upon the introduction of target exosomes, the aptamer selectively binds to its specific target. Consequently, the TPE-TAs/aptamer complexes detach from the GO surface, leading to a distinctive "turn-on" fluorescent signal. The aptasensor proved its efficiency in diagnosing breast tumors and exhibited high sensitivity in discriminating malignant breast cancers. In essence, this aptasensor introduces a promising path for the profiling of tumor-derived exosomal proteins, holding great potential for early diagnosis through liquid biopsy (Li et al. 2020a, b).

1.4.2.4 Upconversion Nanoparticles (UCNPs) Based Quantification of Nanovesicles

This technique employs UCNPs for characterizing surface markers on individual EVs with super-resolution capability. The methodology is rooted in the use of uniformly bright and highly photostable UCNPs, extensively doped with a significant number of lanthanide ions. The EVs are initially captured on a slide coated with CD9 antibodies. Subsequently, they are targeted using biotinylated antibodies against the epithelial cellular adhesion molecule (EpCAM), followed by labeling through streptavidin-UCNPs. The visualization of UCNPs on single small EVs is achieved through a super-resolution microscope based on a single-beam scanning approach, parallel to STED microscopy. In contrast to traditional probes, the utilization of super-resolution microscopy with UCNPs offers additional avenues to reduce nanoparticle size while attaining a higher labeling density. Furthermore, the technique opens doors for employing nanoparticles with diverse emission colors and lifetimes, enabling the targeting of a variety of orthogonal biomarkers (Huang et al. 2022).

1.4.3 Fluorescence Resonance Energy Transfer (FRET)-Based Nanosensors

1.4.3.1 A MoS2–Multiwall Carbon Nanotubes (MWCNT) Based Nanosensor

In this strategy, researchers have introduced a rapid and sensitive exosome detection platform utilizing MoS2–MWCNT as a fluorescence quenching agent. The MoS2–MWCNT surface is covered with anti-CD63-PE. This causes the PE-conjugated antibody's fluorescence to fade because of a process called FRET. This transfer occurs from the antibody to the MoS2–MWCNT structure, resulting in the quenching of fluorescence. This exosome biosensor showcases both high sensitivity and specificity in detecting biomarkers. The application of this MoS2–MWCNT-based fluorometric nanosensor was demonstrated in the analysis of exosomes derived from MCF-7 breast cancer cells. The quantification of CD63 expression was achieved through the recovered fluorescence of the fluorophore. Remarkably, this nanosensing technique enables the quantification of exosomes with varying surface biomarker expressions. The biosensor's versatility is apparent, as it can measure different biomarkers by selecting corresponding antibodies. Additionally, this approach holds promise as a prospective platform for exosome detection, particularly in the context of early cancer detection and therapy. The utilization of MoS2–MWCNT as a fluorescence quencher offers a robust foundation for a versatile and potentially impactful technique (Tayebi et al. 2019).

1.4.3.2 FRET-Based Assay for the Quantification of Extracellular Vesicles

In this study, the researcher introduces an innovative technique for quantifying vesicle content, specifically focusing on the total lipid membrane surface area within a

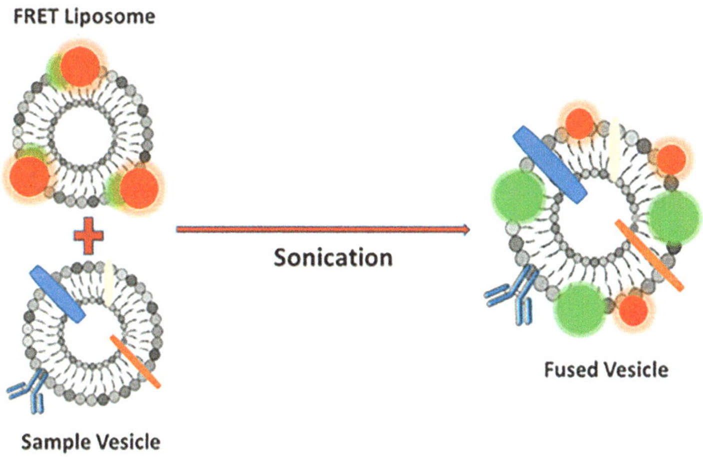

Fig. 1.5 FRET-labeled liposomes (red and green) merge with target vesicles (gray) using sonication. Interaction reduces fluorophore density on vesicles, causing emission changes

given sample. This quantification is achieved using FRET. The method involves inducing vesicle fusion in the target sample through sonication, wherein these vesicles merge with liposomes containing a pair of FRET fluorophores, as shown in Fig. 1.5.

The alteration in emission spectrum resulting from vesicle fusion is directly correlated with the total membrane surface area of the added sample. The incorporation of a calibration curve permits the quantification of diverse vesicle species, encompassing enveloped viruses, bacterial outer membrane vesicles, and mammalian extracellular vesicles. Remarkably, even without intensive optimization of experimental parameters, the technique was capable of quantifying down to approximately 10^9 vesicles mL^{-1}, utilizing merely 60 μL of the sample. The assay's precision closely rivaled that of a commercial nanoparticle tracking analysis system. While its limit of detection was slightly higher, the FRET assay outshines in detecting small vesicles, as its efficacy remains independent of vesicle size. Its versatility and reliability make it a valuable tool for diverse research applications (Thorsteinsson et al. 2020).

1.4.3.3 Ti_3C_2 MXenes Based Ratiometric FRET Sensor

In this study, researchers introduced a novel technique that involved creating a distinctive self-standard ratiometric FRET nanoprobe. This innovative nanoprobe combined a Cy3-labeled CD63 aptamer (Cy3-CD63 aptamer) with Ti3C2 MXenes nanosheets to achieve quantitative detection of exosomes. The binding of Cy3-CD63 aptamer onto Ti3C2 MXene nanosheets was driven by hydrogen bonding and metal chelate interactions, leading to rapid quenching of the fluorescence signal emitted by Cy3-CD63 aptamer due to FRET between Cy3 and MXenes. This newly established self-standard turn-on FRET biosensing platform displayed impressive exosome detection capabilities. Moreover, the fluorescence sensor demonstrated the

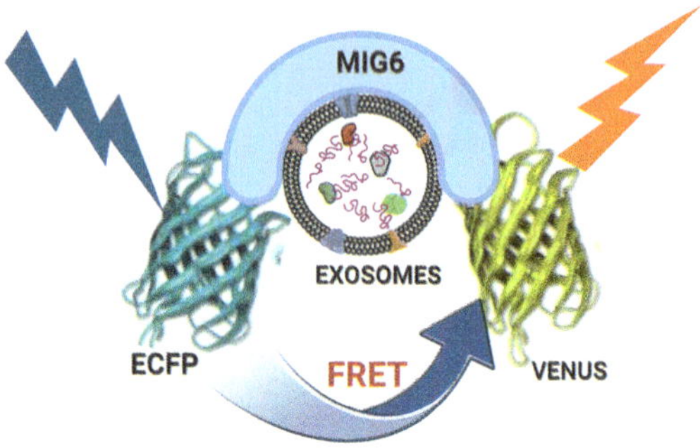

Fig. 1.6 The diagrammatic sensor's concept shows that two fluorescent molecules, ECFP and Venus, come close together due to changes caused by exosomes in the MIG6 protein. This proximity results in the transfer of energy, confirming FRET

ability to identify multiple biomarkers on exosome surfaces and distinguish between various types of exosomes, complemented by fluorescent confocal scanning microscope imaging (Zhang et al. 2018).

1.4.3.4 Real-Time Monitoring of Exosomes via Genetically Encoded FRET Nanosensor

In this study, a group of researchers introduced an innovative FRET-based genetically encoded nanosensor termed "ExoSen" (exosome sensor), designed for quantifying exosomal EGFR concentration both in vitro and in vivo. They utilized a sensing element specific to A549 exosomes, mitogen-inducible gene 6 (MIG6), positioned between the FRET pair ECFP and Venus as depicted in Fig. 1.6.

The interaction of MIG6 with exosomes triggers a structural change, discernible through alterations in the FRET ratio. Furthermore, the developed sensor is expressed in three systems bacterial, yeast, and mammalian cells. Upon the addition of A549 exosomes, the FRET ratio increases, enabling noninvasive quantification of A549 exosomes. ExoSen demonstrates highly sensitive detection of A549 exosomes, even at a low concentration. ExoSen exhibits stability against pH fluctuations and provides accurate, real-time optical readouts in cell-based experiments, facilitated by confocal microscopy. The presented sensor employs a fluorescence-based detection assay to monitor the level of exosomes in real-time and instantly reports them as fluorescence signals from live cells without degrading the biological constituents (Bano et al. 2023).

1.5 Conclusion and Future Perspectives

In conclusion, the area of nanovesicle research stands as a fertile ground for understanding the pivotal role of biomolecule exchange between cells in health and disease. A remarkable advancement in understanding their biogenesis, molecular composition, trafficking mechanisms, and functional attributes has been achieved. This achievement has brought forth an expansive diversity in studying nanovesicles, encompassing various biofluids, diverse biological sources, and innovative biotechnological and therapeutic prospects. Yet, this pursuit for in-depth knowledge has unveiled a necessity to standardize experimental protocols for nanovesicle isolation, identification, and characterization, fostering reproducibility and reliability. Conventional methods often falter in preserving the integrity and purity of nanovesicles while incurring substantial time and cost investments. The spotlight on fluorescent nanosensors coupled with immunoassays, nanoparticles, and FRET as discussed above highlights their cost-effectiveness, time efficiency, and ability to analyze small sample volumes harboring the sought-after nanovesicles. Despite the current compartmentalization of isolation and detection processes in many studies, emerging sensing configurations exhibit an exceptionally low limit of detection, seamlessly integrating isolation and characterization phases within a single device. This contrasts with traditional methods that often entail multi-step approaches, paving the way for more effective and rapid enrichment through miniaturized nanosensors. In the future, specific and sensitive functional outputs will continue to illuminate the intricate biochemical signatures of nanovesicles, shedding light on their complex roles in health and pathology. The ongoing refinement of fluorescent-based platforms, coupled with the integration of isolation and detection, holds immense potential for strengthening the reliability and applicability of nanovesicle-based research. As these innovative technologies converge, the prospect of leveraging nanovesicles as precise disease markers edges closer, imitating new paths in diagnostics, therapeutics, and personalized medicine.

Acknowledgement The first author (RB) kindly acknowledges the University Grants Commission, Govt. of India, for Senior Research Fellowship in the form of Maulana Azad National Fellowship (MANF).

References

Bano R, Soleja N, Mohsin M (2023) Genetically encoded FRET-based nanosensor for real-time monitoring of A549 exosomes: early diagnosis of cancer. Anal Chem 95(13):5738–5746

Boriachek K, Islam MN, Gopalan V, Lam AK, Nguyen NT, Shiddiky MJ (2017) Quantum dot-based sensitive detection of disease specific exosome in serum. Analyst 142(12):2211–2219

Chiriacò MS, Bianco M, Nigro A, Primiceri E, Ferrara F, Romano A, Quattrini A, Furlan R, Arima V, Maruccio G (2018) Lab-on-chip for exosomes and microvesicles detection and characterization. Sensors 18(10):3175

Christ L, Raiborg C, Wenzel EM, Campsteijn C, Stenmark H (2017) Cellular functions and molecular mechanisms of the ESCRT membrane-scission machinery. Trends Biochem Sci 42(1):42–56

Duijvesz D, Versluis CY, Van Der Fels CA, Vredenbregt-van den Berg MS, Leivo J, Peltola MT, Bangma CH, Pettersson KS, Jenster G (2015) Immuno-based detection of extracellular vesicles in urine as diagnostic marker for prostate cancer. Int J Cancer 137(12):2869–2878

Gurunathan S, Kang MH, Jeyaraj M, Qasim M, Kim JH (2019) Review of the isolation, characterization, biological function, and multifarious therapeutic approaches of exosomes. Cells 8(4):307

Henne WM, Buchkovich NJ, Emr SD (2011) The ESCRT pathway. Dev Cell 21(1):77–91

Huang L, Wang DB, Singh N, Yang F, Gu N, Zhang XE (2018) A dual-signal amplification platform for sensitive fluorescence biosensing of leukemia-derived exosomes. Nanoscale 10(43):20289–20295

Huang G, Liu Y, Wang D, Zhu Y, Wen S, Ruan J, Jin D (2022) Upconversion nanoparticles for super-resolution quantification of single small extracellular vesicles. eLight 2(1):20
Kosaka N, Iguchi H, Yoshioka Y, Takeshita F, Matsuki Y, Ochiya T (2010) Secretory mechanisms and intercellular transfer of microRNAs in living cells. J Biol Chem 285(23):17442–17452
Kurian TK, Banik S, Gopal D, Chakrabarti S, Mazumder N (2021) Elucidating methods for isolation and quantification of exosomes: a review. Mol Biotechnol 63:249–266
Lane RE, Korbie D, Hill MM, Trau M (2018) Extracellular vesicles as circulating cancer biomarkers: opportunities and challenges. Clin Transl Med 7(1):14
Li B, Liu C, Pan W, Shen J, Guo J, Luo T, Feng J, Situ B, An T, Zhang Y, Zheng L (2020a) Facile fluorescent aptasensor using aggregation-induced emission luminogens for exosomal proteins profiling towards liquid biopsy. Biosens Bioelectron 168:112520
Li B, Pan W, Liu C, Guo J, Shen J, Feng J, Luo T, Situ B, Zhang Y, An T, Xu C (2020b) Homogenous magneto-fluorescent nanosensor for tumor-derived exosome isolation and analysis. ACS Sens 5(7):2052–2060
Li K, Tu J, Zhang Y, Jin D, Li T, Li J, Ni W, Xiao MM, Zhang ZY, Zhang GJ (2022) Ultrasensitive detection of exosomal miRNA with PMO-graphene quantum dots-functionalized field-effect transistor biosensor. iScience 25(7):104522
Pegtel DM, Cosmopoulos K, Thorley-Lawson DA, van Eijndhoven MA, Hopmans ES, Lindenberg JL, de Gruijl TD, Würdinger T, Middeldorp JM (2010) Functional delivery of viral miRNAs via exosomes. Proc Natl Acad Sci U S A 107(14):6328–6333
Raiborg C, Stenmark H (2009) The ESCRT machinery in endosomal sorting of ubiquitylated membrane proteins. Nature 458(7237):445–452
Ramirez MI, Amorim MG, Gadelha C, Milic I, Welsh JA, Freitas VM, Nawaz M, Akbar N, Couch Y, Makin L, Cooke F (2018) Technical challenges of working with extracellular vesicles. Nanoscale 10(3):881–906
Raposo G, Stoorvogel W (2013) Extracellular vesicles: exosomes, microvesicles, and friends. J Cell Biol 200(4):373–383
Reiner AT, Witwer KW, Van Balkom BW, De Beer J, Brodie C, Corteling RL, Gabrielsson S, Gimona M, Ibrahim AG, De Kleijn D, Lai CP (2017) Concise review: developing best-practice models for the therapeutic use of extracellular vesicles. Stem Cells Transl Med 6(8):1730–1739
Sin ML, Mach KE, Wong PK, Liao JC (2014) Advances and challenges in biosensor-based diagnosis of infectious diseases. Expert Rev Mol Diagn 14(2):225–244
Tayebi M, Yaraki MT, Yang HY, Ai Y (2019) A MoS 2–MWCNT based fluorometric nanosensor for exosome detection and quantification. Nanoscale Adv 1(8):2866–2872
Thorsteinsson K, Olsén E, Schmidt E, Pace H, Bally M (2020) FRET-based assay for the quantification of extracellular vesicles and other vesicles of complex composition. Anal Chem 92(23):15336–15343
Valadi H, Ekström K, Bossios A, Sjöstrand M, Lee JJ, Lötvall JO (2007) Exosome-mediated transfer of mRNAs and microRNAs is a novel mechanism of genetic exchange between cells. Nat Cell Biol 9(6):654–659
Van der Pol E, Böing AN, Gool EL, Nieuwland R (2016) Recent developments in the nomenclature, presence, isolation, detection and clinical impact of extracellular vesicles. J Thromb Haemost 14(1):48–56
Yang Y, Li C, Shi H, Chen T, Wang Z, Li G (2019) A pH-responsive bioassay for paper-based diagnosis of exosomes via mussel-inspired surface chemistry. Talanta 192:325–330
Zhang P, He M, Zeng Y (2016) Ultrasensitive microfluidic analysis of circulating exosomes using a nanostructured graphene oxide/polydopamine coating. Lab Chip 16(16):3033–3042
Zhang Q, Wang F, Zhang H, Zhang Y, Liu M, Liu Y (2018) Universal Ti3C2 MXenes based self-standard ratiometric fluorescence resonance energy transfer platform for highly sensitive detection of exosomes. Anal Chem 90(21):12737–12744

Electrochemical Biosensor: The Most Eligible Diagnostic Tool for the Detection of Various Diseases

2

Mohd. Rahil Hasan, Yashda Singh, Ubaid Mushtaq Naikoo, Shahzad Ahmad, and Jagriti Narang

Abstract

The electrochemical biosensors offer considerable potential in the biomedical field due to their improved specificity, sensitivity, early diagnosis, label-free nature, and cost efficiency for quick point-of-care illness diagnosis at the bedside. An electrochemical biosensor has the potential to reduce the gap between the patient and the treatment as it is an inexpensive, robust, and highly sensitive approach that helps the public manage their disease at an early stage. In this overview, the influence of electrochemical biosensors on effective diagnosis of various diseases was covered and their different strategies with the goal of boosting biosensor sensitivity and employing effective bioreceptors to provide better specificity and easy detection of biomolecules have been thoroughly examined and different electrode setups utilized in electrochemical biosensor are also depicted. This chapter also covers the different components of electrochemical biosensor along with their recent healthcare applications. So, this chapter provides an impactful overview to the reader and they can fulfill their mind with good knowledge about electrochemical biosensor specially those who are new to this field.

Keywords

Electrochemical biosensor · Electrode setups · Methods · Diseases · Diagnosis

M. R. Hasan · Y. Singh · U. M. Naikoo · J. Narang (✉)
Department of Biotechnology, Jamia Hamdard, New Delhi, India

S. Ahmad
Department of Medical Elementology and Toxicology, Jamia Hamdard, New Delhi, India
e-mail: satox@jamiahamdard.ac.in

M. Mohsin, N. Soleja (eds.), *Nanoscale Sensors and their Applications in Biomedical Imaging*, https://doi.org/10.1007/978-981-97-3144-2_2

Highlights

- This chapter covers a recent report on electrochemical biosensor.
- Presents different electrode setup based electrochemical biosensor.
- Different bioreceptors and electrochemical methods are discussed.
- Summarized different applications of electrochemical biosensor based on various diseases.

2.1 Introduction

The diagnosis assumes a pivotal role in the effective management of diverse medical conditions. There are various diagnosis approaches available such as PCR, ELISA, immuno-chromatography assay, microscopy, LAMP, and many more (Krokhine et al. 2021). However, researchers try to advance these approaches as these approaches meet major limitations like these are very expensive, time-consuming, expert-needed, less sensitive, and less specific, so biosensor was employed as an effective diagnostic tool. Biosensor is delineated as a bioanalytical apparatus that integrates a recognition component with a transduction unit. Clark and Lyons in 1962 pioneered the inaugural biosensor, designed for quantification of glucose within biological samples. Their innovation centered on the electrochemical identification of hydrogen and oxygen, accomplished through the utilization of a functionalized electrode with glucose oxidase. Subsequently, there has been remarkable progress both in technological advancements and in the utilization of nano-biosensors, using novel methodologies encompassing electrochemistry, with nickel molybdate nanorods ($NiMoO_4$ NRs) and bioelectronics being incorporated. The chief advantages of biosensors are the easy to practice, low-cost, point-of-care-appeal, on-site detection, technically pro, rapid analysis, and highly sensitive and specific (Anzar et al. 2020). There are various types of biosensors available such as colorimetric biosensor, lateral flow assay-based biosensor, optical biosensor, electrochemical biosensor, piezoelectric biosensor, and many more (Thakur et al. 2023; Hasan et al. 2021a). Of these approaches, the electrochemical biosensor stands out as the most prevalent and dependable choice within the realm of healthcare for the identification of a wide array of perilous medical conditions (Putzbach and Ronkainen 2013). Such biosensors are characterized as "an integrated self-contained gadget capable of providing specific quantitative or semi-quantitative analytical data" (De Girolamo et al. 2018). This technique relies on quantifying alterations in electric-based parameters, such as conductance, current, field effect, or voltage arising from the interaction between the target substance and the bioreceptors situated on the surface of the sensor. Electrochemical biosensors are built on a 3-electrode setup, which includes a working electrode (which functions as the transducing element), a counter electrode (which completes the circuit), and a reference electrode (which is employed to generate a constant potential). Considering that nanomaterial adjustments and surface enhancements are executed on the active electrode, the choice of an appropriate functional electrode assumes paramount importance during the development of an electrochemical biosensor.

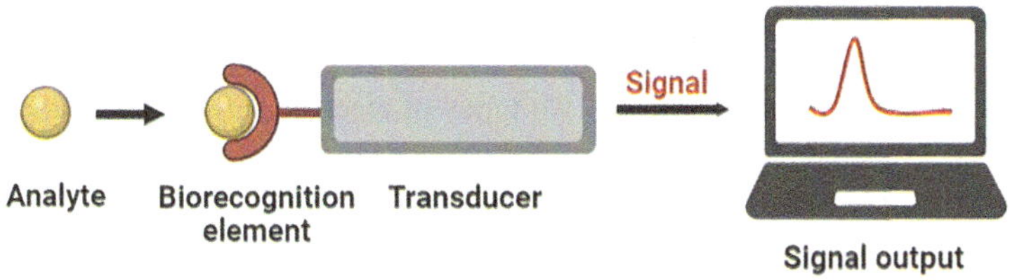

Fig. 2.1 A diagrammatic depiction of an electrochemical biosensor that encompasses distinct components of the sensor, including the specimen, analyte, bio-recognition element, transducers, and electronic system

The fundamental principle underpinning the operational mechanism of electrochemical biosensors is rooted in the kinetics governing the transmission of electrons and ions from the reaction center to the electrode's surface (Purohit et al. 2020). All response measurements are performed using chemical solutions, i.e., electrolytes. These measurements are categorized into three types: impedimetric (change in impedance), amperometric (change in current), and voltammetric (change in potential, such as CV/SWV/DPV) (Hasan et al. 2022a). Electrochemical biosensor comprises bio-analyte, bioreceptors, transducer, and an electronic setup, as shown in Fig. 2.1, which represent all the important parts of the electrochemical biosensor. This chapter outlines the recent report on the electrochemical biosensor. It starts with the general introduction of biosensor, and in the subsequent section, different constituents of the electrochemical biosensor, including the electrode configuration, bio-recognition elements, and methodologies, were delineated. The concluding portion of this chapter is dedicated to exploring the healthcare implementations of the electrochemical sensors, particularly in the identification/diagnosis of various diseases like dengue, HIV, COVID-19, cancer, and cardiovascular disease. This chapter concludes with a concise summary and a contemplation of future prospects in this field.

2.2 Different Components Based Electrochemical Biosensor

There are various components on which the electrochemical is based such as different electrode setups, bioreceptors elements, and electrochemical methods, all these different components were covered in this section.

2.2.1 Electrode Setups

Electrodes are a key component of an electrochemical biosensor in which the detecting process takes place (Cho et al. 2020) and it comes in a large variety of setups such as two-electrode setup, and three-electrode setup, bulky-electrode setup, screen-printed electrode setup, and paper electrode setup, Fig. 2.2. Table 2.1 presents the comparison of different electrode setups.

Typically, electrode configurations are employed within two distinct electrode platforms, specifically a two-electrode arrangement and a three-electrode configuration. The 2-E arrangement had just two types of electrodes (counter and working), but the 3-E setup had three separate electrodes (working, counter, and reference). Researchers discovered that a three-electrode design was more successful than a two-electrode setup in electrochemical biosensors (Hasan et al. 2022b).

(a) *Electrochemical-Bulky electrodes setup*: The electrochemical biosensor's performance has been markedly influenced by the electrodes employed in the electrochemical cell. Almost every electrochemical cell has three electrodes. A reference electrode (Ag/AgCl) is strategically placed distant from the site of the reaction to supply a potential that corresponds to the stable and known solutions. The reference electrode facilitates the normalization of measurements. In contrast, the counter electrode (CE) functions as the current origin, which is later directed toward the WE, i.e., the working electrode (A sensing/redox-electrode) which fulfills the role of the transduction component within the biological process. Both the CE and WE must possess electrical conductivity and chemical stability (Kim et al. 2019; Damiati and Schuster 2020).
(b) *Screen-printed electrode setup*: Diabetes was tested using the first commercially accessible screen-printed electrodes (Pohanka 2020). Screen-printed electrodes are generally made by deposition of patterned coatings over prepared surfaces such as glass. To produce the required electrode shape, the paste material is squeezed through the screen. The screens must be created in accordance with the applications and dimensions given. The paste is often composed of components that can transmit electrons, such as gold, silver, and carbon. Silver is commonly used in the fabrication of tracks and connections. Researchers commercialized screen-printed electrodes to overcome the limits of bulky-

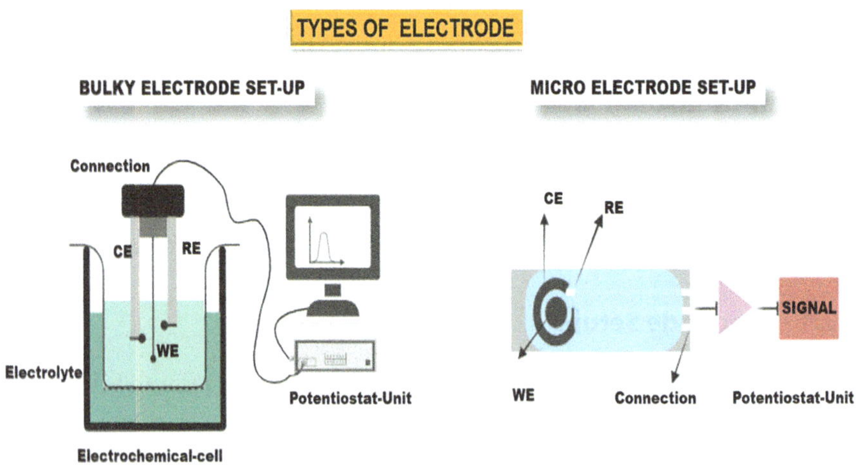

Fig. 2.2 Different electrode setups of electrochemical biosensor

Table 2.1 Comparison of different electrode setups of electrochemical biosensors

S. no	Standards	2-Electrode setup	3-Electrode setup	Bulky-electrode setup	Screen-electrode setup	Paper-electrode setup
1.	Availability	In-house fabrication/ commercial	In-house fabrication/ commercial	Commercial	Commercial	In-house fabrication
2.	Usage	Disposable/re-usable	Disposable/re-usable	Re-usable	Re-usable	Disposable
3.	Substrate	Paper/plastic	Paper/plastic	Glass	Plastic	Paper
4.	Volume/sample	20 μL	20 μL	50–60 μL	20 μL	20–30 μL
5.	Price/set (INR)	3/150	5/150	1000	150–200	3–5
6.	Electrodes comprise	– Working – Reference	– Working (W) – Reference (R) – Counter (R)	W/R/C	W/R/C	W/R/C
7.	Response	Very weak	Very good	Decent	Very accurate	Excellent

electrode setup since screen-printed electrodes provide many benefits such as cheap cost, low amount of sample required, and portable as well as miniaturized configuration. One of the primary benefits of the screen-printed electrode is that it is smaller in size and can be created quickly using a simple printing method, whereas a bulky setup is large, requires multiple costly devices for manufacturing, and requires a lot of time and specialists (Suresh et al. 2021; Alhazmi et al. 2023).

(c) *Paper electrode-based setup*: Commercialized screen electrodes are also expensive, cannot be purchased in bulk, cannot be stored for an extended period, are toxic to the environment because they are made of plastic, and take a long time to purchase from companies, which also does not meet the POC-based devices criteria. To overcome these constraints, scientists created screen-printed electrodes in the lab using a paper substrate rather than a plastic substrate, simply designed as it is based on a wooden screen frame, and the track of electrodes is printed by squeezing the conductive ink onto the cellulose paper substrate. The created electrodes were eco-friendly, economical, could be made in 5 min, easy to handle, could be modified as needed, size could be adjusted, and efficient for the detection of a variety of disorders (Sharma et al. 2023).

These represent key subcategories of electrode configurations that are effectively utilized in the construction of electrochemical biosensors for the diagnosis of diverse analytes.

2.2.2 Bioreceptors (BR) for Specific Electrochemical Biosensor

Fundamentally, the biosensor's selectivity or specificity hinges on the biorecognition element, which possesses the capacity for detection in the presence of an analyte (Islam et al. 2022). The electrochemical biosensor, the selection of biorecognition element is a crucial step as it helps the developed sensor be more selective toward the detection of the desired target. There are various types of BR available such as nucleic acid-BR, antibody-BR, and aptamer-BR. The electrochemical biosensor was also classified on such different BR and names according to the specific baroreceptors as if the electrochemical biosensor utilizes nucleic acid than the sensor called an "electrochemical-genosensor" (Pividori et al. 2000), if the sensor utilizes an antibody then it must be called "electrochemical-immunosensor" (Ramanavicius et al. 2010), and integration of aptamer in electrochemical biosensor was collectively called "electrochemical-aptasensor" (Tabrizi et al. 2015).

Antibody (Abs) is one of the popular bioreceptors utilized in the development of electrochemical immunosensor, in 1987, Vo-Dinh and colleagues demonstrated the application of antibodies in situ for detecting a chemical-based carcinogen using a fiber-optic immunosensor (Tromberg et al. 1987). Abs have subsequently showcased their efficacy as potent instruments for diagnostic purposes. Antibodies are highly suitable as biorecognition elements owing to their remarkably robust affinities and specificities for corresponding antigens. Several critical parameters pertain to the use of antibodies in applications of biosensors. These encompass sensitivity, selectivity, stability, immobilization (and consequently, orientation upon the coated-surface), labeling, and size (which influences the density of the sensor's biomaterial-layer). Recombinant Abs offer a feasible approach for enhancing these factors. Genetic modification primarily eases enhancements in stability and size selectivity, and, additionally, novel antibody fragments of this nature assist in achieving effective immobilization. Enhancements in sensitivity are achieved through the utilization of high-output screening and display libraries, primarily owing to their capability to screen significantly larger recombinant libraries. Leveraging coupling chemistries and/or the genetic incorporation of tags for immobilization proves to be valuable techniques for aiding the orientation and immobilization of innovative recombinant antibodies (Conroy et al. 2009).

But nowadays *aptamers* are emerging as effective alternatives of antibodies and also called artificial antibodies (Ruff et al. 2012). Aptamer was invented in the 1980s, but by 2023, it had advanced to higher development and was being used in a variety of applications, mostly in biosensor technology. The inaugural electrochemical aptasensor was developed in the year 2004, employing a sandwich-based amperometric-biosensor that utilized glucose dehydrogenase-labeled signaling aptamers (Sassolas et al. 2009). Aptamers are short ssDNA/RNA sequences with specific capturing properties to their targets (Sharma and Sharma 2013). The aptamer's molecular weight varies from 10 to 30 kDa, which is much lower than that of antibodies. SELEX is a method for quickly designing and synthesizing aptamers against any target compounds. The word "aptamer" is obtained from the Latin term "Aptus," signifying "to fit." Aptamers have several advantages over conventional bio-recognition elements (antibodies), including higher affinity and specificity, a much smaller size (almost a ten-fold reduction), and the ability to be manufactured in vitro at commercial scale and in a pure state as "chemical antibodies" (https://www.sciencedirect.com/topics/engineering/aptasensor).

2.2.3 Electrochemical Methods

In electrochemical biosensor, electrochemical methods are the main procedure, on which different diagnosis responses are evaluated. These approaches can be categorized into three distinct types: potentiometric, amperometric, and impedance-based methods.

2.2.4 Potentiometric

Potentiometric transduction involves the assessment of alterations in a system's potential, as determined by the Nernst equation (Han et al. 2021a). Variations in potential correspond to the activity of specific ions. This phenomenon occurs when the interaction between a free antigen and an electrode with an immobilized antibody is contrasted with a reference electrode. *Amperometric*, sensors based on such technique quantify the current produced through the action of electrochemical analysis. Most of the targets (analytes) cannot participate as redox couple in the reaction of electrochemical analysis, leading to limited applications for direct amperometric sensors. As a result, electroactive labels are used by amperometric sensors to produce an electric current. In contrast to the RE, where no selective binding is anticipated, the working electrode is where this electric current is generated when an electroactive species undergoes oxidation or reduction. The electroactive species' concentration and the current show a linear relationship. *Impedance (including capacitance/conductance)*, electric-impedance spectroscopy (EIS) identifies the electrical reaction elicited when a periodic, low-amplitude AC current is administered. Alterations in impedimetric transduction occur as a result of interactions between Ags and Abs on the exterior area of electrically conductive polymers (Conroy et al. 2009; Wehmeyer et al. 2021).

2.3 Applications of Electrochemical-Based Biosensor for the Detection of Numerous Diseases

There is a huge need for electrochemical biosensor in the field of heath care to diagnose various dreadful diseases (Fig. 2.3). In this section, different diseases based on electrochemical biosensor were covered such as DENV, COVID-19, HIV, cancer, and cardiovascular disease (Table 2.2).

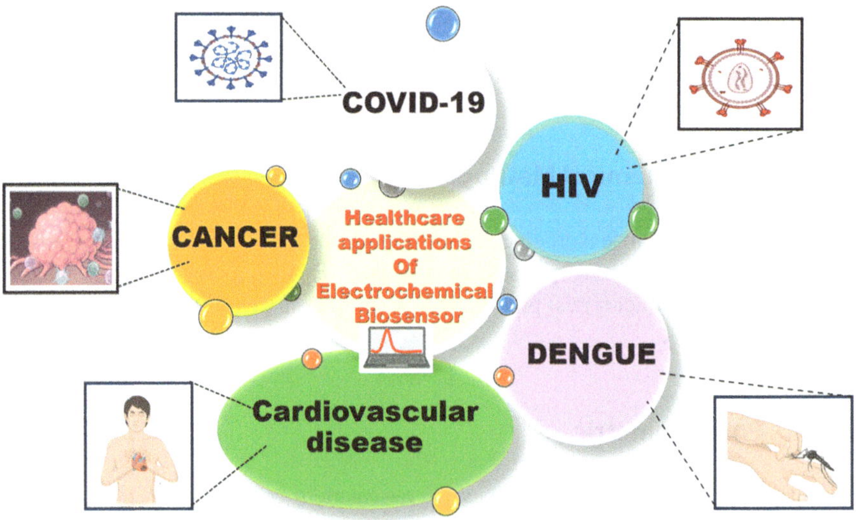

Fig. 2.3 Diagrammatic representation of electrochemical biosensor application in the health care sector comprises different disease diagnoses such as cancer, COVID, HIV, dengue, and cardiovascular diseases

Table 2.2 Electrochemical biosensor-based application for the detection of various diseases

S. no.	Electrochemical biosensor/principle	Analyte of various diseases	Nanomaterial	LOD	References
1.	Electrochemical impedance spectroscopy (EIS) based electrochemical genosensor	Dengue virus RNA	Gold nps	20 plaque-forming units (PFU)/mL	Wu et al. (2020)
	Multiplex-electrochemical paper-based genosensor	All the serotypes (1/2/3/4) of the dengue virus	Graphene oxide-silicon dioxide	100 pM to 100 μM	Singhal et al. (2020)
	Immunosensor electrochemical impedance spectroscopy (EIS)	Dengue virus-based biomarker (NS-1)	Gold nanorod decorated with graphitic carbon nitride	0.09 ng/mL	Ojha et al. (2022)
	Aptasensor based on 2-electrode and 3-electrode setup	Dengue virus antigen	Nanocomposite	0.1 μg/mL	Hasan et al. (2022)
	Immunosensor utilizing a screen-printed carbon electrode	Dengue virus antigen	Nanomaterial	0.16 nM	Alhazmi et al (2023)
2.	An electrochemical biosensor founded on the principles of isothermal amplification	The RNA-dependent RNA polymerase gene of the SARS-CoV-2 virus	Photoresist-Au thin nanofilm	0.972 and 3.925 fg/μL	Kim et al. (2021)
	Label-free electrochemical biosensor based on nature-inspired peptide	Anti-SARS-CoV-2 protein S antibody	Gold nanoparticle	0.2 μg/mL	Castro et al. (2022)
	Antisense oligonucleotide paper-based electrochemical biosensor chip	ssDNA	Gold nanoparticle	6.9 copies/μL	Alafeef et al. (2020)
	Bacterial-cellulose-based electrochemical biosensor modified with GO and human ACE-2	SARS-CoV-2 nucleic acid	GO nanoparticle	0.05 copies/μL	De Lima et al. (2023)
	Electrochemical biosensor based on papain-like cysteine protease for SARS-CoV-2 detection (PLpro biomarker)	PLpro biomarker	Gold nanoparticle	27.18 fM	Liang et al. (2023)

(continued)

Table 2.2 (continued)

S. no.	Electrochemical biosensor/principle	Analyte of various diseases	Nanomaterial	LOD	References
3.	Electrochemical DNA biosensor	HIV-1	Polyaniline/graphene (PAN/GN)	1.0×10^{-16} M	Gong et al. (2019)
	Electrochemical biosensor based on the glassy carbon electrode	HIV gene	Amino-functionalized reduced graphene oxide (NH2-rGO) and beta-cyclodextrin (β-CD)	8.7 fM	Li et al. (2020)
	Electrochemical biosensor based on the single particle	HIV DNA	Platinum nanoparticles	4.86 fM	Luo et al. (2021)
	Electrochemical biosensor based using carbon ink electrodes	HIV RNA	Gold nanoparticles	6 copies/mL	Moco et al. (2021)
	Electrochemical biosensor	HIV DNA	Nanomaterial	0.037 fM	Ma et al. (2023)
4.	Electrochemical biosensor utilizing SWCNT for the detection of prostate cancer	mi-RNAs	Dendritic gold nanostructure	0.01 fmol/L	Sabahi et al (2020)
	Electrochemical biosensor for non-small cell lung cancer	ct-DNA	A nanocomposite comprising nitrogen-doped graphene	7.65 fM	Guo et al. (2022)
	Electrochemical biosensor designed for the detection of the long non-coding RNA MALAT-1 biomarker, utilizing a screen-printed carbon electrode as its foundation	Lnc RNAs	Gold nanocomposite	42.8 fM	Chen et al. (2021)
	KIT electrochemical aptasensor for cancer cell detection	KIT protein	Gold nanoparticle	1.15 pg/mL	Chung et al. (2021)
	Electrochemical biosensor for analysis of tryptophan in breast cancer detection	Tryptophan	Polypyrrole-carbon dot nanocomposite	0.003 µmol/L	Abdel-aa et al. (2023)

5.	Electrochemical biosensor based on multifunctional peptide-oligonucleotide conjugation	Cardiac troponin I	Quantum dots	0.42 pg/mL	Han et al. (2021)
	Dual-aptamer based aptasensor	Cardiac troponin I (cTnI)	Nanomaterial	16 pg/mL	Sun et al. (2019)
	Electrochemical-immunosensor	cTnl	Gold nanoparticles	1.7 pg/mL	Feng et al. (2021)
	Electrochemical aptasensor based on 2D nanomaterials	cTnl	Ruthenium nanoparticles (RuNPs) and nickel molybdate nanorods ($NiMoO_4$ NRs) are employed for the functionalization of porous graphitic carbon nitrides (PCN)	70 pg/mL 50 pg/mL	Khushaim et al. (2022)
	Aptamer-based electrochemical aptasensor	cTnI	Polypyrrole (PPy)-gold nanoparticles	25 pg/mL	Eshlaghi et al. (2023)
	A sandwich-type aptasensor	CD-biomarker	Gold nanoparticles (AuNPs) incorporated into hollow mesoporous carbon spheres that have been modified with polydopamine (HMCS@PDA@AuNPs)	2.3 fg/mL	Wang et al. (2023)

2.3.1 Electrochemical-Based Biosensor for Dengue Diagnosis

Dengue virus (DENV) belongs to the family Flaviviridae and has four different serotypes (DENV-1, 2, 3, and 4) that are transmitted through the bites of Aedes mosquitoes carrying the infection. It leads to a broad range of medical conditions, spanning from mild, asymptomatic cases to severe and possibly fatal conditions such as dengue shock syndrome (DHF/DSS) and dengue hemorrhagic fever (Gupta et al. 2012). It becomes a global concern as it affects millions of lives each year. Despite so much advancement in the field of healthcare still, there is no proper treatment available, only proper diagnosis is the key to managing such stubborn diseases. Various scientists were trying to diagnose the dengue virus by developing effective electrochemical biosensor such as Wu and their co-workers (2020) developed genosensor based on an electrochemical principle designed for the detection of the dengue virus, achieving a commendable LOD at 20 plaque-forming units (PFU) per milliliter. For the electrochemical evaluation of the dengue virus, their study used electrochemical impedance spectroscopy (EIS). The developed sensor is known as "amplification and label-free diagnosis of extracted dengue virus RNA" since the test is more sophisticated and reliable and saves time by doing away with the need for amplification and labelling. To increase sensitivity, the study functionalized the Au electrodes with probe DNA using different doses of the linkers 6-mercapto-1-hexanol (MCH) and 6-mercaptohexanoic acid (MHA). These linkers were chosen for their robust binding affinity to the gold electrode when coupled with immobilized probe molecules. The developed platform (probe-linkers-modified-genosensor) is more sensitive and selective, the study evaluated the specificity of genosensors immobilized with DENV1-specific probe DNA about various DENV-serotypes. Therefore, the absence of labeling and amplification in the diagnosis of dengue RNA contributes to increasing the robustness and sensitivity of the diagnostic process. Electrochemical biosensors demonstrated the capacity to detect different variations of the dengue virus using a unified platform as proved by Singhal and their co-workers (2020) who built a multiplex-electrochemical nanosensor for the diagnosis of all the serotypes (1/2/3/4) of dengue virus and showed a great range of detection, i.e., 100 pM to 100 μM. In this work, they synthesized nanocomposite (graphene oxide-silicon dioxide) which helps in amplifying the response of the sensor as the synthesized nanomaterial is more conducting in nature. The developed multiplexed genosensor exhibited a complete absence of cross-reactivity and demonstrated a notable level of specificity when targeting distinct types of dengue virus. Furthermore, this is the first report of creating a paper based genosensor against dengue disease. Not only genosensor but immunosensor was also constructed for the detection of dengue virus as it provides numerous advantages over genosensor such as antibodies used in immunosensor were more selective toward their targets and provided high specificity to the sensor. Ojha et al. (2022) engineered an immunosensor designed for the diagnosis of a biomarker associated with the dengue virus, specifically NS-1. They quantified the dengue virus serotype 1 by employing an anti-DENV (1–4, and) antibody. The developed immunosensor demonstrated a wide linear range spanning from 0.6 to 216 ng/mL, and it achieved a remarkable LOD, i.e., 0.09 ng/mL. In this study, EIS was employed to analyze the detection response. To enhance the sensor's sensitivity, gold nanorods decorated with graphitic carbon nitride were utilized to functionalize the electrode (glassy carbon). This modification provided the electrode with exceptional electroactivity, rapid electron transfer kinetics, and a heightened catalytic capability. The surface of the electrode was immobilized with an Abs to increase the sensor's specificity. The response was then analyzed in the presence of the target using an electrolyte, specifically $[Fe(CN)_6]^{3-/4-}$ as a redox couple, to monitor the corresponding variations in charge transfer resistance (Rct) linked to the antigen binding with the antibody. The developed sensor performed well in human serum as well. There are various biorecognition elements used in the detection of the dengue virus such as DNA based, antibody-based but other than these another emerging bio-recognition element, i.e., aptamer was used as an effective alternative bio-recognition element as it provides various advantages to the developed sensor like it is very inexpensive and artificially synthesized in the laboratories in a very short time and provides high stability in the ambience environment. Hasan et al. (2022b) were the pioneers in designing an aptamer-based aptasensor for detecting the dengue virus antigen, achieving an impressive LOD 0.1 μg/mL. They created two separate paper-based platforms (2-electrode and 3-electrode setup) for this study and compared the responses of the two in relation to the diagnosis of dengue. The newly designed paper-based aptasensor based on 3 electrode system demonstrated a substantially broader current range (55.53 to 322.21 μA) than the 2-electrode set-up. As indicated by this research, the 3-electrode setup exhibited a substantial current amplification range, approximately 50 times greater than that observed in the 2-electrode setup, which exhibited a relatively feeble current signal.

Accordingly, utilization of the 3-electrode setup has appeared as a promising alternative for highly effective dengue virus detection and holds a possible role for the analysis of various other diseases. In a recent report of Alhazmi et al. (2023), they detected the dengue virus antigen by creating an immunosensor by the help of microelectrodes, i.e., screen-printed electrodes. In this study, DENV antibodies were immobilised using a linker called EDC/NHS i.e., 1-ethyl-3-(3-dimethylaminopropyl) carbodiimide hydrochloride/N-hydroxysulfosuccinimide. The diagnosis response to the target was then obtained through LSV employing a potentiostat and a sensor connector. The devised platform demonstrated outstanding sensitivity in the identification of antigen (DENV) in both phosphate-buffered saline (PBS) at a pH of 7.4 and diluted human plasma, with an LOD in PBS measuring 0.11 nM; conversely, in the case of human plasma, the LOD was determined to be 0.16 nM, and the sensor exhibited the capability to detect the dengue antigen within a remarkably short period of 5 min, thus showcasing rapid detection capabilities.

2.3.2 Electrochemical-Based Biosensor for the Diagnosis of Coronavirus

Originating in late 2019 in Hubei province, China, the global pandemic instigated by the SARS-CoV-2 has resulted in an exceptional scale of over one million fatalities (Bai et al. 2020). This challenging situation needs an accurate testing device for the rapid analysis of patients worldwide. Conventional techniques such as RT-PCR and serological assays were playing the gold standard for detection. However, the limiting drawbacks of this technology, such as high cost, false positive and negative cases in some patients highlighted issues that can be solved by biosensors (Tomichan et al. 2023). The electrochemical biosensor can be integrated with the recombinase polymerase amplification (RPA) for the POC testing using thiol-modified primers immobilized on PR/Au (positive photoresist—PR) modified coplanar electrode. The device demonstrated a detection limit of approximately 0.972 femtograms per microliter (fg/μL) and 3.925 fg/μL for the RdRp and N genes utilized, respectively (Kim et al. 2021). A label-free biosensor based on synthetic peptides for COVID-19 screening was engineered. This biosensor replicates the molecular recognition of the S protein binding site of the anti-SARS-CoV-2 S protein antibody, employing gold nanoparticles and generating a signal output through electrochemical impedance spectroscopy (Castro et al. 2022). An electrochemical biosensor incorporating laser-induced graphene (LIG) decorated with CO_2 infrared laser was established for the diagnosis of antibodies in the serum (human). The reactive sample yielded an average current value of 9.6757 (Oliveira et al. 2022). A highly quantitative, expeditious, and remarkably sensitive sensor chip was created for the rapid detection of COVID within a mere 5 min. This innovative sensor employed an antisense oligonucleotide (ssDNA) designed to target the viral nucleocapsid (N gene), as described in the work by Alafeef et al. in 2020. The gold nanoparticle was used for the signal amplification which was capped with ssDNA. The sensor exhibited an elevated electrical signal response, characterized by a sensitivity of 231 $(\text{copies}/\mu L)^{-1}$ and an LOD of 6.9 copies/μL. Notably, this sensitivity level obviates the need for additional amplification processes (Alafeef et al. 2020). Environmentally friendly technology was employed in the creation of an electrochemical biosensor, which incorporates bacterial cellulose (BC) as a biodegradable substrate. Additionally, carbon-printed electrodes were coated via PEG and graphene oxide in the sensor's construction. In less than 10 min, this biosensor detected SARS-CoV-2 with an LOD 4.26×10^{-18} *g/mL SP* and provided excellent performance in detection (de Lima et al. 2023). Using papain-like cysteine protease, an electrochemical biosensor was designed which uses the SARS-CoV-2 PLpro biomarker and fabricated a novel PLpro-activated peptide-DNA smart nanoprobe. They successfully attained a linear detection span spanning from 1 picogram per milliliter (pg/mL) to 10 micrograms per milliliter (μg/mL), accompanied by a detection threshold of 27.18 femtomolar (fM), as represented by the equation $I = 1.63 + 4.44 * \log(C)$. Currently, it is feasible to swiftly conduct inhibitor screening for the SARS-CoV-2 PLpro enzyme. As a result, the creation of new drugs and the fight against COVID-19 have a lot to gain by using this electrochemical biosensor (Liang et al. 2023).

2.3.3 Electrochemical Biosensor for the Diagnosis of Human Immunodeficiency Virus (HIV)

This ailment is categorized among the life-threatening conditions (Shao and Shao 2023). This pertains to the Retroviridae family, falling under the sub-family Orthoretrovirinae and the genus Lentivirus. HIV-1 and HIV-2 are the taxonomic classifications of the human immunodeficiency virus (HIV). The HIV genome consists of two identical single-stranded RNA (ssRNA) molecules that are tethered to the core of the viral particle. The proviral genome, alternatively referred to as proviral DNA, is formed through the process of reverse transcription of the viral RNA genome into DNA. This results in the integration of double-stranded DNA (belonging to HIV) into the human genome, accompanied by the degradation of RNA. The hallmark of the initial phases of infection is the presence of intricate protein–protein interactions. Macrophages, T helper cells, astrocytes, and dendritic cells are all susceptible to HIV infection and are characterized by the presence of CD4 receptors on their surfaces. The introduction of proviral DNA into the cell concludes the process of HIV infection and initiates the establishment of a persistent, chronic infection (Thakur et al. 2023). Thorough preventive measures and prompt diagnostic interventions have been demonstrated by numerous researchers to transform this life-threatening condition into a manageable disease such as Gong et al. (2019), who engineered an electrochemical biosensor designed for the detection of HIV, which exhibited a commendable LOD of 1.0×10^{-16} M. With the use of reverse-phase polymerization, they created a nanomaterial known as polyaniline/graphene (PAN/GN) in this study. The synthesized nanocomposites were then dropped onto the surface of the electrode (glassy carbon electrode). The process increases the sensor's sensitivity, and the probe molecules were immobilized on an electrode that had been treated with nanomaterials for increased specificity. Through π–π* stacking interactions, the HIV-1 phosphate backbone's negative charge adhered to the altered electrode surface. Double-stranded DNA (dsDNA) was produced as a result of the interaction between the HIV-1 target and the single-stranded DNA (ssDNA) probe. The electrode's electron transfer resistance was then measured using impedimetric experiments with a $[Fe(CN)_6]^{3-/4-}$ redox pair. The platform that was developed exhibited remarkable discriminatory ability in distinguishing non-complementary or mismatched sequences and demonstrated exceptional selectivity and sensitivity in detecting the HIV-1 gene. Therefore, this biosensor offers a novel approach to the detection of fragments of the HIV gene. In another report, Li et al. (2020), an electrochemical biosensor was fabricated to detect the HIV gene, achieving a limit of detection at 8.7 fM. In this study, the modification of the glassy carbon electrode involved a process wherein amino-reduced graphene oxide (NH2-rGO) and β-cyclodextrin (β-CD) were self-assembled onto the electrode surface. Subsequently, single-stranded DNA (ssDNA) was employed, with a sequence that could hybridize with the HIV gene. The electrode surface was initially functionalized with the attachment of single-stranded DNA (ssDNA), followed by the interaction of methylene blue (MB) with the ssDNA. The ensuing interaction between the probe and the target was subsequently observed and recorded utilizing the DPV-electrochemical analysis technique. The developed platform featuring the modified electrode exhibited notable reproducibility, exceptionally high sensitivity, and outstanding selectivity. This electrochemical biosensor was effectively employed to detect the presence of the HIV gene in human serum samples. Luo et al. (2021), A single-particle-based electrochemical biosensor was developed for the diagnosis of HIV, demonstrating the capability to detect 4.86 fM of HIV DNA. In this study, a magnetic separation method and DNA walker amplification process were employed as integral components of the experimental approach, magnetic nanobeads (MBs) possess the capacity to selectively capture and isolate target entities from intricate sample matrices, this not only guarantees the method's resistance to interference but also mitigates the issue of platinum nanoparticle aggregation induced by the presence of numerous coexisting substances. A minimal quantity of targets can induce the liberation of a greater number of Pt NPs, thereby triggering an increased occurrence of collision current transients and facilitating cyclic amplification. In contrast to basic hybridization techniques, the utilization of a DNA walker can elevate collision frequencies approximately threefold, substantially enhancing detection sensitivity. A correlation between collision frequency and target concentration is then employed to achieve quantification. The developed platform demonstrates that this distinctive approach can attain an impressively low sensor threshold at the femtomolar level. This achievement opens up new prospects for the effective detection of HIV DNA in real-world samples and the advancement of innovative single-entity biosensors. In a separate study, as reported by Moco et al. (2021) who engineered an electrochemical biosensor designed for the detection of HIV RNA, achieving a remarkable limit of detection at 6 copies/mL. In this work, they constructed a screen-printed electrode (SPE) with the help of carbon-conductive ink and functionalized these electrodes with gold nanoparticles, which helps to enhance the signal of the sensor. The detection response of the target and probe complex was then evaluated with the aid of EtBr as a duplex formation indicator after the probes were attached to the surface of a screen-printed electrode that had undergone alteration. The biosensor that was developed possesses the capability to distinguish between negative and positive samples, thereby enabling the quantification of viral load. This capacity is pivotal in the advancement of biosensors geared toward the early detection of HIV. Recently, Ma et al. (2023), The electrochemical biosensor was engineered to identify HIV DNA within a linear range spanning from 0.1 fM to 10 nM, achieving a notably lower limit of detection (LOD) at 0.037 fM. In this procedure, an abundant quantity of probe molecules was generated through the application of exonuclease III, which fragmented the cyclic target DNA. Subsequently, the Zr^{4+}-modified probe DNA, featuring multiple branched DNA structures, was effectively captured. Subsequently, the initiator was attached via an esterification process, specifically α-bromophenylacetic acid, with the branched chains of the probe DNA. A multitude of electroactive monomers, specifically ferrocenyl methyl methacrylate (FMMA), are synthesized at the termini of the branched chains by an electrochemically facilitated atomic transfer radical polymerization process. The signal originating from FMMA experiences a significant amplification, primarily attributable to the presence of a substantial quantity of probe clusters loaded onto branched chains, which in turn contain a considerable number of FMMA moieties. The sensor that has been engineered exhibits exceptional sensitivity and specificity. Furthermore, it demonstrates robust performance when applied to actual samples, underscoring its capability likely for application in the primary clinical diagnosis of AIDS.

2.3.4 Electrochemical Biosensor for the Diagnosis of Cancer

Timely detection has a crucial part in combatting one of the most severe and potentially fatal illnesses, cancer. Biosensors offer expeditious, highly sensitive, selective, and cost-effective cancer detection in its nascent phase, addressing a historical limitation of conventional technologies (Zhang et al. 2019). Diverse categories of electrochemical-based biosensors have been engineered for the diagnosis of cancer biomarkers such as microRNAs (miRNAs), p53 gene mutation (Afsharan et al. 2016), prostate-specific antigen (PSA), cancer antigen 15-3 (CA15-3), CD 44, and carcinoembryonic antigen (CEA) (Hasan et al. 2021b). Several of these cancer biomarkers possess the potential for early cancer diagnosis, preempting the onset of symptoms in patients. Cancer biomarker mi-RNA 21 for the diagnosis of prostate cancer was detected using an electrochemical-based nano-genosensor made using dendritic gold nanostructures and cadmium for signal amplification and labeling. The concentration range, spanning from 0.01 fmol/L to 1 μmol/L, was determined by employing differential pulse voltammetry (DPV) techniques to measure the signal (Sabahi et al. 2020). At present, there is an ongoing effort to advance the utilization of metal-organic frameworks (MOFs) in enhancing the detection of biomarkers. This stems from their notable attributes, which include remarkable porosity, substantial regenerative capacities, the presence of specific functional sites, and the adjustability of their framework (Zhang et al. 2021; Mohan et al. 2022). Recently, a pioneering electrochemical biosensor was created using a composite of polyethyleneimine (PEI) and nitrogen-doped graphene, specifically the covalent organic framework (NG-PEI-COFTAPB-TFPB). This biosensor system was designed for the detection of tumor DNA associated with non-small cell lung cancer (NSCLC), achieving an impressively low detection limit of 7.65 fM (Guo et al. 2022). An innovative biosensor was engineered to exploit long non-coding RNAs (lnc RNAs) MALAT-1 as a diagnostic biomarker, enabling the highly sensitive diagnosis of NSCLC. This biosensor uses gold nanocage, which is coupled with a carbon nanotube with multiple layers decorated on screen-printed electrode. This biosensor system gives a low limit of detection 42.8 fM with high accuracy and selectivity (Chen et al. 2021). An electrochemical aptasensor was developed using ferrocene-labeled aptamer which is conjugated with the gold nanoparticle decorated electrode for detecting soluble KIT protein in the human serum. This sensor exhibited a detection time of 35 min, offering an LOR spanning from 0.010 to 100 ng/mL, having a limit of detection of 1.15 pg/mL. It can be functional for both in vivo and in vitro monitoring (Chung et al. 2021). Tryptophan catabolism can be monitored to detect breast cancer. To do this, a polypyrrole/carbon dot nanocomposite biosensor was created, and it demonstrated outstanding electrochemical catalytic activity. This biosensor may be utilised for evaluating tryptophan levels (Abdel-aal et al. 2023).

2.3.5 Electrochemical-Based Biosensor for the Diagnosis of Cardiovascular Disease (CD)

Cardiovascular ailments constitute the predominant factor leading to mortality worldwide (Reyes-Retana and Duque-Ossa 2021). Heart arrest and myocardial infarction (MI) are closely related conditions that cause death and organ damage from insufficient oxygen delivery, damaging the heart muscle irreversibly. In the realm of medical diagnosis, electrocardiogram (ECG) analysis has traditionally served as the primary technique, but its efficacy is limited. Consequently, methodologies such as assessing biomarker concentrations in plasma or saliva have gained prominence as a widely adopted approach for prognosticating acute myocardial infarction (AMI), particularly because certain biomarkers exhibit specific associations with heart attacks. Numerous investigations are ongoing to ascertain alterations in biological markers. However, these examinations frequently exhibit drawbacks associated with protracted time requirements and labor-intensive procedures. To address these challenges, researchers worldwide have been devising various methods to augment detection capabilities by harnessing biosensors (Hui et al. 2023). Cardiac troponin I, or cTnI, is a common blood biomarker used to diagnose acute myocardial infarction. Precisely detecting this condition is crucial to the effective treatment of this condition. Numerous investigators have employed cardiac troponin as a prospective biomarker for the early identification of cardiovascular ailments as proved by Han et al. (2021b), yielded an electrochemical biosensing device designed for the quantification of the cardiovascular disease biomarker, cardiac troponin I, with a very low LOD of 0.42 pg/mL. In this study, a multifunctional peptide-oligonucleotide conjugation was employed. This conjugation is structured to include a peptide component with a high specificity for binding to cTnI, alongside a DNA segment that facilitates the controlled development of quantum dots (QDs), thereby acting as per the signal basis. The presence of the target cTnI results in the attraction of quantum dots (QDs) onto magnetic beads, consequently generating a notable electrochemical reaction when applied to a screen-printed electrode. The established platform demonstrates considerable efficacy in the detection of cTnI, holding substantial promise for its application in both cell biology research and clinical diagnostic settings. In another research, Sun et al. (2019) established an electrochemical biosensing apparatus utilizing aptamers to study the quantification of the protein cardiac troponin I (cTnI), achieving a limit of detection which was as low as 16 pg/mL. In this study, magnetic-metal organic framework (MMOF) nano-catalysts were employed in conjunction with DNA nanotetrahedron (NTH)-based dual-aptamer probes used for the nonenzymatic detection of protein cardiac troponin I (cTnI). These components were immobilized onto the surface of a screen-printed gold electrode, followed by the introduction of the biomarker to diagnose cardiovascular disease. The sensor that was created demonstrated exceptional sensitivity and specificity for cTnI in the form of an aptasensor, thereby showcasing substantial potential for utilization in clinical diagnostic scenarios. In addition to the aptasensor, immunosensors were also employed for the detection of CD. Feng et al. (2021) conducted the diagnosis of troponin I (cTnI) through the development of an immunosensing device, achieving an impressively low limit of detection, specifically at 1.7 pg/mL. In this study, a multifunctional nanocomposite based on a covalent organic framework (COF) was utilized, with its composition including horseradish peroxidase (HRP), Ab2 (secondary antibody), and gold (Au). This nanocomposite served as the probe for signal amplification. The gold nanoparticles incorporated onto the COF surface serve a dual purpose: they offer a solid foundation for the add-on of the secondary antibody (Ab2) and simultaneously enhance the electric conduction of the COF material. The probes that were devised not only exhibit the precise target recognition properties of Ab2 toward cTnI but also enhance the sensitivity through signal amplification, attributed to the outstanding enzymatic capabilities of HRP. Furthermore, this study presented a straightforward methodology for the quantitative assessment of cTnI within real-world samples, thereby expanding the potential applications of the COF-based nanocomposite within the domain of electrochemical immunosensors. 2D nanomaterial is one of the trending nanomaterials due to its unique physiochemical properties over other nanomaterials. Using a wet chemical synthesis technique, Khushaim et al. (2022) synthesised nickel molybdate nanorods ($NiMoO_4$ NRs) and ruthenium nanoparticles (RuNPs) to functionalize porous graphitic carbon nitrides (PCN). These modified materials were then used to fabricate the biosensors with an electrochemical nature for the detection of cardiac troponin I (cTn-I), achieving impressive LOD of 70.0 pg/mL and 50.0 pg/mL, respectively. In this study, 2D nanomaterials were used for the variation of the working electrodes composed of laser-scribed graphene electrodes. Subsequently, these modified electrodes were immobilized with thiol-functionalized aptamers specific to cTn-I, demonstrating a pronounced affinity for the target molecule. Furthermore, the biosensors exhibited exceptional selectivity and practical utility. The conducted research suggests that 2D materials based on PCN serve as highly promising candidates for the development of biosensors targeting cardiac biomarkers. Eshlaghi et al. (2023) devised an aptamer-based aptasensing system for measuring the quantity of cTnI, achieving an LOD of 25 pg/mL. In this work, they employed a polypyrrole (PPy)-gold nanoparticles to modify the electrode surface (screen-printed carbon electrode), modification was also confirmed by the help of various characterization techniques (SEM, AFM, EDX CV, SWV). Following the modification of the electrode, aptamers were affixed to the surface of the modified electrode. Subsequently, the target analyte was introduced, and its detection range was recorded using square wave voltammetry (SWV) in conjunction with an appropriate electrolyte. The aptasensor presented in this study displayed noteworthy selectivity and sensitivity suitable for clinical diagnostic applications, especially when compared to the conventional enzyme-linked immunosorbent assay (ELISA) analysis. This aptasensor enables rapid, quantitative, and label-free detection of cardiac markers. Recently, Wang et al. (2023) also made a sandwich-type aptasensor design that can be used for the identification of the cardiovascular disease biomarker, achieving an impressive LOD of 2.3 fg/mL. In this study, dendritic Pt-Cu alloy NPs were employed and laden onto graphene-like copper oxide and titanium dioxide mesoporous ultrathin nanosheets, designated as platinum-copper DNs/MUN-CuO-TiO_2, to serve as labeling substances. Additionally, gold nanoparticles were loaded onto polydopamine-modified hollow mesoporous carbon spheres, designated as

HMCS@PDA@AuNPs, to function as electrode-altered material. These materials were utilized to expedite electron transfer processes. Aptamers were attached to the outward of the altered electrode, and the target analyte was employed to assess the electrochemical functionality of the biosensing device. The suggested sensor has been employed for the examination of human serum samples, yielding satisfactory outcomes, thus demonstrating its viability for potential clinical deployment.

2.4 Conclusion

Among the numerous contemporary biosensor measurement methodologies, electrochemical techniques stand out as particularly well-suited for biosensing applications due to their ability to fulfill all the essential criteria for genuine biosensor advancement. Furthermore, electrochemical techniques offer several advantages, such as swift response times, heightened sensitivity and specificity, inherent capacity for miniaturization, user-friendly operation, and portability. These attributes collectively render these sensors exceptionally suitable for the creation of biosensors tailored to specific diseases. Consequently, numerous electrochemical biosensors have garnered substantial attention in recent times for prospective applications. Several crucial prerequisites must be satisfied to advance the development of electrochemical biosensors. These prerequisites encompass exceptional sensitivity, the ability to enhance the substrate's conductivity, and the judicious selection of an electrode configuration, all of which offer additional advantages to the resultant sensor. In summary, this chapter offers valuable insights into the evolution of electrochemical biosensors and presents innovative guidance and inspiration for the creation of precise, dependable, and functional electrochemical biosensors designed for the detection of a spectrum of serious diseases.

References

Abdel-aal FAM, Kamel RM, Abdeltawab AA, Mohamed FA, Mohamed A-MI (2023) Polypyrrole/carbon dot nanocomposite as an electrochemical biosensor for liquid biopsy analysis of tryptophan in the human serum of normal and breast cancer women. Anal Bioanal Chem 415(20):4985–5001. https://doi.org/10.1007/s00216-023-04784-7

Afsharan H, Navaeipour F, Khalilzadeh B, Tajalli H, Mollabashi M, Ahar MJ, Rashidi M-R (2016) Highly sensitive electrochemiluminescence detection of p53 protein using functionalized Ru–silica nanoporous@gold nanocomposite. Biosens Bioelectron 80:146–153. https://doi.org/10.1016/j.bios.2016.01.030

Alafeef M, Dighe K, Moitra P, Pan D (2020) Rapid, ultrasensitive, and quantitative detection of SARS-CoV-2 using antisense oligonucleotides directed electrochemical biosensor chip. ACS Nano 14(12):17028–17045. https://doi.org/10.1021/acsnano.0c06392

Alhazmi HA, Ahsan W, Taha MM, Albratty M, Najmi A, Farasani A, Abdulhaq AA, Darwish IA (2023) Development of screen-printed carbon electrode-based immunosensors for the electrochemical detection of dengue virus antigen. J King Saud Univ Sci 35(3):102568

Anzar N, Hasan MR, Akram M, Yadav N, Narang J (2020) Systematic and validated techniques for the detection of ovarian cancer emphasizing the electro-analytical approach. Process Biochem 94:126–135

Bai Y, Yao L, Wei T, Tian F, Jin DY, Chen L, Wang M (2020) Presumed asymptomatic carrier transmission of COVID-19. JAMA 323(14):1406–1407. https://doi.org/10.1001/jama.2020.2565

Castro ACH, Bezerra ÍRS, Pascon AM, da Silva GH, Philot EA, de Oliveira VL, Mancini RSN, Schleder GR, Castro CE, de Carvalho LRS, Fernandes BHV, Cilli EM, Sanches PRS, Santhiago M, Charlie-Silva I, Martinez DST, Scott AL, Alves WA, Lima RS (2022) Modular label-free electrochemical biosensor loading nature-inspired peptide toward the widespread use of COVID-19 antibody tests. ACS Nano 16(9):14239–14253. https://doi.org/10.1021/acsnano.2c04364

Chen M, Wu D, Tu S, Yang C, Chen D, Xu Y (2021) A novel biosensor for the ultrasensitive detection of the lncRNA biomarker MALAT1 in non-small cell lung cancer. Sci Rep 11(1):3666. https://doi.org/10.1038/s41598-021-83244-7

Cho IH, Kim DH, Park S (2020) Electrochemical biosensors: perspective on functional nanomaterials for on-site analysis. Biomater Res 24(1):1–2

Chung S, Sicklick JK, Ray P, Hall DA (2021) Development of a soluble KIT electrochemical aptasensor for cancer theranostics. ACS Sens 6(5):1971–1979. https://doi.org/10.1021/acssensors.1c00535

Conroy PJ, Hearty S, Leonard P, O'Kennedy RJ (2009) Antibody production, design and use for biosensor-based applications. Semin Cell Dev Biol 20(1):10–26

Damiati S, Schuster B (2020) Electrochemical biosensors based on S-layer proteins. Sensors 20(6):1721

De Girolamo A, McKeague M, Pascale M, Cortese M, DeRosa MC (2018) Immobilization of aptamers on substrates. In: Aptamers for analytical applications: affinity acquisition and method design, pp 85–126

de Lima LF, Ferreira AL, Ranjan I, Collman RG, de Araujo WR, de la Fuente-Nunez C (2023) A bacterial cellulose-based and low-cost electrochemical biosensor for ultrasensitive detection of SARS-CoV-2. Cell Rep Phys Sci 4(8):101476. https://doi.org/10.1016/j.xcrp.2023.101476

Eshlaghi SN, Syedmoradi L, Amini A, Omidfar K (2023) A label-free electrochemical aptasensor based on screen printed carbon electrodes with gold nanoparticles-polypyrrole composite for detection of cardiac troponin I. IEEE Sens J 23(4):3439–3445

Feng S, Yan M, Xue Y, Huang J, Yang X (2021) Electrochemical immunosensor for cardiac troponin I detection based on covalent organic framework and enzyme-catalyzed signal amplification. Anal Chem 93(40):13572–13579

Gong Q, Han H, Yang H, Zhang M, Sun X, Liang Y, Liu Z, Zhang W, Qiao J (2019) Sensitive electrochemical DNA sensor for the detection of HIV based on a polyaniline/graphene nanocomposite. J Mater 5(2):313–319

Guo L, Mu Z, Yan B, Wang J, Zhou J, Bai L (2022) A novel electrochemical biosensor for sensitive detection of non-small cell lung cancer ctDNA using NG-PEI-COFTAPB-TFPB as sensing platform and Fe-MOF for signal enhancement. Sens Actuators B Chem 350:130874. https://doi.org/10.1016/j.snb.2021.130874

Gupta N, Srivastava S, Jain A, Chaturvedi UC (2012) Dengue in India. Indian J Med Res 136(3):373

Han T, Mattinen U, Mousavi Z, Bobacka J (2021a) Coulometric response of solid-contact anion-sensitive electrodes. Electrochim Acta 367:137566

Han K, Li G, Tian L, Li L, Shi Y, Huang T, Li Y, Xu Q (2021b) Multifunctional peptide-oligonucleotide conjugate promoted sensitive electrochemical biosensing of cardiac troponin I. Biochem Eng J 174:108104

Hasan MR, Sharma P, Anzar N, Pundir CS, Pilloton R, Narang J, Shetti NP (2021a) Analytical methods for detection of human cytomegalovirus clinched biosensor a cutting-edge diagnostic tool. Biomed Eng Adv 1:100006

Hasan MR, Ahommed MS, Daizy M, Bacchu MS, Ali MR, Al-Mamun MR, Saad Aly MA, Khan MZH, Hossain SI (2021b) Recent development in electrochemical biosensors for cancer biomarkers detection. Biosens Bioelectron X 8:100075. https://doi.org/10.1016/j.biosx.2021.100075

Hasan MR, Anzar N, Sharma P, Singh S, Hassan H, Rawat C, Narang J (2022a) Mycobacterium tuberculosis diagnosis from conventional to biosensor—a systematic review. Int J Environ Anal Chem:1–6
Hasan MR, Sharma P, Shaikh S, Singh S, Pilloton R, Narang J (2022b) Electrochemical aptasensor developed using two-electrode setup and three-electrode setup: comprising their current range in context of dengue virus determination. Biosensors 13(1):1
Hui H, Gopinath SC, Ismail ZH, Chen Y, Pandian K, Velusamy P (2023) Cardiovascular biomarker troponin I biosensor: aptamer-gold-antibody hybrid on a metal oxide surface. Biotechnol Appl Biochem 70(2):581–591
Islam MA, Karim A, Ethiraj B, Raihan T, Kadier A (2022) Antimicrobial peptides: promising alternatives over conventional capture ligands for biosensor-based detection of pathogenic bacteria. Biotechnol Adv 55:107901
Khushaim W, Mani V, Peramaiya K, Huang KW, Salama KN (2022) Ruthenium and nickel molybdate-decorated 2D porous graphitic carbon nitrides for highly sensitive cardiac troponin biosensor. Biosensors 12(10):783
Kim JH, Cho CH, Ryu MY, Kim JG, Lee SJ, Park TJ, Park JP (2019) Development of peptide biosensor for the detection of dengue fever biomarker, nonstructural 1. PLoS One 14(9):e0222144
Kim HE, Schuck A, Lee SH, Lee Y, Kang M, Kim YS (2021) Sensitive electrochemical biosensor combined with isothermal amplification for point-of-care COVID-19 tests. Biosens Bioelectron 182:113168. https://doi.org/10.1016/j.bios.2021.113168
Krokhine S, Torabi H, Doostmohammadi A, Rezai P (2021) Conventional and microfluidic methods for airborne virus isolation and detection. Colloids Surf B: Biointerfaces 206:111962
Li J, Jin X, Feng M, Huang S, Feng J (2020) Ultrasensitive and highly selective electrochemical biosensor for HIV gene detection based on amino-reduced graphene oxide and β-cyclodextrin modified glassy carbon electrode. Int J Electrochem Sci 15(3):2727–2738
Liang Q, Huang Y, Wang M, Kuang D, Yang J, Yi Y, Shi H, Li J, Yang J, Li G (2023) An electrochemical biosensor for SARS-CoV-2 detection via its papain-like cysteine protease and the protease inhibitor screening. Chem Eng J 452:139646. https://doi.org/10.1016/j.cej.2022.139646
Luo F, Chen F, Xiong Y, Wu Z, Zhang X, Wen W, Wang S (2021) Single-particle electrochemical biosensor with DNA walker amplification for ultrasensitive HIV-DNA counting. Anal Chem 93(10):4506–4512
Ma M, He L, Shi X, Wang Y, Hai H, Wei X (2023) Ultrasensitive detection of HIV DNA using an electrochemical biosensor with branched DNA amplification. Int J Electrochem Sci 18(10):100286
Moco AC, Neto JA, de Moraes DD, Guedes PH, Brussasco JG, Flauzino JM, Luz LF, Soares MM, Madurro JM, Brito-Madurro AG (2021) Carbon ink-based electrodes modified with nanocomposite as a platform for electrochemical detection of HIV RNA. Microchem J 170:106739
Mohan B, Kumar S, Xi H, Ma S, Tao Z, Xing T, You H, Zhang Y, Ren P (2022) Fabricated metal-organic frameworks (MOFs) as luminescent and electrochemical biosensors for cancer biomarkers detection. Biosens Bioelectron 197:113738. https://doi.org/10.1016/j.bios.2021.113738
Ojha RP, Singh P, Azad UP, Prakash R (2022) Impedimetric immunosensor for the NS1 dengue biomarker based on the gold nanorod decorated graphitic carbon nitride modified electrode. Electrochim Acta 411:140069
Oliveira ME, Lopes BV, Rossato JHH, Maron GK, Gallo BB, La Rosa AB, Balboni RDC, Alves MLF, Ferreira MRA, da Silva Pinto L, Conceição FR, Piva E, de Pereira CMP, Escote MT, Carreño NLV (2022) Electrochemical biosensor based on laser-induced graphene for COVID-19 diagnosing: rapid and low-cost detection of SARS-CoV-2 biomarker antibodies. Surfaces 5(1):187–201. https://doi.org/10.3390/surfaces5010012
Pividori MI, Merkoci A, Alegret S (2000) Electrochemical genosensor design: immobilisation of oligonucleotides onto transducer surfaces and detection methods. Biosens Bioelectron 15(5–6):291–303
Pohanka M (2020) Screen printed electrodes in biosensors and bioassays. A review. Int J Electrochem Sci 15:11024–11035

Purohit B, Vernekar PR, Shetti NP, Chandra P (2020) Biosensor nanoengineering: design, operation, and implementation for biomolecular analysis. Sens Int 1:100040

Putzbach W, Ronkainen NJ (2013) Immobilization techniques in the fabrication of nanomaterial-based electrochemical biosensors: a review. Sensors 13(4):4811–4840

Ramanavicius A, Finkelsteinas A, Cesiulis H, Ramanaviciene A (2010) Electrochemical impedance spectroscopy of polypyrrole based electrochemical immunosensor. Bioelectrochemistry 79(1):11–16

Reyes-Retana JA, Duque-Ossa LC (2021) Acute myocardial infarction biosensor: a review from bottom up. Curr Probl Cardiol 46(3):100739

Ruff P, Pai RB, Storici F (2012) Real-time PCR-coupled CE-SELEX for DNA aptamer selection. ISRN Mol Biol 2012:939083

Sabahi A, Salahandish R, Ghaffarinejad A, Omidinia E (2020) Electrochemical nano-genosensor for highly sensitive detection of miR-21 biomarker based on SWCNT-grafted dendritic Au nanostructure for early detection of prostate cancer. Talanta 209:120595. https://doi.org/10.1016/j.talanta.2019.120595

Sassolas A, Blum LJ, Leca-Bouvier BD (2009) Electrochemical aptasensors. Electroanalysis 21(11):1237–1250

Shao S, Shao Y (2023) From a deadly disease to a manageable chronic disease, HIV/AIDS remains a challenge for mankind. Infect Microbes Dis 5(2):41–43

Sharma A, Sharma RK (2013) Aptamers—a promising approach for sensing of biothreats using different bioinformatics tools. Soft Nanosci Lett 03(04):1–5

Sharma P, Hassan H, Hasan MR, Fatima T, Mohan H, Khanuja M, Kaushik S, Narang J (2023) PBIS-based system integrated with zinc–silver nanocomposite for the detection of chikungunya virus. Biosens Bioelectron X 13:100303

Singhal C, Shukla SK, Jain A, Pundir C, Khanuja M, Narang J, Shetti NP (2020) Electrochemical multiplexed paper nanosensor for specific dengue serotype detection predicting pervasiveness of DHF/DSS. ACS Biomater Sci Eng 6(10):5886–5894

Sun D, Luo Z, Lu J, Zhang S, Che T, Chen Z, Zhang L (2019) Electrochemical dual-aptamer-based biosensor for nonenzymatic detection of cardiac troponin I by nanohybrid electrocatalysts labeling combined with DNA nanotetrahedron structure. Biosens Bioelectron 134:49–56

Suresh RR, Lakshmanakumar M, Arockia Jayalatha JB, Rajan KS, Sethuraman S, Krishnan UM, Rayappan JB (2021) Fabrication of screen-printed electrodes: opportunities and challenges. J Mater Sci 56:8951–9006

Tabrizi MA, Shamsipur M, Farzin L (2015) A high sensitive electrochemical aptasensor for the determination of VEGF165 in serum of lung cancer patient. Biosens Bioelectron 74:764–769

Thakur D, Fatima T, Sharma P, Hasan MR, Malhotra N, Khanuja M, Shukla SK, Narang J (2023) High-performance biosensing systems for diagnostics of sexually transmitted disease—a strategic review. Process Biochem 126:223–237

Tomichan R, Sharma A, Akash K, Siddiqui AA, Dubey A, Upadhyay TK, Kumar D, Pandey S, Nagraik R (2023) Insight of smart biosensors for COVID-19: a review. Luminescence 38(7):1102–1110. https://doi.org/10.1002/bio.4430

Tromberg BJ, Sepaniak MJ, Vo-Dinh T, Griffin GD (1987) Fiber-optic chemical sensors for competitive binding fluoroimmunoassay. Anal Chem 59(8):1226–1230

Wang Z, Zhao H, Chen K, Cao S, Lan M (2023) Sandwich-type electrochemical aptasensor based on HMCS@PDA@AuNPs and PtCu DNs/MUN-CuO-TiO_2 for ultrasensitive detection of cardiac troponin I. Sensors Actuators B Chem 393:134275

Wehmeyer KR, White RJ, Kissinger PT, Heineman WR (2021) Electrochemical affinity assays/sensors: brief history and current status. Annu Rev Anal Chem 14:109–131

Wu CC, Yen HY, Lai LT, Perng GC, Lee CR, Wu SJ (2020) A label-free impedimetric genosensor for the nucleic acid amplification-free detection of extracted RNA of dengue virus. Sensors 20(13):3728

Zhang Y, Li M, Gao X, Chen Y, Liu T (2019) Nanotechnology in cancer diagnosis: progress, challenges and opportunities. J Hematol Oncol 12(1):137. https://doi.org/10.1186/s13045-019-0833-3

Zhang S, Rong F, Guo C, Duan F, He L, Wang M, Zhang Z, Kang M, Du M (2021) Metal–organic frameworks (MOFs) based electrochemical biosensors for early cancer diagnosis in vitro. Coord Chem Rev 439:213948. https://doi.org/10.1016/j.ccr.2021.213948

3 Golden Threads of Detection: Unveiling Electrochemical DNA Sensors Empowered by Gold Nanoparticles

Parul Narwal and Neelam Prabha Negi

Abstract

In recent years, the intersection of nanotechnology and biosensing has sparked remarkable advancements in the field of DNA detection. Electrochemical DNA sensors, distinguished for their simplicity, accuracy, and cost-effectiveness, have emerged as indispensable tools in this pursuit. This chapter delves into the multifaceted role of gold nanoparticles (AuNPs) in advancing DNA detection methodologies.

The discussion centres around the manifold functions of AuNPs, encompassing DNA immobilization, electron transfer facilitation, and signal amplification. Emphasis is placed on elucidating their effectiveness as labels in electrochemical sensors, concurrently accentuating their potential for surface modification to augment detection limits. The catalytic and conductive attributes of gold are harnessed to underscore the rationale behind the enhanced performance of AuNP-empowered sensors. Furthermore, this chapter provides insights into the practical applications of these sensors, focusing on their roles in clinical diagnostics, environmental monitoring, and genetic research. The culmination of this exploration exemplifies the potential of gold nanoparticle-based electrochemical DNA sensors to revolutionize molecular diagnostics and biosensing applications.

Keywords

Gold nanoparticles · Electrochemical sensor · DNA sensor · Biosensors · Electron transfer mediator

---

P. Narwal · N. P. Negi (✉)
University Institute of Biotechnology, Chandigarh University, Mohali, Punjab, India

M. Mohsin, N. Soleja (eds.), *Nanoscale Sensors and their Applications in Biomedical Imaging*, https://doi.org/10.1007/978-981-97-3144-2_3

3.1 Introduction

A biosensor is a device that utilizes biological molecules to detect and analyse specific substances (Negi and Choephel 2020). The demand for an affordable and highly sensitive DNA detection method for early-stage genetic disease diagnosis is substantial (He et al. 2005). DNA biosensors, integrating sequence-specific probes with transducers, offer a promising solution (Paleček et al. 2002). Nano-biosensors, specifically electrochemical sensors, combine nanotechnology and biosensing principles to detect and quantify biological molecules by measuring changes in electrical properties. These sensors utilize nanoparticles to enhance sensitivity, selectivity, and speed of detection. They have the potential to revolutionize healthcare, environmental monitoring, and various industries. Ongoing research in nanotechnology and biosensing is likely to lead to further advancements in sensor design, sensitivity, and specificity, expanding their applications and impact (Javaid et al. 2021). For instance, electrochemical biosensors can be designed to measure substances like glucose, lactate, catecholamines, nucleic acids, uric acid, and many others. These biosensors work by exploiting the unique electrochemical properties of the target molecules, often relying on enzymes or other biorecognition elements to enhance specificity and sensitivity. The advancement of electrochemical biosensors has yielded substantial progress in diverse domains such as medical diagnostics and environmental monitoring. For instance, electrochemical biosensors have been utilized for real-time glucose monitoring in diabetes management, enabling patients to monitor their blood glucose levels continuously and adjust their treatment accordingly (Hammond et al. 2016). This real-time monitoring capability has significantly improved the quality of life for diabetic patients and enhanced their overall health management. Furthermore, the portability and simplicity of electrochemical biosensors have enabled their deployment in remote and resource-limited settings. An illustrative example is the utilization of electrochemical biosensors to check the water quality monitoring in developing regions. These biosensors can rapidly detect contaminants and pollutants in water sources, providing valuable information for ensuring safe drinking water and preventing waterborne diseases (Murray et al. 2018). The study conducted by Kilic et al. (2023) further emphasizes the potential of electrochemical biosensors in environmental applications. Their research illustrates how a portable electrochemical biosensor could possibly be designed to detect heavy metals in water samples. The biosensor demonstrated great specificity and sensitivity, which make it a potential device for on-site environmental monitoring.

Electrochemical DNA detection methods provide an economical and straightforward approach, yielding direct electronic signals without the need for expensive transduction equipment. These electrochemical DNA sensors serve as robust tools in molecular diagnostics and biosensing, facilitating rapid, sensitive, and accurate identification of specific DNA sequences. Their applications are pivotal in clinical diagnostics, to environmental monitoring and genetic research (Grabowska et al. 2014; Liu et al. 2008; Wang 2002). The straightforward immobilization of single-stranded DNA (ssDNA) probes onto versatile electrode substrates adds to their appeal (Feng et al. 2015; Jayakumar et al. 2012; Zhu et al. 2015).

The incorporation of nanomaterials in electrochemical DNA sensors enhances their performance, allowing for improved detection limits and selectivity (Kurkina 2012; Rasheed and Sandhyarani 2017). These nanomaterials, including magnetic nanoparticles (MNPs), quantum dots (QDs), carbon-based nanomaterials, metal nanoparticles (MNPs), and polymeric NPs, contribute to the development of more efficient sensors (Kurkina 2012; Rasheed and Sandhyarani 2017). By harnessing the principles of electrochemistry, these sensors facilitate the direct transduction of DNA hybridization events into measurable electrical signals. This conversion of biological recognition events into quantifiable electrochemical responses has paved the way for innovative sensing platforms with versatile applications. There are several methods used in electrochemical DNA sensing, including direct as well as indirect DNA-specific redox indicator detection, DNA electrochemistry, nanoparticle-based electrochemical signal amplification, and electron-mediated charge transport (Drummond et al. 2003). Among the commonly employed techniques are chronoamperometry, differential pulse voltammetry (DPV), cyclic voltammetry, and electrochemical impedance spectroscopy (EIS), each offering unique advantages for DNA detection. Cyclic voltammograms reveal distinctive peaks that indicate DNA hybridization, while DPV enhances the resolution of these peaks. Additionally, chronoamperometry provides insight into current values over defined time intervals, and impedimentary enables the assessment of impedance variations across different frequency ranges. These strategies collectively contribute to the robustness and versatility of electrochemical DNA sensors, enabling them to fulfil various analytical needs.

To overcome the challenge of irreversible oxidation and reduction of nucleic acids at electrodes, various labels have been employed to generate electrochemical signals reflecting DNA concentration (Wang 2006). Nanomaterial-based labels, particularly gold nanoparticles (AuNPs), have garnered attention for their superior performance (Cao et al. 2011). Gold nanoparticles possess unique physicochemical properties that make them highly attractive for sensor development. Their remarkable optical, electronic, and catalytic properties, combined with their facile functionalization, have made them extensively used as signal-enhancing elements in DNA sensing platforms. DNA hybridization detection frequently uses AuNPs as oligonucleotide labels due to their ability to enhance signal sensitivity and specificity (Saha et al. 2012). Electrode substrates, notably glassy carbon electrodes (GCE) and gold electrodes, significantly influence sensor responses (Lucarelli). Utilizing nanomaterials like AuNPs, carbon nanotubes (CNTs), graphene and their derivatives enhances the modification of these substrates, improving detection limits and sensor stability (Shi et al. 2014; Zhang et al. 2012). The integration of gold nanoparticles into electrochemical DNA sensors has not only improved the sensitivity and selectivity of detection but has also enabled the development of miniaturized and portable sensing devices for point-of-care applications.

The utilization of gold nanoparticles offers several distinct advantages in the design of electrochemical DNA sensors. These nanoparticles provide a large surface area for immobilizing DNA probes, facilitating the efficient capture of target DNA molecules. Furthermore, their unique optical properties, such as surface plasmon resonance, enable label-free detection and real-time monitoring of DNA hybridization events. The inherent biocompatibility of gold nanoparticles further ensures minimal interference with biological samples, making them suitable for various clinical and biomedical applications.

This chapter aims to provide an in-depth exploration of electrochemical DNA sensors that leverage the capabilities of gold nanoparticles. We will delve into the principles underlying electrochemical DNA sensing, highlighting the role of electrode interfaces and DNA hybridization mechanisms. Additionally, we will examine the distinct attributes of gold nanoparticles that contribute to their significance in the development of advanced DNA sensing platforms.

3.2 Gold Nanoparticle in Biosensing

Gold nanoparticles (AuNPs) have garnered substantial attention in the field of biosensing due to their remarkable physicochemical properties and versatile applications. These nanoparticles, typically ranging from 1 to 100 nm in size, exhibit unique optical, electronic, and catalytic characteristics that make them particularly well-suited for sensor development (Dreaden et al. 2012; Sperling and Parak 2010). Their tuneable optical properties, notably the phenomenon of localized surface plasmon resonance (LSPR), enable them to interact with light in a size-dependent manner, resulting in distinct colour changes or shifts in absorbance spectra (Haes et al. 2004). This property has been harnessed for label-free detection and quantification of biomolecules, including DNA. AuNPs also possess a high surface area-to-volume ratio, facilitating efficient functionalization and immobilization of biomolecules like DNA probes (Rosi and Mirkin 2005). Their unique electronic properties allow for facile electron transfer, crucial in electrochemical detection and catalytic reactions, making them attractive elements for enhancing sensor performance (Saldan et al. 2018). Additionally, AuNPs can be easily functionalized with various ligands, including thiolate DNA, peptides, and polymers, enabling tailored interaction with biomolecular targets (Alkilany and Murphy 2010). The applications of AuNPs in biosensing are diverse and span across multiple domains. They find utility in medical diagnostics, environmental monitoring, and food safety assessment (Huang et al. 2007). In medical diagnostics, AuNPs have been employed for the detection of biomarkers associated with diseases such as cancer, cardiovascular disorders, and infectious diseases (Cordeiro et al. 2016). In environmental monitoring, AuNPs have been utilized for the detection of pollutants, heavy metals, and pathogens (Li et al. 2020). Additionally, in food safety assessment, AuNPs have been integrated into sensors for the detection of harmful additives and contaminants in food products (Han et al. 2016).

Integration of gold nanoparticles into DNA sensors has led to significant improvements in the sensitivity, specificity, and versatility of detection platforms. The high surface area of AuNPs allows for effective immobilization of DNA probes, enhancing the capture of target DNA sequences (Pividori et al. 2000). Moreover, AuNPs can be functionalized with DNA strands to create nanoprobes that exhibit complementary base pairing with target sequences, leading to the formation of stable duplexes (Mirkin et al. 2020). The use of AuNPs in DNA sensors often involves exploiting their aggregation behaviour. Upon target DNA binding, the interparticle distances change, leading to aggregation and shifts in optical properties, such as colour changes or altered absorbance spectra (Zhang et al. 2006a, b). This aggregation-induced signal change can be easily detected and quantified, enabling sensitive DNA detection (Ren et al. 2013). In addition to colorimetric changes, AuNPs have been integrated into electrochemical DNA sensors. Their excellent electronic properties facilitate efficient charge transfer, which can be exploited for transducing DNA hybridization events into measurable electrical signals (Albrecht et al. 2003). The presence of AuNPs enhances the conductivity of the sensor surface, leading to improved signal-to-noise ratios and lower limits of detection (Zhang et al. 2009). The integration of AuNPs into DNA sensors has broadened their applicability in various fields. For instance, in medical diagnostics, AuNP-enhanced DNA sensors have enabled rapid and sensitive detection of disease-related genetic mutations. In environmental monitoring, these sensors have facilitated the identification of specific DNA sequences associated with pathogens or pollutants (Dai et al. 2021). Their use in food safety assessment has allowed for the detection of genetically modified organisms and pathogens in food products (Soni and Jayaprakash 2023).

3.3 Gold Nanoparticle-Based Electrochemical Sensors

Gold nanoparticles are made of gold that have dimensions on the nanoscale. These materials exhibit unique properties due to their small size and the surface plasmon resonance phenomenon, which makes them highly versatile and valuable in various applications across fields such as nanotechnology, materials science, medicine, agricultural science, and electronics (Roy et al. 2022; Mir et al. 2023). Some common types of gold nanoparticles include gold nanoparticles, gold nanowires, and gold nanoclusters (Corti et al. 2002).

Gold nanoparticle-based electrochemical sensors are a type of analytical device that utilizes nanoscale gold materials to detect and quantify various chemical or biological species through electrochemical reactions. These sensors take advantage of the unique properties of gold nanoparticles (NPs) and other gold-based nanoparticles to enhance the sensitivity, selectivity, and overall performance of the sensor (Neri et al. 2019; Pingarrón et al. 2008; Yanez-Sedeno and Pingarron 2005).

3.3.1 Gold Nanoparticle-Based Electrochemical Immunosensors

Gold nanoparticle (AuNP)-based electrochemical immunosensors are specialized devices that utilize the unique properties of gold nanoparticles to detect specific target molecules, such as antigens or antibodies, through electrochemical signals. These sensors play a significant role in a variety of fields, including medical diagnostics, environmental monitoring, and food safety testing (Rastogi et al. 2022).

AuNPs have distinctive optical and electrochemical properties due to their small size and unique electronic structure. When AuNPs are functionalized with specific biomolecules like antibodies or antigens, they can bind to corresponding target molecules. This binding event induces changes in the properties of the AuNPs, such as their conductivity, surface charge, and redox properties (Pingarrón et al. 2008; Rasheed and Sandhyarani 2017). These changes can be measured using electrochemical techniques to quantitatively determine the presence and concentration of the target molecules. Also, the surface of AuNPs can be functionalized with biomolecules to create specific recognition sites for target molecules. For example, antibodies can be immobilized onto the AuNPs' surface to form antibody-functionalized AuNPs, allowing the immunosensor to selectively detect antigens (Hammond et al. 2016; Prinith and Manjunatha 2022). In a typical electrochemical immunosensor setup, the target antigen in the sample binds to the antibody-functionalized AuNPs, forming antigen–antibody complexes. The extent of binding is proportional to the concentration of the target antigen in the sample. Also, changes in the electrochemical properties of the AuNPs due to antigen–antibody binding are measured using techniques such as cyclic voltammetry, chronoamperometry, or impedance spectroscopy. These changes are converted into measurable electrical signals, which can be correlated to the concentration of the target antigen in the sample. AuNP-based electrochemical immunosensors offer several advantages, including high sensitivity, selectivity, and the ability to detect multiple targets simultaneously. The unique optical properties of AuNPs can also be exploited for colorimetric readouts, enabling visual detection in addition to electrochemical measurements (Brett 2022). AuNP-based electrochemical immunosensors combine the specific binding capabilities of antibodies with the unique properties of gold nanoparticles to provide highly sensitive and selective detection of target molecules. These sensors contribute to advancements in diagnostics and monitoring, offering rapid and accurate results for a wide range of applications.

3.3.2 Gold Nanoparticle-Based Electrochemical Enzyme Sensors

Enzymes were the first particular molecules utilized as biosensors, and they continue to remain an foundational element in this field today (Kilic et al. 2023). Enzymes can be selected based on their ability to interact with a specific analyte and catalyse a reaction that generates a measurable signal. The choice of enzyme depends on the target analyte and the desired application of the biosensor (Aldewachi et al. 2018; Musa et al. 2023). Enzymes are indeed essential molecules in living organisms, and they play a crucial role in accelerating various biochemical processes. Enzymatic biosensors are specialized devices that utilize enzymes to detect and quantify specific analytes in samples. These biosensors work by integrating an enzyme with a transducer, which is a device that converts a biological signal into an electronic or optical signal. The enzyme acts as a bioreceptor, interacting with the target analyte present in the sample. When the enzyme catalyses a reaction involving the analyte, a detectable signal is generated by the transducer, and this signal is proportional to the amount of analyte present in the sample. This allows for the accurate measurement of analyte concentrations. These biosensors are also used in various fields, including medical diagnostics, environmental monitoring, and food analysis. Their specificity and sensitivity make them valuable tools for detecting a wide range of substances in complex samples (Kilic et al. 2023).

Additionally, electrochemical analysis is a method to determine the types of enzymes used in a biosensor and whether they function through oxidation or reduction reactions. This refers to the electrochemical processes involved in the catalytic reactions facilitated by enzymes in biosensors. By understanding the electrochemical behaviour of the enzyme–substrate interaction, researchers can optimize the biosensor's performance and sensitivity (Chen et al. 2008).

Enzyme-based electrochemical biosensors indeed benefit from the high catalytic activity and selectivity of enzymes. These attributes, combined with the availability of commercially purified enzymes, contribute to the success and advancement of these bioanalytical devices. In electrochemical enzyme biosensors, the detection principle relies on observing electroactive species produced as a result of redox reactions (Neri et al. 2019). These reactions occur either directly at the electrode surface or with the assistance of a mediator. This approach has led to the creation of notable devices like the enzyme immunoassay system and the widely recognized glucose biosensor. These devices have found use in self-monitoring blood glucose levels and are often designed in disposable formats for ease of use. Many variations of such devices are currently available on the market.

Enzyme biosensors offer several benefits. These advantages include the ability to amplify the biosensor's response by controlling enzyme activity in response to a specific analyte, a wide array of commercially accessible enzymes with established properties, the potential to modify enzyme catalytic properties or substrate specificity using genetic engineering techniques, and the availability of materials for consistent supply. For enzyme-based biosensors to detect the analyte accurately, presence of the target analyte in the system is must, leading to observable changes such as alterations in colour, mass, light absorption, composition, or emission. These changes subsequently lead to variations in the optical or electrical signals emitted by the biosensor. The intensity of these signals is directly proportional to the amount of the target analyte (Chang et al. 2019). The development of affordable enzyme electrodes and small meters for at-home blood glucose monitoring has been significantly aided by the introduction of screen-printed enzyme electrodes This innovation offers the advantages of simplicity, portability, and continuous operation (Lee et al. 2022). Overall, enzymes and enzymatic biosensors are integral components of modern analytical techniques, enabling rapid and accurate detection of various substances in a wide range of samples.

3.4 Fabrication of Electrochemical DNA Sensors

Fabrication of electrochemical DNA sensors involves the integration of sequence-specific probes with transducers, creating a powerful platform for the detection of DNA sequences (Palecek and Fojta 2005). These sensors offer distinct advantages in terms of simplicity, sensitivity, and direct electronic signal generation, making them valuable tools in molecular diagnostics and biosensing applications (Salamifar and Lai 2014). One of the critical components of electrochemical DNA sensors is the design of the sensing interface, which enables the selective capture and recognition of target DNA molecules.

The process of fabricating electrochemical DNA sensors typically includes several key steps:

3.4.1 Probe Immobilization

Probe immobilization is a fundamental step in the fabrication of electrochemical DNA sensors, playing a critical role in ensuring the accurate and selective recognition of complementary DNA sequences (Marchuk and Willets 2014). This process involves anchoring single-stranded DNA (ssDNA) probes onto the sensor surface, forming a stable interface for target DNA hybridization. Several techniques have been employed for probe immobilization, each offering distinct advantages based on the specific sensor design and application:

3.4.1.1 Self-Assembly

Self-assembly relies on the spontaneous adsorption of ssDNA probes onto the sensor surface through weak interactions, such as van der Waals forces and hydrogen bonding. This technique is relatively simple and does not require extensive chemical modifications, making it suitable for rapid sensor development (Paleček et al. 1998).

3.4.1.2 Covalent Attachment

Covalent attachment involves chemically linking ssDNA probes to the sensor surface using reactive functional groups. This approach provides strong and stable attachment, minimizing probe detachment during hybridization events. Techniques like salinization or functionalization of gold surfaces with thiol groups are commonly used for covalent immobilization (Nguyen et al. 2015).

3.4.1.3 Hybridization with Capture Probes

In this method, complementary capture probes are pre-immobilized on the sensor surface. These capture probes selectively hybridize with target DNA sequences, ensuring specific recognition. This approach enhances the binding efficiency and minimizes nonspecific interactions (Rashid and Yusof 2017).

3.4.1.4 Biotin-Avidin Binding

Biotinylated DNA probes can be immobilized onto sensor surfaces functionalized with avidin. The strong biotin–avidin interaction provides stable attachment, allowing for efficient target DNA recognition (Wan et al. 2013).

3.4.1.5 Layer-by-Layer Assembly

This technique involves the sequential deposition of alternating layers of ssDNA and polycations on the sensor surface. Each layer contributes to enhanced stability and specific binding, making it suitable for multi-layered sensor architectures (Zhang et al. 2006a, b).

3.4.2 Signal Amplification

Signal amplification is a pivotal aspect of electrochemical DNA sensor fabrication aimed at enhancing sensitivity in the detection process. By integrating various amplification strategies into the sensor design, the electrochemical signal generated upon DNA hybridization can be significantly amplified, enabling the detection of target DNA molecules even at low concentrations (Kumar et al. 2019). Several amplification methods have been incorporated into electrochemical DNA sensor designs:

3.4.2.1 Enzyme Labels

Enzymes conjugated to DNA probes serve as amplification labels. Upon hybridization, the enzymatic reaction catalysed by the enzyme label generates a detectable electrochemical signal. This approach enhances the signal intensity and enables the detection of even trace amounts of target DNA (Rashid and Yusof 2017).

3.4.2.2 Nanoparticles

Nanoparticles, such as gold nanoparticles (AuNPs) or magnetic nanoparticles (MNPs), are functionalized with DNA probes and introduced into the sensor system. Upon DNA hybridization, the nanoparticles aggregate, leading to an amplified signal due to the increased electrode surface area for electrochemical detection (Wang 2002).

3.4.2.3 Catalytic Reactions

Catalytic reactions involving redox-active species, such as the oxidation of a substrate by an enzyme or the reduction of metal ions, are utilized to amplify the electrochemical signal. The catalytic process leads to a higher concentration of electroactive species, resulting in enhanced current signals (Popovich and Thorp 2002).

3.4.2.4 Rolling Circle Amplification (RCA)

RCA is a DNA amplification technique where a circular DNA template is enzymatically replicated in a highly targeted manner. This amplifies the number of DNA sequences complementary to the target, resulting in a proportional increase in the electrochemical signal (Goda et al. 2015).

3.4.2.5 Hybridization Chain Reaction (HCR)

HCR involves the sequential hybridization of two DNA probes that trigger a chain reaction of hybridization events. This results in the formation of long DNA structures, amplifying the electrochemical signal upon target DNA binding (Borum and Jokerst 2021).

3.4.3 Transduction Mechanism

The transduction mechanism serves as a crucial bridge between the DNA hybridization event and the measurable electrochemical signal in the fabrication of electrochemical DNA sensors (Palecek and Fojta 2005). This process involves translating the specific recognition of target DNA into quantifiable changes in redox properties, impedance, or conductivity at the sensor surface. Several common transduction methods are employed in electrochemical DNA sensor design:

3.4.3.1 Redox Property Changes

Changes in redox properties occur due to the electrochemical activity of redox-active species attached to DNA probes or involved in the hybridization process. Upon target DNA binding, changes in the electrochemical response of redox species are detected, providing an indication of the hybridization event (Paleček et al. 1998).

3.4.3.2 Impedance Spectroscopy

Impedance spectroscopy involves measuring changes in the impedance of the electrode–solution interface caused by DNA hybridization. The binding of target DNA alters the charge transfer resistance, capacitance, and frequency-dependent behaviour of the system, which can be analysed to deduce the presence of the target DNA (Park and Park 2009).

3.4.3.3 Conductivity Changes

Changes in the electrical conductivity of the sensor surface are observed upon DNA hybridization. This can be attributed to the change in charge transport properties due to the binding of target DNA molecules. Monitoring these conductivity changes enables the detection of DNA hybridization events (Brett 2022).

3.4.3.4 Field-Effect Transistors (FETs)

FET-based transduction involves integrating DNA probes onto the gate of a field-effect transistor. Upon hybridization, the presence of target DNA modulates the conductance of the transistor channel, leading to changes in the drain current, which serves as the electrochemical signal (Stern et al. 2007).

3.4.3.5 Optical Readout

Optical transduction methods involve using labelled DNA probes that emit fluorescence, chemiluminescence, or absorb light upon hybridization. The intensity of the emitted or absorbed light is proportional to the concentration of the bound target DNA (Xu et al. 2018).

3.4.4 Nanomaterial Integration

The integration of nanomaterials is a pivotal aspect of electrochemical DNA sensor fabrication, where materials like gold nanoparticles (AuNPs), carbon nanotubes (CNTs), and graphene are seamlessly integrated into the sensor design to enhance sensitivity, selectivity, and stability (Lohcharoenkal et al. 2021). These nanomaterials play a vital role in improving the electrochemical properties of the sensor surface, thereby contributing to enhanced signal-to-noise ratios and overall sensor performance. Nanomaterial integration offers several key advantages:

3.4.4.1 Enhanced Electrochemical Properties

The unique physicochemical properties of nanomaterials, such as high surface area and excellent conductivity, synergistically enhance the electrochemical behaviour of the sensor. This results in improved charge transfer kinetics and increased sensitivity for detecting DNA hybridization events (Malik et al. 2021).

3.4.4.2 Signal Amplification

Nanomaterials serve as excellent platforms for immobilizing DNA probes and amplifying the electrochemical signal. Their large surface area allows for higher probe loading, leading to greater target DNA capture and stronger signal generation upon hybridization (Cajigas and Orozco 2020).

3.4.4.3 Improved Stability

The incorporation of nanomaterials provides a stable scaffold for the attachment of DNA probes, preventing their degradation and ensuring long-term sensor stability. This is crucial for consistent and reliable sensing performance (Ibrahim et al. 2021).

3.4.4.4 Enhanced Selectivity

Nanomaterials can be functionalized with specific recognition elements, such as aptamers or ligands, to impart high selectivity for the target DNA sequence. This reduces nonspecific binding and enhances the accuracy of DNA detection (Malik et al. 2021).

3.4.4.5 Multi-Modal Sensing

Nanomaterials can serve as versatile platforms that enable multi-modal sensing, combining different transduction mechanisms for improved analytical capabilities. For example, AuNPs can be integrated into a sensor design to enhance both redox-based and optical sensing modalities (Jeerapan and Poorahong 2020).

3.4.5 Platform Detection

The determination of the detection platform is contingent upon the particular transduction mechanism employed and the type of electrochemical measurement utilized in the fabrication of electrochemical DNA sensors (Sadighbayan et al. 2019). Several techniques, including cyclic voltammetry, differential pulse voltammetry, chronoamperometry, and electrochemical impedance spectroscopy, are frequently employed for the quantitative evaluation of DNA hybridization events and their corresponding electrochemical signals (Rahman et al. 2015). These detection platforms offer distinct advantages:

3.4.5.1 Cyclic Voltammetry (CV)

CV involves applying a potential sweep to the electrode surface and measuring the resulting current response. The distinctive peaks observed in the cyclic voltammogram provide insights into the electrochemical behaviour of redox-active species involved in DNA hybridization, facilitating quantitative analysis (Taha et al. 2020).

3.4.5.2 Differential Pulse Voltammetry (DPV)

DPV enhances the sensitivity of cyclic voltammetry by employing a series of potential pulses at the electrode. This technique improves the resolution of current peaks, leading to improved signal-to-noise ratios and precise quantification of target DNA concentrations (Prinith and Manjunatha 2022).

3.4.5.3 Chronoamperometry

Chronoamperometry measures the current at a fixed potential over time, enabling real-time monitoring of DNA hybridization events. The change in current over time is indicative of the kinetics of DNA binding, contributing to quantitative analysis (Teles and Fonseca 2008).

3.4.5.4 Electrochemical Impedance Spectroscopy (EIS)

EIS involves measuring the impedance of the electrode–solution interface across a range of frequencies. The changes in impedance caused by DNA hybridization offer insights into the charge transfer resistance and capacitance changes, enabling the determination of target DNA concentration (Magar et al. 2021).

3.5 Application of Gold Nanoparticle-Based Nanosensors

Gold nanoparticle-based electrochemical DNA sensors have found diverse applications in various fields due to their exceptional sensitivity, selectivity, and versatility. By leveraging the unique properties of gold nanoparticles, these sensors offer promising solutions for detecting specific DNA sequences in clinical diagnostics, environmental monitoring, food safety, quality control, drug safety and development, and genetic research (Fig. 3.1). Some of the prominent applications of these sensors are as follows:

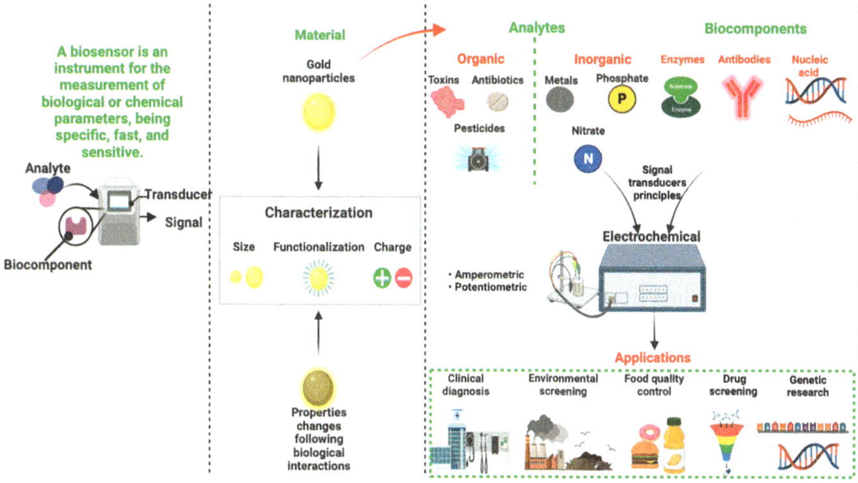

Fig. 3.1 Gold nanoparticle-based electrochemical sensors and applications

3.5.1 Clinical Diagnostics

Gold nanoparticles can be functionalized with biomolecules like antibodies, enzymes, or DNA to make extremely sensitive and precise biosensors. These sensors are capable of detecting biomarkers linked to disorders including diabetes, cancer, and infectious diseases. The relation between the biomolecules and the target analytes induces changes in the electrochemical properties, allowing for the detection and quantification of the analytes (Aldewachi et al. 2018; Rabbani et al. 2020).

AuNPs indeed offer a range of unique properties that make them incredibly attractive for use in diagnostics and therapeutics. AuNPs can be easily synthesized and their surfaces can be functionalized with various molecules, such as ligands, antibodies, and peptides. This allows for precise tuning of their properties and enables them to interact with specific targets in biological systems (Chandra et al. 2013). It also exhibits excellent biocompatibility, meaning they can be introduced into biological environments without causing significant harm. They are also highly sensitive to changes in their surroundings, which makes them suitable for sensing and detecting biomolecules.

AuNPs are being explored as versatile nanoplatforms for diagnostics and therapeutics. They can be functionalized with specific ligands to target and detect biomolecules with high sensitivity. Additionally, their ability to carry drugs or genetic material allows for targeted drug delivery and gene therapy applications. Despite their potential, there are still challenges to overcome. The behaviour of AuNPs in biological systems needs a deeper understanding to ensure their safe and effective use. Moreover, efforts to engineer novel AuNPs with enhanced properties are ongoing, aiming to address limitations of existing AuNPs (Chandra et al. 2013; Larguinho and Baptista 2012). Future developments in AuNPs could lead to enhanced biodistribution within the body, improving their efficiency in diagnostics and therapeutics. This could involve optimizing their size, shape, and surface properties for better interactions with biological systems. Advances in their electrical and optical characteristics could further enhance their performance in clinical applications (Aldewachi et al. 2018).

Tessaro has focused on using gold nanoparticles (AuNPs) for the rapid detection of Influenza viruses. Such a study would indeed be valuable in advancing our ability to detect and diagnose Influenza viruses quickly and accurately. Gold nanoparticles have unique optical properties that can be leveraged for biosensing applications, and their use in conjunction with biosensors holds great promise for improving the speed and efficiency of detection compared to traditional methods like ELISA and PCR.

3.5.2 Environmental Monitoring

Gold nanoparticle-based sensors can be used to detect pollutants, heavy metals, and toxins in the environment. They can be functionalized with receptors that specifically bind to the target analytes, leading to changes in the electrochemical signal. These sensors are particularly valuable for water quality monitoring and air pollution control (Rabbani et al. 2020).

Huber and Leopold present a comprehensive exploration of analytical strategies for the trace analysis of mercury in environmental and drinking water samples. They focused on utilizing nanomaterials (NMs) to enhance the efficiency and sensitivity of analytical methods and cover a range of techniques, encompassing both sample preparation and detection. These techniques involve nanomaterial-based sample preparation as well as optical and electrochemical detection methods. By evaluating the sensitivity, selectivity, real-water matrix compatibility, simplicity, and analysis duration, they assesses the suitability of these approaches for monitoring purposes.

In essence, Jessica Huber's highlights the remarkable potential of nanomaterials in revolutionizing the field of mercury tracer analysis. By offering improved efficiency, sensitivity, and simplicity in both sample preparation and detection, these approaches hold promise for enhancing our ability to monitor mercury levels in diverse water sources and ensuring the safety of drinking water and environmental quality.

3.5.3 Food Safety and Quality Control

Gold nanoparticle-based sensors are employed to detect contaminants and adulterants in food products. They can detect allergens, pathogens, and chemical contaminants, ensuring food safety and quality. These sensors are especially important for the food industry to comply with regulations and maintain consumer trust (Li and Sheng 2014; Rastogi et al. 2022).

Food additives serve as substances used in food processing to enhance quality, extend shelf life, and control colour and flavour. The presence of illegal additives, which are toxic and prohibited in food products, remains a concern, posing health risks to consumers. To address this issue, the development of sensors for detecting such harmful additives is imperative. This advanced sensing system is based on gold nanoparticles (AuNPs) for detecting melamine, preservatives, and food dyes (Zhang et al. 2018).

Melamine, a cyanamide trimer, is commonly used in various industries but has been illicitly added to animal feeds and dairy products to inflate protein content. Consumption of melamine-contaminated items can lead to severe kidney and reproductive system damage. Reliable detection methods are crucial to identify melamine. Its structure, featuring three amino groups, allows it to bind with 2–3 AuNPs through interactions with coating agents like citrate, thymine, and 1,4-dithiothreitol (DTT), leading to crosslinking aggregation. Some strategies combine crosslinking and deprotection techniques for colorimetric assays. To enhance sensitivity, innovative approaches have been proposed (Zhang et al. 2018).

For instance, Zhang et al. (2016) devised a "turn-off" FRET sensor by utilizing AuNP aggregation and conjugated polymer NPs as donors. Fluorescence signal was quenched without melamine, but its introduction caused AuNP aggregation, leading to fluorescence recovery with a low LOD of 1.7 nM. To broaden the dynamic detection range, Chen et al. (2006) stabilized AuNPs with polyethylene glycol to prevent colour fading due to rapid aggregation upon melamine introduction. This method covered a linear range from 1.05 nM to 1 mM.

Preservatives have a pivotal role in enhancing the shelf life and transportability of food products, yet illicit ones pose serious health risks. The consumption of toxic illegal preservatives, such as nitrite and formaldehyde, can lead to ailments like gastric cancer and leukaemia. While research on optical sensors for detecting food preservatives remains limited compared to electrochemical methods, several noteworthy developments have emerged. Martinez-Aquino et al. (2019) utilized resorcinol-functionalized AuNPs for formaldehyde detection, displaying a low LOD of 0.05 ppm. Nie et al. (2021) achieved formaldehyde detection via non-functionalized AuNPs in a SERS sensor, demonstrating an LOD of 1 × 10–4 μg mL^{-1} in real samples like duck blood and rice flour. To address the presence of thiabendazole, an unapproved EU food preservative and antifungal agent, Alsammarraie et al. (2018) designed a SERS sensor with enhanced SERS signal using Au nanorods as substrates. This sensor exhibited LODs of 149, 216, and 179 μg L^{-1} in lemon, carrot, and mango juice, respectively.

Food dyes impart appealing colours to various products, but their toxicity has led to the prohibition of several types, including Sudan dyes, Malachite Green, Allura Red, and Sunset Yellow. Malachite Green, functioning both as a dye and antimicrobial in aquaculture, requires detection in fish and seafood. Jia et al. (2022) harnessed RNA aptamer affinity to MG, utilizing aptamer-labelled AuNPs for MG detection through an anti-aggregation strategy. In the presence of MG, RNA aptamer preferentially bound to MG over protecting AuNPs, causing aggregation under high NaCl concentrations and changing the solution colour from red to blue. Zhao et al. (2019)

exploited a similar affinity, developing an indirect colorimetric sensor that employed CTAB as inhibitors for the peroxidase-like activity of AuNPs. The system displayed three colour states, correlating with varying levels of RNA aptamer and MG presence. These innovations offer potential for highly sensitive and specific detection, as evidenced by their LODs of 15.95 and 1.8 nM, respectively. The development of AuNP-based sensing systems addresses the critical need for detecting harmful food additives like melamine, preservatives, and food dyes. These methods offer enhanced sensitivity and expanded detection ranges, demonstrating their potential for improving food safety and quality control.

3.5.4 Drug Screening and Development

Gold nanoparticle-based sensors can be utilized to screen the interactions between drugs and their target molecules. This is crucial in drug discovery and development, as it helps identify potential drug candidates and assess their binding affinities. These sensors can also be used to study the kinetics of enzymatic reactions involved in drug metabolism (Aldewachi et al. 2018; Selli et al. 2022; Vo-Dinh et al. 2001).

The study conducted by Wu et al. (2020) described the development of colorimetric aptasensor, which provides a rapid and simple method for the simultaneous detection of two antibiotics: chloramphenicol (CAP) and tetracycline (TET), based on the unique aggregation responses of AuNPs induced by the specific binding of the multifunctional aptamer to these antibiotics. This type of sensor has potential applications in food safety testing and environmental monitoring where the presence of antibiotics in various samples needs to be assessed. In the presence of one antibiotic, a specific fragment of the aptamer recognized antibiotic and detached from the AuNP surface. However, the nonspecific fragment of the aptamer remained attached and controlled the aggregation of AuNPs. Different antibiotics caused distinct states of AuNP aggregation, leading to observable colour changes in the solutions. This system successfully enabled the separate detection of two different antibiotics, achieving low limits of detection (LODs) of 32.9 nM for TET and 7.0 nM for CAP.

Yan's group developed an even more intricate system for the detection of five types of aminoglycoside antibiotics. This system involved using two kinds of AuNPs, each labelled with a distinct aptamer. The distinct binding sites of the different antibiotics resulted in specific binding events that generated distinguishable signals. By utilizing the different responses of the aptamer-functionalized AuNPs, this system was capable of separately detecting five different types of antibiotics. These systems utilize the unique binding properties of aptamers and the aggregation behaviour of gold nanoparticles (AuNPs) to enable the detection of multiple antibiotics in a single assay.

However, both of these described sensing systems have limitations when it comes to detecting two or more antibiotics simultaneously. They were designed for separate detection of multiple antibiotics in a single assay. Detecting multiple antibiotics simultaneously in a single reaction is more challenging due to the need to distinguish different colorimetric responses or other signals from the various aptamer–antibiotic interactions.

Developing a system that can simultaneously detect multiple antibiotics in a single reaction requires careful design to ensure that the signals generated by each interaction are distinguishable from one another. This could involve differentiating the responses using varying nanoparticle sizes, aptamer sequences, or other innovative strategies. While the studies we have mentioned have made significant strides in multiplex antibiotic detection, simultaneously detecting multiple antibiotics in a single assay remains a complex challenge.

3.5.5 Genetic Research

Gold nanoparticles can be used in DNA and RNA detection through techniques like electrochemical DNA hybridization assays. These assays rely on the specific binding between DNA probes and target sequences, leading to changes in the electrochemical signal. This technology has applications in genetic testing, forensic analysis, and pathogen detection (Girigoswami and Girigoswami 2021; Zhou et al. 2019).

AuNPs exhibit a phenomenon called localized surface plasmon resonance (LSPR), which results in the absorption and scattering of light at specific wavelengths. The LSPR properties of AuNPs can be tuned by changing their size, shape, and the surrounding environment. This property forms the basis for their use in DNA/RNA detection (Rasheed and Sandhyarani 2017). To detect DNA/RNA, AuNPs are often functionalized with complementary DNA or RNA sequences. These sequences are designed to hybridize with the target DNA/RNA of interest. The binding of the target nucleic acid to the functionalized AuNPs induces changes in the LSPR properties, leading to detectable shifts in the absorption or scattering spectra of light. One common method for detecting DNA/RNA involves the colorimetric change of AuNPs. In this approach, when the target DNA/RNA binds to the functionalized AuNPs, the nanoparticles aggregate due to hybridization between the complementary sequences. This aggregation causes a shift in the LSPR band, resulting in a visible colour change from red to blue. This colour change can be observed with the naked eye or measured using spectrophotometry (Song et al. 2022).

AuNPs can also enhance the fluorescence of nearby fluorophores due to plasmon-enhanced electromagnetic fields. This property is utilized in fluorescence-based DNA/RNA detection assays. Fluorescently labelled DNA probes are designed to bind to the target DNA/RNA, and the fluorescence signal is enhanced when the probes are in close proximity to the AuNPs. AuNPs provide a platform for signal amplification in DNA/RNA detection. Multiple DNA/RNA targets can bind to a single AuNP, leading to the formation of nanoparticle clusters and further enhancing the signal change. This property allows for increased sensitivity in detecting low concentrations of nucleic acids (Paleček et al. 2002; Song et al. 2022).

Overall, the combination of gold nanoparticles' unique properties, the ease of functionalization, and their compatibility with electrochemical techniques makes them valuable tools for a wide range of applications in fields such as healthcare, environmental monitoring, and materials science.

3.6 Conclusion and Future Prospective

DNA/RNA sensors play a pivotal role in medical and clinical fields, driving the exploration of highly sensitive and selective DNA detection methods. Among these methods, electrochemical sensors stand out for their cost-effectiveness, sensitivity, and specificity in detecting specific DNA sequences. These sensors leverage nanoscale interactions between target sequences, recognition layers, and electrode surfaces to achieve their remarkable performance. The sensitivity of electrochemical sensors relies on effective electron transfer between the analyte and the electrode surface. Gold nanoparticles (AuNPs) have emerged as key players, enhancing sensitivity by acting as immobilization platforms or electrochemical labels. This review underscores the significant contribution of AuNPs in improving detection limits and sensor stability. Combining multiple nanomaterials, like graphene and AuNPs, has the potential to further boost sensing performance. While carbon nanostructures and AuNPs offer zeptomolar detection, alternative strategies such as self-assembled monolayers hold promise for multi-target detection.

Future development hinges on novel nanomaterials with well-defined structures and superior properties. Integration of 2D layered materials like graphene with AuNPs is expected to elevate sensor sensitivity. Beyond achieving single-molecule detection, the focus should shift to enhancing selectivity. This involves using electrochemical mediators that selectively bind to DNA bases in conjunction with nano systems.

Two primary strategies for AuNPs in electrochemical DNA sensing are adsorption onto electrode surfaces and functionalized use as labels. Future progress involves optimizing chemical interactions, conjugation with DNA, and simplifying fabrication procedures while minimizing toxic materials and complex detection methods. A significant challenge remains in detecting specific sequences from mixed populations. Developing sensors capable of detecting mutations, even at low ratios within wild-type sequences, holds great potential. Integration of sensor arrays and individual nanomaterial-based sensors could enable single-cell and molecule-level bio-detection, along with point-of-care diagnostics. Envisioning nanomaterial-based DNA sensors for in vivo studies is a tantalizing prospect. In conclusion, the integration of nanotechnology into DNA sensors, especially those empowered by AuNPs, opens avenues for revolutionizing diagnostics. Advances should focus on improving sensitivity, selectivity, miniaturization, and speed while harnessing nanotechnology for portable and user-friendly devices.

References

Albrecht C, Blank K, Lalic-Multhaler M, Hirler S, Mai T, Gilbert I, Schiffmann S, Bayer T, Clausen-Schaumann H, Gaub HE (2003) DNA: a programmable force sensor. Science 301(5631):367–370

Aldewachi H, Chalati T, Woodroofe MN, Bricklebank N, Sharrack B, Gardiner P (2018) Gold nanoparticle-based colorimetric biosensors. Nanoscale 10(1):18–33

Alkilany AM, Murphy CJ (2010) Toxicity and cellular uptake of gold nanoparticles: what we have learned so far? J Nanopart Res 12:2313–2333

Alsammarraie FK, Lin M, Mustapha A, Lin H, Chen X, Chen Y, Wang H, Huang M (2018) Rapid determination of thiabendazole in juice by SERS coupled with novel gold nanosubstrates. Food Chem 259:219–225

Borum RM, Jokerst JV (2021) Hybridizing clinical translatability with enzyme-free DNA signal amplifiers: recent advances in nucleic acid detection and imaging. Biomater Sci 9(2):347–366

Brett CMA (2022) Electrochemical impedance spectroscopy in the characterisation and application of modified electrodes for electrochemical sensors and biosensors. Molecules 27(5):1497

Cajigas S, Orozco J (2020) Nanobioconjugates for signal amplification in electrochemical biosensing. Molecules 25(15):3542

Cao X, Ye Y, Liu S (2011) Gold nanoparticle-based signal amplification for biosensing. Anal Biochem 417(1):1–16

Chandra P, Singh J, Singh A, Srivastava A, Goyal RN, Shim YB (2013) Gold nanoparticles and nanocomposites in clinical diagnostics using electrochemical methods. J Nanopart 2013:1–12

Chang C-C, Chen C-P, Wu T-H, Yang C-H, Lin C-W, Chen C-Y (2019) Gold nanoparticle-based colorimetric strategies for chemical and biological sensing applications. Nanomaterials 9(6):861

Chen S, Guo C, Hu G-H, Wang J, Ma J-H, Liang X-F, Zheng L, Liu H-Z (2006) Effect of hydrophobicity inside PEO-PPO-PEO block copolymer micelles on the stabilization of gold nanoparticles: experiments. Langmuir 22(23):9704–9711

Chen PC, Mwakwari SC, Oyelere AK (2008) Gold nanoparticles: from nanomedicine to nanosensing. Nanotechnol Sci Appl 1:45–65

Cordeiro M, Ferreira Carlos F, Pedrosa P, Lopez A, Baptista PV (2016) Gold nanoparticles for diagnostics: advances towards points of care. Diagnostics 6(4):43

Corti CW, Holliday RJ, Thompson DT (2002) Developing new industrial applications for gold: gold nanotechnology. Gold Bull 35:111–117

Dai X, Ren Q, Aydın S, Yardımcı MY, Lesage K, De Schutter G (2021) Enhancing thixotropy and structural build-up of alkali-activated slag/fly ash pastes with nano clay. Mater Struct 54(4):163

Dreaden EC, Alkilany AM, Huang X, Murphy CJ, El-Sayed MA (2012) The golden age: gold nanoparticles for biomedicine. Chem Soc Rev 41(7):2740–2779

Drummond TG, Hill MG, Barton JK (2003) Electrochemical DNA sensors. Nat Biotechnol 21(10):1192–1199

Feng X, Cheng H, Pan Y, Zheng H (2015) Development of glucose biosensors based on nanostructured graphene-conducting polyaniline composite. Biosens Bioelectron 70:411–417

Girigoswami K, Girigoswami A (2021) A review on the role of nanosensors in detecting cellular miRNA expression in colorectal cancer. Endocr Metab Immune Disord Drug Targets 21(1):12–26

Goda T, Tabata M, Miyahara Y (2015) Electrical and electrochemical monitoring of nucleic acid amplification. Front Bioeng Biotechnol 3:29

Grabowska I, Malecka K, Jarocka U, Radecki J, Radecka H (2014) Electrochemical biosensors for detection of avian influenza virus—current status and future trends. Acta Biochim Pol 61(3):471–478

Haes AJ, Stuart DA, Nie S, Van Duyne RP (2004) Using solution-phase nanoparticles, surface-confined nanoparticle arrays and single nanoparticles as biological sensing platforms. J Fluoresc 14:355–367

Hammond JL, Formisano N, Estrela P, Carrara S, Tkac J (2016) Electrochemical biosensors and nanobiosensors. Essays Biochem 60(1):69–80
Han G, Wang X, Hamel J, Zhu H, Sun R (2016) Lignin-AuNPs liquid marble for remotely-controllable detection of Pb^{2+}. Sci Rep 6(1):38164
He P, Xu Y, Fang Y (2005) A review: electrochemical DNA biosensors for sequence recognition. Anal Lett 38(15):2597–2623
Huang X, Jain PK, El-Sayed IH, El-Sayed MA (2007) Gold nanoparticles: interesting optical properties and recent applications in cancer diagnostics and therapy. Nanomedicine (Lond) 2(5):681–693
Ibrahim N, Jamaluddin ND, Tan LL, Mohd Yusof NY (2021) A review on the development of gold and silver nanoparticles-based biosensor as a detection strategy of emerging and pathogenic RNA virus. Sensors 21(15):5114
Javaid M, Haleem A, Singh RP, Rab S, Suman R (2021) Exploring the potential of nanosensors: a brief overview. Sens Int 2:100130
Jayakumar K, Rajesh R, Dharuman V, Venkatasan R, Hahn JH, Pandian SK (2012) Gold nano particle decorated graphene core first generation PAMAM dendrimer for label free electrochemical DNA hybridization sensing. Biosens Bioelectron 31(1):406–412
Jeerapan I, Poorahong S (2020) Flexible and stretchable electrochemical sensing systems: materials, energy sources, and integrations. J Electrochem Soc 167(3):37573
Jia H-R, Zhang Z, Fang X, Jiang M, Chen M, Chen S, Gu K, Luo Z, Wu F-G, Tan W (2022) Recent advances of cell surface modification based on aptamers. Mater Today Nano 18:100188
Kilic NM, Singh S, Keles G, Cinti S, Kurbanoglu S, Odaci D (2023) Novel approaches to enzyme-based electrochemical nanobiosensors. Biosensors 13(6):622
Kumar A, Purohit B, Maurya PK, Pandey LM, Chandra P (2019) Engineered nanomaterial assisted signal-amplification strategies for enhancing analytical performance of electrochemical biosensors. Electroanalysis 31(9):1615–1629
Kurkina T (2012) Label-free electrical biosensing based on electrochemically functionalized carbon nanostructures. Thesis-Universitätsbibliothek Giessen
Larguinho M, Baptista PV (2012) Gold and silver nanoparticles for clinical diagnostics—from genomics to proteomics. J Proteomics 75(10):2811–2823
Lee S-E, Jeong S-E, Hong J-S, Im H, Hwang S-Y, Oh JK, Kim S-E (2022) Gold-nanoparticle-coated magnetic beads for ALP-enzyme-based electrochemical immunosensing in human plasma. Materials 15(19):6875
Li Z, Sheng C (2014) Nanosensors for food safety. J Nanosci Nanotechnol 14(1):905–912
Li J, Yang L, Ruan Y, Chu S, Wang H, Li Z, Jiang C, Liu B, Yang L, Zhang Z (2020) Dual-mode optical nanosensor based on gold nanoparticles and carbon dots for visible detection of As (III) in water. ACS Appl Nano Mater 3(8):8224–8231
Liu X, Shi L, Niu W, Li H, Xu G (2008) Amperometric glucose biosensor based on single-walled carbon nanohorns. Biosens Bioelectron 23(12):1887–1890
Lohcharoenkal W, Abbas Z, Rojanasakul Y (2021) Advances in nanotechnology-based biosensing of immunoregulatory cytokines. Biosensors 11(10):364
Magar HS, Hassan RYA, Mulchandani A (2021) Electrochemical impedance spectroscopy (EIS): principles, construction, and biosensing applications. Sensors 21(19):6578
Malik P, Gupta R, Malik V, Ameta RK (2021) Emerging nanomaterials for improved biosensing. Meas Sens 16:100050
Marchuk K, Willets KA (2014) Localized surface plasmons and hot electrons. Chem Phys 445:95–104
Martinez-Aquino C, Costero AM, Gil S, Gaviña P (2019) Resorcinol functionalized gold nanoparticles for formaldehyde colorimetric detection. Nanomaterials 9(2):302
Mir MA, Bansal N, Sharma S, Negi NP (2023) Nanoparticle application for sustainable agriculture and post-harvest technology: advances and future perspective. Indian J Agric Biochem 36(1):10–25

Mirkin CA, Letsinger RL, Mucic RC, Storhoff JJ (2020) A DNA-based method for rationally assembling nanoparticles into macroscopic materials. In: Spherical nucleic acids. Jenny Stanford Publishing, pp 3–11

Murray JLK, Marques DFP, Cameron RL, Potts A, Bishop J, von Wissmann B, William N, Reynolds AJ, Robertson C, McMenamin J (2018) Moving epidemic method (MEM) applied to virology data as a novel real time tool to predict peak in seasonal influenza healthcare utilisation. The Scottish experience of the 2017/18 season to date. Euro Surveill 23(11):18–79

Musa AM, Kiely J, Luxton R, Honeychurch KC (2023) An electrochemical screen-printed sensor based on gold-nanoparticle-decorated reduced graphene oxide–carbon nanotubes composites for the determination of 17-β estradiol. Biosensors 13(4):491

Negi NP, Choephel T (2020) Biosensor: an approach towards a sustainable environment. In: Nanobiosensors for agricultural, medical and environmental applications, pp 43–62

Neri G, Fazio E, Mineo PG, Scala A, Piperno A (2019) SERS sensing properties of new graphene/gold nanocomposite. Nanomaterials 9(9):1236

Nguyen HH, Park J, Kang S, Kim M (2015) Surface plasmon resonance: a versatile technique for biosensor applications. Sensors 15(5):10481–10510

Nie X, Chen Z, Tian Y, Chen S, Qu L, Fan M (2021) Rapid detection of trace formaldehyde in food based on surface-enhanced Raman scattering coupled with assembled purge trap. Food Chem 340:127930

Palecek E, Fojta M (2005) Electrochemical DNA sensors. In: Bioelectronics: from theory to applications, pp 127–192

Paleček E, Fojta M, Tomschik M, Wang J (1998) Electrochemical biosensors for DNA hybridization and DNA damage. Biosens Bioelectron 13(6):621–628

Paleček E, Fojta M, Jelen F (2002) New approaches in the development of DNA sensors: hybridization and electrochemical detection of DNA and RNA at two different surfaces. Bioelectrochemistry 56(1–2):85–90

Park J-Y, Park S-M (2009) DNA hybridization sensors based on electrochemical impedance spectroscopy as a detection tool. Sensors 9(12):9513–9532

Pingarrón JM, Yanez-Sedeno P, González-Cortés A (2008) Gold nanoparticle-based electrochemical biosensors. Electrochim Acta 53(19):5848–5866

Pividori MI, Merkoci A, Alegret S (2000) Electrochemical genosensor design: immobilisation of oligonucleotides onto transducer surfaces and detection methods. Biosens Bioelectron 15(5–6):291–303

Popovich N, Thorp H (2002) New strategies for electrochemical nucleic acid detection. Electrochem Soc Interface 11(4):30

Prinith NS, Manjunatha JG (2022) Electrochemical analysis of L-tryptophan at highly sensitive poly (glycine) modified carbon nanotube paste sensor. Mater Res Innov 26(3):134–143

Rabbani M, Hoque ME, Mahbub ZB (2020) Nanosensors in biomedical and environmental applications: perspectives and prospects. In: Nanofabrication for smart nanosensor applications, pp 163–186

Rahman MM, Li X-B, Lopa NS, Ahn SJ, Lee J-J (2015) Electrochemical DNA hybridization sensors based on conducting polymers. Sensors 15(2):3801–3829

Rasheed PA, Sandhyarani N (2017) Electrochemical DNA sensors based on the use of gold nanoparticles: a review on recent developments. Microchim Acta 184:981–1000

Rashid JIA, Yusof NA (2017) The strategies of DNA immobilization and hybridization detection mechanism in the construction of electrochemical DNA sensor: a review. Sens Biosens Res 16:19–31

Rastogi S, Kumari V, Sharma V, Ahmad FJ (2022) Gold nanoparticle-based sensors in food safety applications. Food Anal Methods 15:1–17

Ren W, Liu H, Yang W, Fan Y, Yang L, Wang Y, Liu C, Li Z (2013) A cytometric bead assay for sensitive DNA detection based on enzyme-free signal amplification of hybridization chain reaction. Biosens Bioelectron 49:380–386

Rosi NL, Mirkin CA (2005) Nanostructures in biodiagnostics. Chem Rev 105(4):1547–1562

Roy A, Pandit C, Gacem A, Alqahtani MS, Bilal M, Islam S, Hossain MJ, Jameel M (2022) Biologically derived gold nanoparticles and their applications. Bioinorg Chem Appl 2022:8184217
Sadighbayan D, Sadighbayan K, Khosroushahi AY, Hasanzadeh M (2019) Recent advances on the DNA-based electrochemical biosensing of cancer biomarkers: analytical approach. TrAC Trends Anal Chem 119:115609
Saha K, Agasti SS, Kim C, Li X, Rotello VM (2012) Gold nanoparticles in chemical and biological sensing. Chem Rev 112(5):2739–2779
Salamifar SE, Lai RY (2014) Fabrication of electrochemical DNA sensors on gold-modified recessed platinum nanoelectrodes. Anal Chem 86(6):2849–2852
Saldan I, Dobrovetska O, Sus L, Makota O, Pereviznyk O, Kuntyi O, Reshetnyak O (2018) Electrochemical synthesis and properties of gold nanomaterials. J Solid State Electrochem 22:637–656
Selli GI, Bonatto AET, Bonatto FT, Anzanello MJ, Bergmann CP (2022) Nanosensors in forensic sciences. Environ Appl Nanomater 15:239–253
Shi Y, Pan Y, Zhang H, Zhang Z, Li M-J, Yi C, Yang M (2014) A dual-mode nanosensor based on carbon quantum dots and gold nanoparticles for discriminative detection of glutathione in human plasma. Biosens Bioelectron 56:39–45
Song J, Mou H-Z, Li X-Q, Liu Y, Yang X-J, Chen H-Y, Xu J-J (2022) Self-assembled DNA/RNA nanospheres with cascade signal amplification for intracellular MicroRNA imaging. Sens Actuators B Chem 360:131644
Soni I, Jayaprakash GK (2023) A short review on the analysis of the adsorptive behavior of surfactants at carbon paste electrodes for electrochemical sensing. J Mol Liq 388:122737
Sperling RA, Parak WJ (2010) Surface modification, functionalization and bioconjugation of colloidal inorganic nanoparticles. Philos Trans R Soc A Math Phys Eng Sci 368(1915):1333–1383
Stern E, Wagner R, Sigworth FJ, Breaker R, Fahmy TM, Reed MA (2007) Importance of the Debye screening length on nanowire field effect transistor sensors. Nano Lett 7(11):3405–3409
Taha BA, Al Mashhadany Y, Hafiz Mokhtar MH, Dzulkefly Bin Zan MS, Arsad N (2020) An analysis review of detection coronavirus disease 2019 (COVID-19) based on biosensor application. Sensors 20(23):6764
Teles FRR, Fonseca LP (2008) Trends in DNA biosensors. Talanta 77(2):606–623
Vo-Dinh T, Cullum BM, Stokes DL (2001) Nanosensors and biochips: frontiers in biomolecular diagnostics. Sens Actuators B Chem 74(1–3):2–11
Wan Y, Xu H, Su Y, Zhu X, Song S, Fan C (2013) A surface-initiated enzymatic polymerization strategy for electrochemical DNA sensors. Biosens Bioelectron 41:526–531
Wang J (2002) Electrochemical nucleic acid biosensors. Anal Chim Acta 469(1):63–71
Wang J (2006) Electrochemical biosensors: towards point-of-care cancer diagnostics. Biosens Bioelectron 21(10):1887–1892
Wu Y-Y, Huang P, Wu F-Y (2020) A label-free colorimetric aptasensor based on controllable aggregation of AuNPs for the detection of multiplex antibiotics. Food Chem 304:125377
Xu C, Liu K, Lei M, Yang A, Li Y, Hughes TR, Min J (2018) DNA sequence recognition of human CXXC domains and their structural determinants. Structure 26(1):85–95
Yanez-Sedeno P, Pingarron JM (2005) Gold nanoparticle-based electrochemical biosensors. Anal Bioanal Chem 382:884–886
Zhang J, Lang H-P, Huber F, Bietsch A, Grange W, Certa U, McKendry R, Güntherodt H-J, Hegner M, Gerber C (2006a) Rapid and label-free nanomechanical detection of biomarker transcripts in human RNA. Nat Nanotechnol 1(3):214–220
Zhang J, Liu Y, Ke Y, Yan H (2006b) Periodic square-like gold nanoparticle arrays templated by self-assembled 2D DNA nanogrids on a surface. Nano Lett 6(2):248–251
Zhang W, Yang T, Zhuang X, Guo Z, Jiao K (2009) An ionic liquid supported CeO_2 nanoshuttles—carbon nanotubes composite as a platform for impedance DNA hybridization sensing. Biosens Bioelectron 24(8):2417–2422

Zhang X, Servos MR, Liu J (2012) Instantaneous and quantitative functionalization of gold nanoparticles with thiolated DNA using a pH-assisted and surfactant-free route. J Am Chem Soc 134(17):7266–7269
Zhang JJ, Cheng F, Li J, Zhu J-J, Lu Y (2016) Fluorescent nanoprobes for sensing and imaging of metal ions: recent advances and future perspectives. Nano Today 11(3):309–329
Zhang C, Gao Z, Wang Q, Zhang X, Yao J, Qiao C, Liu Q (2018) Highly sensitive detection of melamine based on the fluorescence resonance energy transfer between conjugated polymer nanoparticles and gold nanoparticles. Polymers 10(8):873
Zhao C, Hong C, Lin Z, Chen X, Huang Z-Y (2019) Detection of Malachite Green using a colorimetric aptasensor based on the inhibition of the peroxidase-like activity of gold nanoparticles by cetyltrimethylammonium ions. Microchim Acta 186:1–8
Zhou H, Liu J, Xu J-J, Zhang S, Chen H-Y (2019) Advances in DNA/RNA detection using nanotechnology. Adv Clin Chem 91:31–98
Zhu C, Yang G, Li H, Du D, Lin Y (2015) Electrochemical sensors and biosensors based on nanomaterials and nanostructures. Anal Chem 87(1):230–249

FRET-Based Tools for Detection and Real-Time Quantification of Vitamins and Their Bioimaging

4

Irfan and Mohd. Mohsin

Abstract

Vitamins play a crucial role in facilitating proper metabolism, cellular development, and functioning of other biomolecules, thereby ensuring the proper operation of various body systems. These essential micronutrients enter the body through food and supplements, thus it is important to quantify their concentrations in food, medicines, and biological fluids. Various methods are employed to measure the amount of these organic micronutrients, with selectivity, sensitivity, and limit of detection playing essential performance parameters for analytical techniques. To detect and continuously measure essential micronutrients in source samples and living cells for therapeutic purposes, a simple, reliable, sensitive, and non-invasive technique is required. Metabolite sensors detect metabolite changes and report their abundance within biological networks. Among the emerging options, FRET-based biosensors are gaining popularity due to their capability to visualize the spatiotemporal fluctuations in metabolite dynamics within living cells. FRET occurs within a range of 1–10 nm (Forster radius) between a bright donor molecule and an energy acceptor fluorescent molecule. FRET-based sensor requires a protein that exhibits a high affinity and specificity for the target molecule, along with a significant conformational change upon binding. These nanosensors have the capability to detect alterations in vitamin content in experimental samples in real-time.

Keywords

Vitamins · Biomolecules · Metabolite · Biosensors · Affinity · FRET · Non-invasive

Irfan · M. Mohsin (✉)
Department of Biosciences, Jamia Millia Islamia, New Delhi, Delhi, India
e-mail: mmohsin1@jmi.ac.in

M. Mohsin, N. Soleja (eds.), *Nanoscale Sensors and their Applications in Biomedical Imaging*, https://doi.org/10.1007/978-981-97-3144-2_4

Abbreviations

CaM	Calmodulin
CD	Carbon dots
EGFP	Enhanced green fluorescent protein
FLIP-α	Fluorescence indicator for α-tocopherol
FLIPT	Fluorescent indicator protein for thiamine
FRET	Fluorescence resonance energy transfer
Ln-MOFs	Lanthanide (III)-based metal-organic frameworks
OSenB	Optical sensor for biotin
PtdIns(4,5)P2	Phosphatidylinositol 4,5-bisphosphate
QDs	Quantum dots
RF	Riboflavin
SenVitAL	Sensor for vitamin anemia linked
TRITC	Tetramethyl-rhodamine iso-thiocyanate

4.1 Introduction

Vitamins are micronutrients obtained from dietary sources, and these organic compounds are vital for sustaining and development of cells, tissues, and organs and also fortifying the immune system. They aid in the transformation of food into energy-rich compounds, help in decreasing the risk of illness, and advocate for a healthy lifestyle. Vitamins also play key roles in development and metabolism, and deviations from their appropriate levels can signify health issues and disruptions in normal metabolic processes. Due to their typical daily requirement of less than 100 mg, vitamins are categorized as micronutrients (Sami et al. 2014). These compounds exhibit diverse chemical, structural, and physical properties and are known for their function in energy production and normal functioning of muscle, skin, eye, hair, and liver health.

There are 13 essential vitamins, divided into fat-soluble and water-soluble categories that are necessary for proper bodily function. Insufficient levels of these vitamins can lead to a number of serious health problems, including night blindness, scurvy, and rickets (Saenger et al. 2006). Several vitamins, including those in the B-complex and vitamin C, are water-soluble. Vitamins like thiamine (B1), riboflavin (B2), niacin (B3), pantothenic acid (B5), pyridoxine (B6), biotin (B8), folate (B9), cyanocobalamin (B12) are all part of the B-group vitamins. Several B-group vitamins serve crucial functions in energy-releasing mechanisms and other metabolic processes by primarily functioning as coenzymes in lipid, carbohydrate, and protein metabolic enzymes (Helliwell 2017). The global focus on food safety, high-quality nutrition, and making healthy and safe dietary choices is closely linked to understanding the nutritive value for a healthier society. Nutritional science research has identified bioactive components, such as vitamins, in various foods. There is ongoing curiosity about vitamin's function in energy metabolism. Notably, B-complex vitamins such as B6, B12, and folate play crucial roles in the regulation of mitochondrial enzymes responsible for essential reactions in the electron transfer chain reaction, a process vital for generating energy-rich compounds like ATP molecules (Depeint et al. 2006). Current nutrition science studies attempt to identify bioactive components such as vitamins in food (Asensi-Fabado and Munne-Bosch 2010). Numerous analytical methods, including thermogravimetry, spectrophotometry, high-performance liquid chromatography (HPLC) coupled with mass spectrometry, colorimetry, fluorescence, electrophoresis, and others, have been applied to determine the actual vitamin content in natural sources. However, in the majority of the aforementioned techniques, additional procedures must be taken before their quantification (Fatima et al. 2019; Chawla and Kvarnberg 2014).

For the accurate measurement of clinically essential micronutrients in source samples and living cells, a simple, reliable, specific, non-invasive approach that enables the detection and real-time quantification of these molecules at minimal concentrations is required. Metabolite sensors connect biological networks and the environment in metabolite-sensing and signaling pathways. The term "sensor" refers to an electronic component that can sense and report changes in its surrounding environment (Looger et al. 2003). These metabolite sensors are biological molecules that can detect changes or the presence of specific metabolites and convey information about their abundance in biological networks. The process of metabolite sensing occurs at the intracellular and extracellular boundaries of a biological network. For the proper functioning of these sensors, specificity, dynamicity, and functionality are critical criteria (Wang and Lei 2018). Nanomaterials have broadened the selection of powerful sorbents for purifying, bio-sensing, and bio-imaging of biological analytes. The enhanced sensing capabilities of nanosensors make them promising analytical instruments for elucidating biological properties and the concentration of elements in the given samples (Sargazi et al. 2022).

4.2 Biosensors

The term "biosensors" refers to sensors that can detect and respond to target ligands in a highly precise and sensitive manner using biological products such as proteins, cells, or cell organelles. They can be used in a variety of applications, including drug discovery and environmental monitoring (Dixon et al. 2021). Due to their impressive performance, biosensors have found widespread application across a variety of scientific fields. In the medical field, biosensors can be utilized for the sensitive and specific detection of a wide variety of targets, including tumors, infections, high blood glucose levels in diabetic patients, and various toxins (Agunloye and Oboh 2021). The affordability, reasonable sensitivity, outstanding accuracy, and user-friendly operation of biosensors make them promising tools for detecting various biological substances (Luong et al. 2008). The most popular sensing methods are optical and electrochemical. Spectrophotometers and pH electrodes are used for optical sensing, whereas conduction ions are used for electrochemical sensing (Steinegger et al. 2020). These conventional techniques require specialized biosensing equipment, leading to higher costs and non-specificity for many applications (Rajska et al. 2021). Nanotechnology-based nanomolecules in biosensors improve reliability, validity, stability, sensitivity, and ability to achieve lower detection limits (Malekzad et al. 2017). Based on the types of signal transduction mechanisms, biosensors can be categorized into the following types:

- Voltammetric and potentiometric sensors
- Calorimetric and enzymatic sensors
- DNA and protein-based sensors
- Optical sensors

Among the aforementioned sensors, optical sensors are widely employed for detecting analytes within living cells. Optical sensor fibers detect analytes through light scattering, fluorescence, or absorption. Optical sensors are versatile due to their signal transmission capabilities, which can adjust for variations in light polarity, time, wavelength, wave propagation, spectral distribution, and intensity (Fig. 4.1) (Long et al. 2013; Peltomaa et al. 2018). Fluorescence-based optical biosensors rely on labeled optical phenomena to detect analytes or molecules of interest. Due to their excellent selectivity, sensitivity, and rapid response time, these types of biosensors have received significant attention for applications in medical diagnostics, monitoring of environment-degrading substances, and testing food quality. These biosensors make use of several different fluorescent substances like fluorescent organic dyes, quantum dots (QDs), lanthanides, and biological fluorescent molecules (Fig. 4.2) (Malhotra and Ali 2018; Fang et al. 2023; Demchenko et al. 2009).

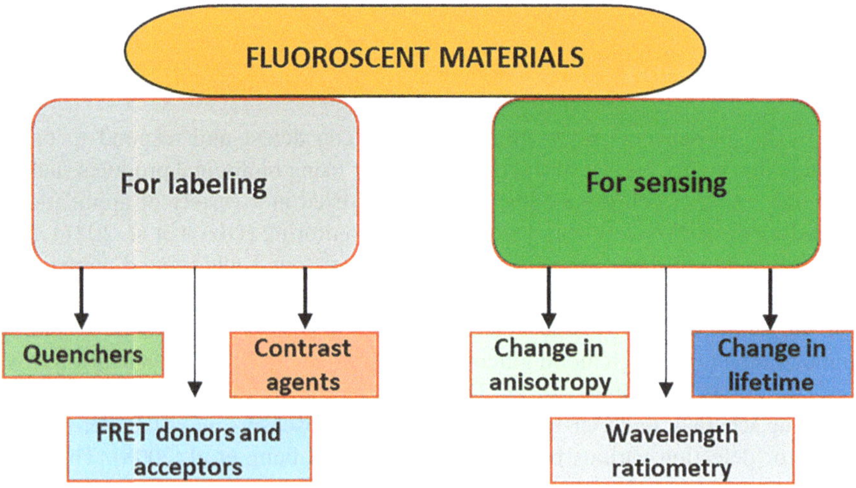

Fig. 4.1 Fluorescent sensing and imaging trends

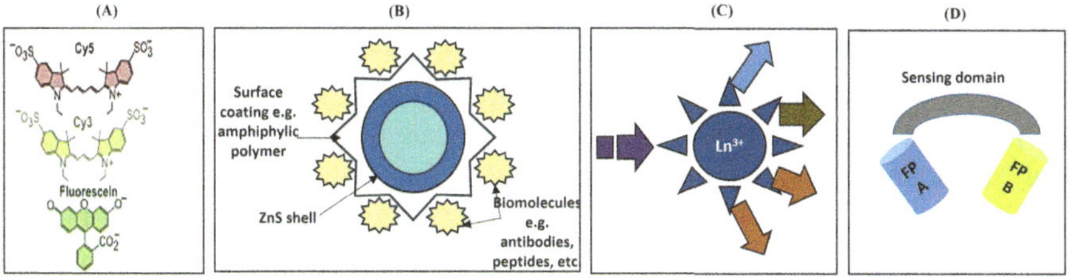

Fig. 4.2 Fluorescent nanomaterials: (**a**) fluorescent organic dye, (**b**) quantum dots, (**c**) lanthanide probe, (**d**) green fluorescent proteins

4.2.1 Fluorescent Organic Dyes

The optical properties of fluorescent organic dyes are determined by the nature of their electronic transition. Specifically, organic dyes emit light through a delocalized optical transition that encompasses the chromophore, and they are referred to as resonant dyes due to their resonant emission (Resch-Genger et al. 2008). Popular members of this class include the fluorescent dye fluorescein and the rhodamine derivative tetramethyl-rhodamine iso-thiocyanate (TRITC) (Barone et al. 2011). However, because dyes often exhibit mono-exponential decay kinetics, it is possible to easily identify dyes based on measurements of their fluorescence lifetimes. This characteristic makes dyes suitable for applications that require lifetime measurements (Resch-Genger et al. 2008).

4.2.2 Fluorescent Quantum Dots

Quantum dots (QDs) possess optical properties that are determined by factors such as their composition, particle size and size distribution, and surface chemistry, particularly the number of electrons (n) available to each photon. To improve fluorescence quantum yields, decay kinetics, and stability, the surfaces of these particles are often modified using inorganic passivation layers or organic capping ligands (Dabbousi et al. 1997). By varying the semiconductor material and nanocrystal size during synthesis, a QD-based probe can be tuned to emit light at a specific wavelength, making it useful for various experimental applications (Rosenthal 2001).

4.2.3 Lanthanides

Rare earth elements called lanthanides are frequently utilized in conjunction with regular organic dyes as FRET donors and acceptors (Pilch-Wrobel et al. 2022; Qiu et al. 2022). Due to their unique characteristics, such as a wide excitation band, a narrow emission band, a long duration of fluorescence, and a significant stokes shift in the fluorescence spectra, lanthanides are of particular significance (Li et al. 2022). These features make them excellent candidates as luminophores in sensing devices. Samarium, europium, terbium, dysprosium, and actinium are the most widely used lanthanides (Fang et al. 2023).

4.2.4 Biological Fluorescent Molecules

During the early 1960s, researchers investigating the bioluminescent properties of the *Aequorea victoria* jellyfish made two significant discoveries. First, they identified a bioluminescent protein called aequorin, which emitted blue light. Simultaneously, adjacent to this, they also observed and examined another protein known as the green fluorescent protein (GFP) (Shimomura et al. 1962). Within the light organs of *A. victoria*, the proteins aequorin and GFP work together to convert Ca^{2+}-induced luminescent signals into the distinctive green fluorescence emission from GFP (Chalfie and Kain 2006). As soon as it was cloned, GFP was put to use in monitoring gene expression sensory neurons of the nematode *C. elegans* and bacterial cells. Since then, a plethora of blue, cyan, and yellow variants of jellyfish-derived GFP have been created and the fluorescent proteins from a wide range of other species were discovered, allowing for the expansion of the color palette into the orange, red, and far-red spectral range. Collectively, these powerful genetically encoded probes are known as fluorescent proteins (Kremers et al. 2011).

4.2.5 Principle of Fluorescence

Due to their sensitivity and selectivity, fluorescent technologies are widely employed in food safety, medical diagnostics, the detection of various analytes, and environmental monitoring (Fang et al. 2023). As a result, they have garnered attention from scientists across disciplines, including biology and chemistry. Fluorescent sensors offer a novel method for detecting these analytes, which can then be used to learn more about their clinical and physiological functions (Pickup et al. 2005). Upon being exposed to light, the lower energy ground state electrons in fluorescent substances absorb the kinetic energy of photons and transition to a higher energy excited state in about 10^{-15} s. Excited electrons are unstable and can return to the ground state through nonradiative and radiative decay. Radiative decay is accompanied by photon emission, leading to fluorescence (Algar et al. 2019).

4.2.6 The FRET Phenomenon

Originally established theoretically in 1948, Forster resonance energy transfer (FRET) is a well-described photophysical phenomenon in which energy is transferred from a donor fluorophore to an acceptor fluorophore (Forster 1948). FRET is superior to other methods in several scientific fields, including molecular interactions and biomolecule structural and dynamic changes. FRET is used to investigate molecular interactions over time due to its non-invasive nature. In addition to its improved sensitivity, short observation interval in nanoseconds, working distance over which most biomolecular processes occur, ease of experimentation, and applicability to diluted sample solutions, this method has several other benefits (Juzeliunas and Andrews 2000; Hof et al. 2004; Lakowicz 2006; Valeur 2001). Fluorescence resonance energy transfer (FRET) is a radiation-free mechanism where a fluorophore transfers energy to another fluorophore via dipole–dipole coupling. This whole mechanism describing the transfer of energy from an excited donor fluorophore to an acceptor fluorophore in a non-radiative manner is FRET.

The FRET efficiency (*E*) is influenced not only by the physical separation and orientation of the two dipoles but also by the spectral characteristics of each. This encompasses the acceptor's extinction coefficient, the donor's quantum yield, and the overlap (*J*) between the two spectra. The FRET efficiency is also affected by the refractive index (*n*), which is the ratio of the speed of light in a vacuum to the speed of light in the medium in which the energy transfer occurs (Kuzmenkina et al. 2006).

$$\mathrm{FRET}(E) = \frac{R_0^6}{R_0^6 + R^6}$$

where R_0, also known as the Forster radius is the critical distance at which FRET efficiency is 50% and varies depending on the FRET pair. At distances below R_0, FRET efficiency is nearly maximal, whereas, at distances above R_0, it is at its lowest. Since FRET allows for more accurate measurements and interpretations of data at close ranges to R_0, it is used as a molecular comparator (Fig. 4.3) (Clegg 1996; Stryer 1978; Pelosse et al. 2016).

4.2.7 Fluorescent Nanosensors

Fluorescence imaging is employed for real-time and in situ identification of different biomolecules owing to its high sensitivity, selectivity, rapid feedback, and exceptional spatiotemporal resolution. There has been a proliferation of fluorescent probes for imaging and bioassay applications in recent years. Fluorescence bioanalysis and imaging using novel probes have gained extensive acceptance in the

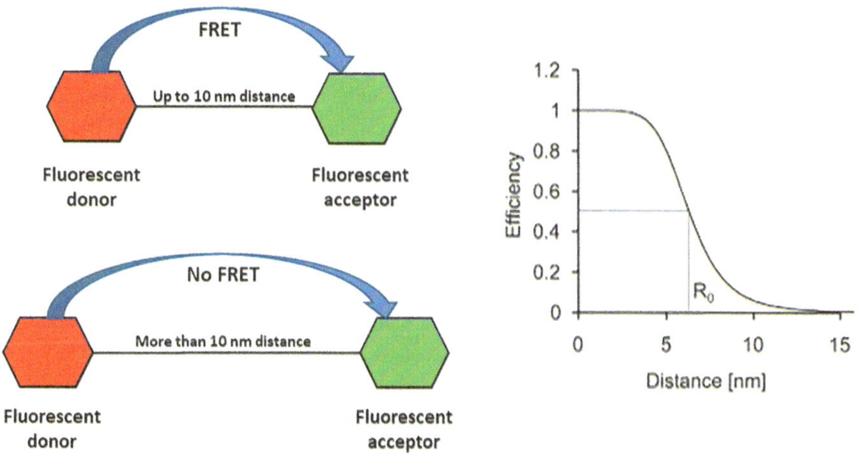

Fig. 4.3 Distance dependence of FRET efficiency

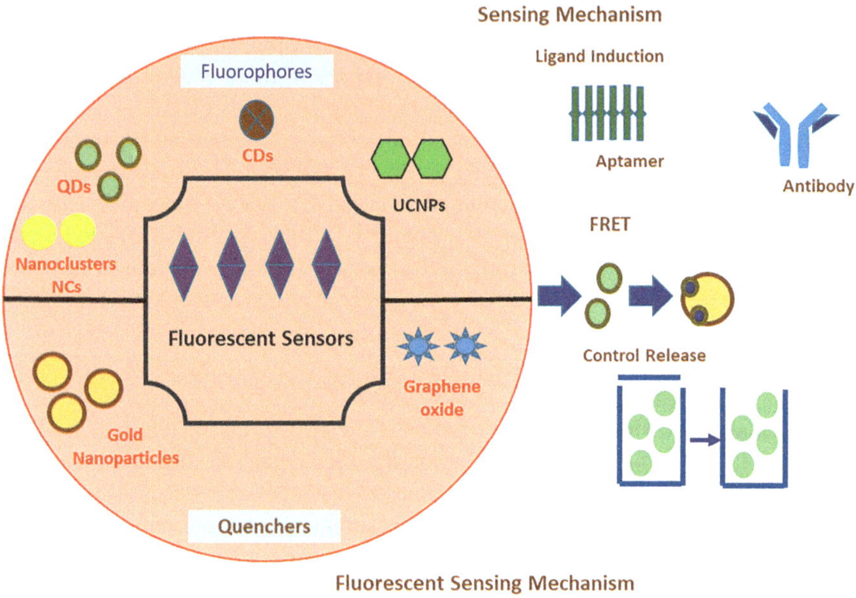

Fig. 4.4 Fluorescence methods to understand nanomaterial fluorophores and quenchers

fields of biomedicine, preclinical studies, diagnostics, and various life science processes due to their remarkable attributes, including high sensitivity, selectivity, resolution, simplicity, and the ability to provide non-invasive and real-time monitoring (Fig. 4.4) (Sargazi et al. 2022). Translocation-based probes, single FP probes: intensity-based, single FP probes: ratiometric-based, and FRET-based sensors are the types of fluorescent nanosensors (Fig. 4.5) (Palmer et al. 2011).

4.2.7.1 Translocation-Based Probes

To measure phosphatidylinositol 4,5-bisphosphate (PtdIns(4,5)P2) in the plasma membrane using a translocation-based probe, phospholipase C Delta1 (PLC-d1) is employed, with GFP fused to its pH domain, which specifically recognizes PtdIns(4,5)P2. Reduced levels of plasma membrane PtdIns(4,5)P2 cause the sensor to translocate to the cytosol, leading to an increase in its fluorescence.

4.2.7.2 Single FP Probes: Intensity-Based

An intensity-based single FP probe, GCaMP2, comprises myosin light chain kinase M13, enhanced green fluorescent protein (EGFP), and calmodulin. As the ionic concentration of Ca^{2+} rises, M13 binds to CaM, changing the protonation state of the chromophore and increasing the fluorescence intensity. This mechanism is used in various sensors, including those for Cl^- and H^+.

4.2.7.3 Single FP Probes: Ratiometric-Based

A single-FP-based ratiometric redox sensor, known as reduction–oxidation sensitive green fluorescent protein (roGFP1's), relies on its redox state to influence the relative fluorescence intensity of its two excitation peaks. In a reduced state, the

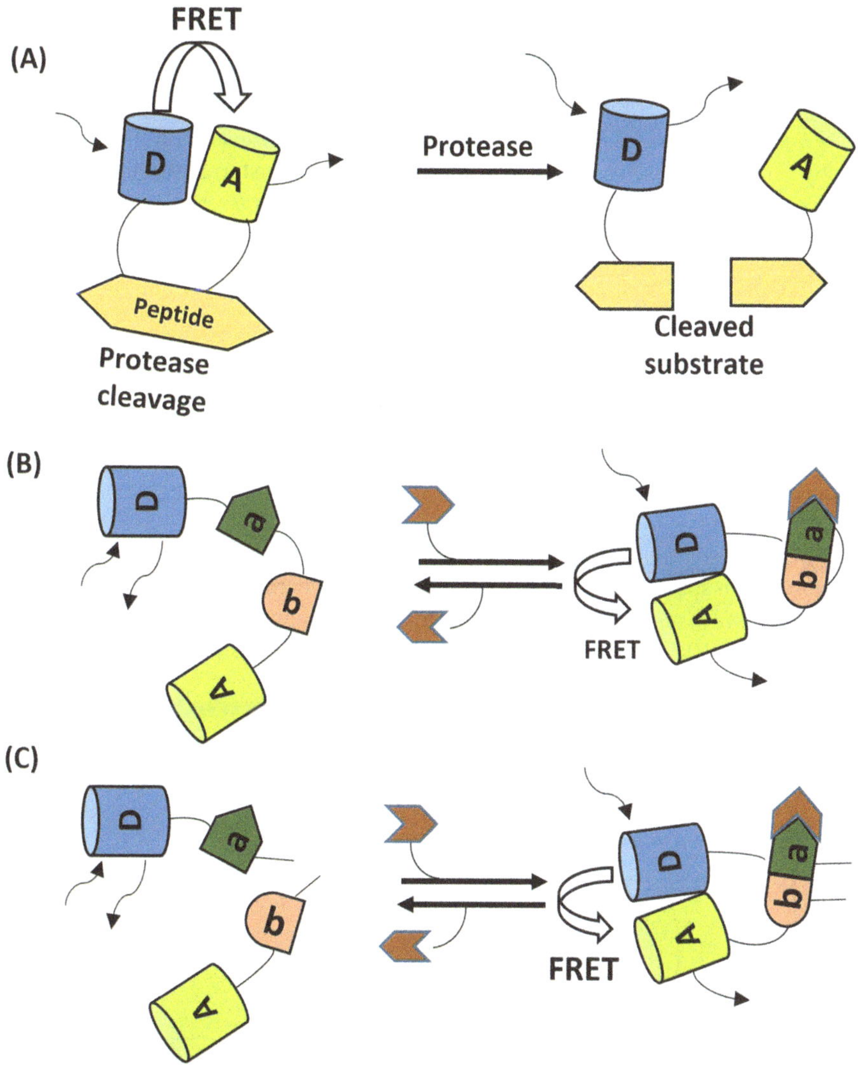

Fig. 4.5 Types of FRET-based sensors include: (A) Protease-activated FRET-based sensors, (B) sensors when metabolite binding occurs, and (C) sensors for protein-protein interactions, where D denotes the donor, A the acceptor, and a and b the protein domains

excitation at 400 nm is decreased while the excitation at 480 nm is increased. Similar principles are utilized in other probes, including the ratiometric H^+ sensor pHlorin, as well as sensors for cGMP and membrane potential.

4.2.7.4 FRET-Based Nanosensors

Biosensors that work on the FRET principle are used in a ratiometric fashion to detect intracellular metabolite levels. This type of sensor was first published in 1997 and is now widely applicable to various ligands, including metal ions, sugars, amino acids, and vitamins (Steffen et al. 2016).

Types of FRET-Based Nanosensors

FRET-based sensors and their working mechanism can be subdivided into the following types (Zeug et al. 2014):

1. *Protease-activated FRET sensors*: Cleavage of the biosensor on the linker side by certain proteases reduces the amount of energy transferred through FRET.
2. *Intramolecular or conformational change FRET sensor*: The interaction of a ligand induces a conformational change in an active unit, and this change in FRET can be monitored using an intramolecular FRET-based biosensor.
3. *Intermolecular FRET-based biosensor*: Upon ligand binding, both parts of the biosensor dimerize, triggering FRET. It is more difficult to quantify the signals because of the unknown amount of free donor and acceptor components.

Required Conditions for FRET

Several parameters contribute to the FRET phenomena, and the most important ones are outlined here (Fig. 4.6):

1. *Integral spectrum overlap (J)*: Both the emission spectrum of the donor and the absorption spectrum of the acceptor need to considerably overlap.
2. *FRET pair distance (R)*: The distance between donor and acceptor fluorophores must be 1–10 nm. The sixth power of *R* inversely affects FRET efficiency. Thus, even a small change in *R* can influence FRET signals.

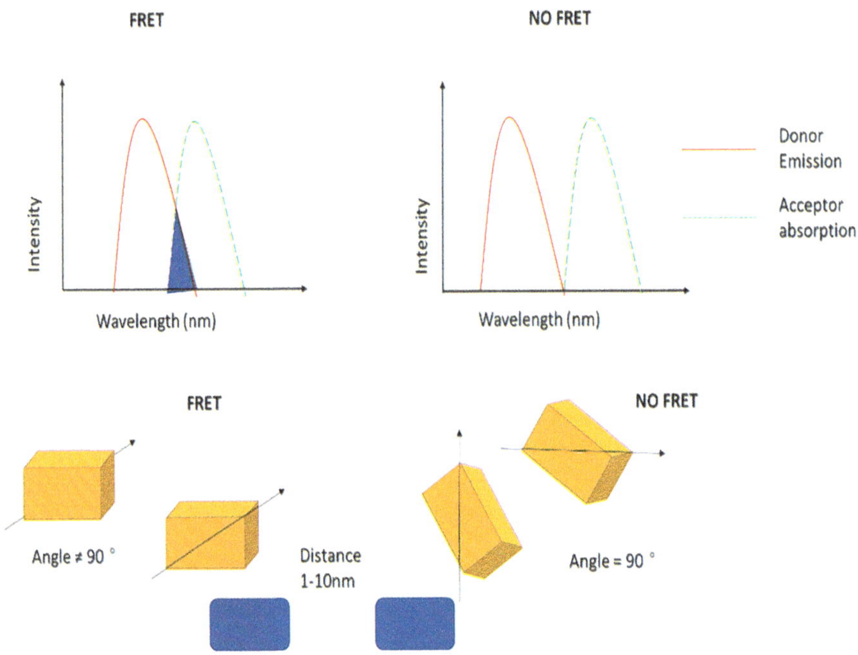

Fig. 4.6 Three prerequisites for FRET

3. *Interaction between dipoles (k^2)*: The dipole moment of emission from the donor, the absorption from the acceptor, and the separation vectors between the two must be in a parallel orientation.

4.2.7.5 Measurements of FRET

There have been various approaches employed for measuring FRET. FRET measurement methodologies can be divided into four primary categories (Jares-Erijman and Jovin 2003; Sekar and Periasamy 2003):

1. Techniques for tracking alterations in donor fluorescence
2. Methods for measuring changes in acceptor fluorescence
3. The use of spectral imaging to assess donor and acceptor fluorescence variations in real-time
4. Strategies for monitoring fluorophore orientation changes

Quantifying FRET efficiency is most conveniently accomplished by observing variations in donor fluorescence in both conditions when the acceptor is present and absent. FLIM-FRET, a technique reliant on fluorescence lifetime imaging, identifies FRET by assessing alterations in the fluorescent lifetime of a donor. This measurement indicates how rapidly a donor fluorophore emits light after a short excitation pulse. Donor fluorescence intensity was examined before and after acceptors were photobleached and this technique is known as acceptor bleaching (Vogel et al. 2006). If FRET occurs, then the donor's fluorescence longevity should decrease. The FRET efficiency can be determined by comparing the donor's lifetime in the presence and absence of acceptors.

Acceptor emission upon donor excitation is the most common method used to quantify FRET (Gordon et al. 1998). This method's key benefit is that, like acceptor bleaching, it can be carried out using inexpensive, widely used wide-field fluorescence microscopes in cell research laboratories. Because it requires different combinations of fluorescent filters for taking images with different wavelengths, this approach is often called the "three-cube method." To generate a FRET image, it is necessary to establish an optimized filter set that excites the donor fluorophore while capturing the emission from the acceptor. The obtained FRET image is subsequently processed using appropriate software that corrects it (Vogel et al. 2006).

Recently, a photon-efficient FRET method was developed, using spectral imaging to simultaneously track changes in donor and acceptor fluorophore intensities. Spectral microscopy measures an entire emission spectrum for each pixel (Zimmermann et al. 2003). The spectrum contains data about the number of donors and acceptors present, as well as the obtained FRET efficiency. Spectral imaging has been described as a theoretical method for measuring FRET (Neher and Neher 2004). This method utilizes two-photon microscopy, which employs two simultaneous infrared photons, requiring half the energy to excite a fluorophore (Denk et al. 1990). Besides FRET efficiency, spectral imaging assesses the abundance of both donors and acceptors. Tunable lasers are used for two-photon excitation to stimulate donors and acceptors. Optimizing excitation wavelengths enhances donor and acceptor signal-to-noise ratios in specific samples (Thaler et al. 2005).

Anisotropy-based FRET techniques examine the correlation between the orientation of molecules that are excited by linearly polarized light and fluorescent molecules emitting fluorescence in response to excitation. If FRET occurs, the molecule that gets excited may differ from the one emitting fluorescence, diminishing the correlation between their orientations. In samples featuring distinct fluorophores for donors and acceptors, anisotropy has been used to monitor FRET (Rizzo and Piston 2005).

4.3 Bioimaging Probes for the Detection of Vitamins

Carbon Dots for Vitamin B2 Riboflavin (RF) is a water-soluble vitamin that supports nucleic acid repair, apoptosis, and electron transfer in the respiratory chain within cells. Vitamin B2 is needed for lipid, ketone body, carbohydrate, and protein metabolism, making it essential to the human diet (Foraker et al. 2003; Kumar et al. 2004). Carbon dots (CDs) have garnered attention as a bioimaging probe due to their remarkable aqueous stability, consistent fluorescence, and simple synthesis. Following 24-h incubation with CDs, the cells maintained their typical morphology, indicating strong biocompatibility.

Lanthanides for Vitamin B6 Lanthanide (III)-based metal-organic frameworks (Ln-MOFs) feature a unique spectrum, substantial Stokes shift, and long fluorescence lifetime, making them suitable for optical and biological applications (Yu et al. 2021; Wu et al. 2020). One of the most important biological coenzymes, vitamin B6 (VB6), is involved in over 100 metabolic reactions involving amino acids, lipids, and glucose. It supports red blood cell formation, normal functioning of immunological and neurological systems, and brain development during pregnancy and infancy. Few MOFs have been tested to detect VB6 luminescence. Chemical sensors for VB6 can be made using europium (III)-and terbium (III)-MOFs with narrow and characteristic fluorescence emissions. Complex Tb-MOFs, utilizing the rigid ligand H5DDB, serve as fluorescent-sensing materials for vitamins in phosphate buffer solutions, exhibiting excellent luminescence and remarkable water stability. A titration experiment was conducted, wherein a VB6 solution was added to a suspension of Tb-MOFs, and its luminous intensity was measured. The luminous peak at 551 nm was decreased with increasing VB6 concentrations, and it nearly disappeared at 0.65 mM VB6 concentration (Wang et al. 2021).

4.4 The Developed Fluorescent Protein Sensors for Vitamins and Their Bioimaging

FLIP-α Fluorescence indicator for α-tocopherol (FLIP-α) nanosensor is based on FRET and is designed to track α-tocopherol (vitamin E) flux in both prokaryotic and eukaryote cells. The separation distance of the FRET fluorophore pair from the ligand-binding protein is crucial to the success of the sensor's development. Using a ligand-binding protein that undergoes structural changes to bring the fluorophores within the requisite distance of less than 10 nm greatly assisted in sensor development. Conformational changes induced by α-tocopherol in the human α-tocopherol transfer protein, which acts as the α-tocopherol-binding protein in the sensor construct, resulted in the close proximity of the fluorophores ECFP and Venus, leading to the FRET signal. The α-tocopherol flow was monitored with great accuracy using FLIP-α. It was determined that the sensor is pH-insensitive and therefore suitable for use in physiological conditions and shows an affinity (K_d) of about 100 μM. FLIP-α, being non-invasive, offers excellent temporal and spatial resolutions, making it invaluable for bioimaging of highly compartmentalized metabolic pathways (Kausar et al. 2021).

SenVitAL FRET-based nanosensor is designed for the real-time monitoring of vitamin B12 levels in bacterial, yeast, and mammalian cells. The development of this nanosensor involved the utilization of the vitamin B12 binding protein (BtuF) in conjunction with cyan (CFP) and yellow (YFP) variants of the green fluorescent protein. Various expression vectors were used to introduce this recombinant construct into the cells. The resulting FRET sensor, named SenVitAL (short for "sensor for vitamin anemia linked"), demonstrated a remarkable level of selectivity for vitamin B12. Notably, SenVitAL is insensitive to changes in pH and is capable of detecting vitamin B12 in a concentration-dependent manner, with a K_d value of 157 μM (Ahmad et al. 2018).

FLIPT A FRET-based genetically encoded nanosensor named "fluorescent indicator protein for thiamine" (FLIPT) has been developed to enable the real-time monitoring of thiamine levels within living cells. The thiamine binding protein (thiB) is placed in between the FRET pair of CFP and YFP (yellow fluorescent protein) to create the nanosensor. The sensory domain thiB undergoes conformational changes in response to changes in thiamine concentration. This novel nanosensor exhibited pH stability within the physiological range and demonstrated high specificity for thiamine. FLIPT has an estimated K_d of 529 nM. The sensor effectively identified and visualized the influx of thiamine into the cytoplasm of these biological systems, demonstrating its potential to monitor thiamine concentrations in both prokaryotic and eukaryotic cellular cytoplasm (Manzoor et al. 2019).

OSenB The OSenB nanosensor, employing the FRET pair ECFP and Venus linked to avidin protein's N- and C-termini, enables real-time monitoring and quantification of biotin. The eukaryotic avidin protein derived from *Gallus gallus* serves as a recognition element within biotin sensors, alongside two provided GFP variants. OSenB exhibits exceptional specificity and selectivity in detecting biotin, providing rapid responses while maintaining pH resistance. It offers a precise FRET signal within a detectable range spanning from 12 pM to 165 nM. Since the OSenB can detect and monitor biotin in extremely low concentrations, it has significant medical applications. This sensor measures biotin in subcellular organelles in vitro and in vivo by monitoring energy transfer between donor and acceptor fluorophores (Irfan et al. 2023).

4.5 Advantages of FRET Sensor Over Conventional Fluorescent Techniques

FRET offers significant advantages in practical applications over traditional fluorescent approaches (Fang et al. 2023):

Dynamic Visualization Real-time monitoring of changes in metabolite concentration and donor and acceptor fluorescence intensity is possible within living cells.

Highly Sensitive In contrast to conventional fluorescence methods, the donors and acceptors involved in FRET are highly sensitive to distance changes.

Anti-interference Ability The FRET system's anti-interference ability is greater than that of conventional fluorescence methods, and the FRET efficiency is maximized only under light with a specified excitation wavelength.

Magnification Effect The use of FRET in nucleic acid hybridization systems can result in a significant increase in the fluorescence of the acceptor. Conventional fluorescent technologies are incapable of achieving this effect since only one fluorescence is used.

4.6 Conclusions

Adjacent cells or subcellular compartments may have different compositions. To understand complex biological phenomena, metabolites must be identified at the cellular or subcellular levels. As coenzymes in lipid, carbohydrate, and protein metabolism, certain B-group vitamins play critical roles in energy production and other metabolic processes. The nutritive value of foods is closely linked to food safety, quality, and healthy diets. Vitamins' role in energy metabolism is significant. To effectively identify and measure these therapeutically critical micronutrients in source samples and living cells in real-time, there is a demand for a technique that is simple, reliable, specific, sensitive, and non-invasive. Nanosensors are extremely sensitive, making them useful analytical tools for demonstrating biological characteristics. Quantifying vitamins in biological samples with a FRET-based sensor is easier, non-invasive, selective, and more sensitive than current approaches. These genetically encoded nanosensors can be incorporated into plasmids or other vectors to express non-toxic chimeric sensor proteins. These proteins emit fluorescence, offering a non-invasive method to monitor time-dependent variations in vitamin concentrations. This approach is applicable to both prokaryotic and eukaryotic systems, and it enables tracking the real-time changes in the vitamin content of individual cells.

Acknowledgments I, as the first author, want to express my gratitude to the Council of Scientific and Industrial Research (CSIR), Government of India, for granting me a Junior and Senior Research Fellowship.

References

Agunloye OM, Oboh G (2021) Effect of diet supplemented with P. ostreatus and L. subnudus on memory index and key enzymes linked with Alzheimer's disease in streptozotocin-induced diabetes rats. J Food Biochem 45(3):e13355

Ahmad M, Mohsin M, Iqrar S, Manzoor O, Siddiqi TO, Ahmad A (2018) Live cell imaging of vitamin B12 dynamics by genetically encoded fluorescent nanosensor. Sens Actuators B Chem 257:866–874

Algar WR, Hildebrandt N, Vogel SS, Medintz IL (2019) FRET as a biomolecular research tool—understanding its potential while avoiding pitfalls. Nat Methods 16(9):815–829

Asensi-Fabado MA, Munne-Bosch S (2010) Vitamins in plants: occurrence, biosynthesis and antioxidant function. Trends Plant Sci 15(10):582–592

Barone V, Bloino J, Monti S, Pedone A, Prampolini G (2011) Fluorescence spectra of organic dyes in solution: a time dependent multilevel approach. Phys Chem Chem Phys 13(6):2160–2166

Chalfie M, Kain S (2006) Green fluorescent protein: properties, applications, and protocols. Wiley-Interscience, Hoboken, NJ

Chawla J, Kvarnberg D (2014) Hydrosoluble vitamins. Handb Clin Neurol 120:891–914

Clegg RM (1996) Fluorescence resonance energy transfer. Fluoresc Imaging Spectrosc Microsc 137:179–251

Dabbousi BO, Rodriguez-Viejo J, Mikulec FV, Heine JR, Mattoussi H, Ober R, Jensen KF, Bawendi MG (1997) (CdSe) ZnS core−shell quantum dots: synthesis and characterization of a size series of highly luminescent nanocrystallites. J Phys Chem B 101(46):9463–9475

Demchenko AP, Mély Y, Duportail G, Klymchenko AS (2009) Monitoring biophysical properties of lipid membranes by environment-sensitive fluorescent probes. Biophys J 96(9):3461–3470

Denk W, Strickler JH, Webb WW (1990) Two-photon laser scanning fluorescence microscopy. Science 248(4951):73–76

Depeint F, Bruce WR, Shangari N, Mehta R, O'Brien PJ (2006) Mitochondrial function and toxicity: role of the B vitamin family on mitochondrial energy metabolism. Chem Biol Interact 163(1–2):94–112

Dixon TA, Williams TC, Pretorius IS (2021) Sensing the future of bio-informational engineering. Nat Commun 12(1):388

Fang C, Huang Y, Zhao Y (2023) Review of FRET biosensing and its application in biomolecular detection. Am J Transl Res 15(2):694

Fatima Z, Jin X, Zou Y, Kaw HY, Quinto M, Li D (2019) Recent trends in analytical methods for water-soluble vitamins. J Chromatogr A 1606:360245

Foraker AB, Khantwal CM, Swaan PW (2003) Current perspectives on the cellular uptake and trafficking of riboflavin. Adv Drug Deliv Rev 55(11):1467–1483

Forster T (1948) Intermolecular energy transfer and fluorescence. Ann Phys Leipzig 2:55–75

Gordon GW, Berry G, Liang XH, Levine B, Herman B (1998) Quantitative fluorescence resonance energy transfer measurements using fluorescence microscopy. Biophys J 74(5):2702–2713
Helliwell KE (2017) The roles of B vitamins in phytoplankton nutrition: new perspectives and prospects. New Phytol 216(1):62–68
Hof M, Hutterer R, Fidler V (eds) (2004) Fluorescence spectroscopy in biology: advanced methods and their applications to membranes, proteins, DNA, and cells. Springer, Berlin
Irfan, Soleja N, Mohsin M (2023) Designing of nanosensor and its applications for spatiotemporal visualization and real time quantification of biotin in biological samples. Biochem Eng J 195:108899
Jares-Erijman EA, Jovin TM (2003) FRET imaging. Nat Biotechnol 21(11):1387–1395
Juzeliunas G, Andrews D (2000) Quantum electrodynamics of resonance energy transfer. Adv Chem Phys 112:357–410
Kausar H, Ambrin G, Okla MK, Alamri SA, Soufan WH, Ibrahim EI, Abdel-Maksoud MA, Ahmad A (2021) FRET-based genetically encoded nanosensor for real-time monitoring of the flux of α-tocopherol in living cells. ACS Omega 6(13):9020–9027
Kremers GJ, Gilbert SG, Cranfill PJ, Davidson MW, Piston DW (2011) Fluorescent proteins at a glance. J Cell Sci 124(2):157–160
Kumar V, Lockerble O, Kell SD, Ruane PH, Platz MS, Martin CB, Ravanat JL, Cadet J, Goodrich RP (2004) Riboflavin and UV-light based pathogen reduction: extent and consequence of DNA damage at the molecular level. Photochem Photobiol 80(1):15–21
Kuzmenkina EV, Heyes CD, Nienhaus GU (2006) Single-molecule FRET study of denaturant induced unfolding of RNase H. J Mol Biol 357(1):313–324
Lakowicz JR (ed) (2006) Principles of fluorescence spectroscopy. Springer, Boston, MA
Li L, Li T, Hu Y, Cai C, Li Y, Zhang X, Liang B, Yang Y, Qiu J (2022) Mechanism of the trivalent lanthanides' persistent luminescence in wide bandgap materials. Light Sci Appl 11(1):51
Long F, Zhu A, Shi H (2013) Recent advances in optical biosensors for environmental monitoring and early warning. Sensors 13(10):13928–13948
Looger LL, Dwyer MA, Smith JJ, Hellinga HW (2003) Computational design of receptor and sensor proteins with novel functions. Nature 423(6936):185–190
Luong JH, Male KB, Glennon JD (2008) Biosensor technology: technology push versus market pull. Biotechnol Adv 26(5):492–500
Malekzad H, Zangabad PS, Mirshekari H, Karimi M, Hamblin MR (2017) Noble metal nanoparticles in biosensors: recent studies and applications. Nanotechnol Rev 6(3):301–329
Malhotra BD, Ali MA (2018) Nanomaterials in biosensors: fundamentals and applications. In: Nanomaterials for biosensors, p 1
Manzoor O, Soleja N, Khan P, Hassan MI, Mohsin M (2019) Visualization of thiamine in living cells using genetically encoded fluorescent nanosensor. Biochem Eng J 146:170–178
Neher RA, Neher E (2004) Applying spectral fingerprinting to the analysis of FRET images. Microsc Res Tech 64(2):185–195
Palmer AE, Qin Y, Park JG, McCombs JE (2011) Design and application of genetically encoded biosensors. Trends Biotechnol 29(3):144–152
Pelosse M, Cottet-Rousselle C, Grichine A, Berger I, Schlattner U (2016) Genetically encoded fluorescent biosensors to explore AMPK signaling and energy metabolism. In: AMP-activated protein kinase, pp 491–523
Peltomaa R, Glahn-Martínez B, Benito-Peña E, Moreno-Bondi MC (2018) Optical biosensors for label-free detection of small molecules. Sensors 18(12):4126
Pickup JC, Hussain F, Evans ND, Rolinski OJ, Birch DJ (2005) Fluorescence-based glucose sensors. Biosens Bioelectron 20(12):2555–2565
Pilch-Wrobel A, Kotulska AM, Lahtinen S, Soukka T, Bednarkiewicz A (2022) Engineering the compositional architecture of core-shell upconverting lanthanide-doped nanoparticles for optimal luminescent donor in resonance energy transfer: the effects of energy migration and storage. Small 18(18):2200464

Qiu X, Xu J, Cardoso Dos Santos M, Hildebrandt N (2022) Multiplexed biosensing and bioimaging using lanthanide-based time-gated Forster resonance energy transfer. Acc Chem Res 55(4):551–564
Rajska D, Brzózka A, Marciszko-Wiąckowska M, Marzec MM, Chlebda D, Hnida-Gut KE, Sulka GD (2021) Optimization of synthesis conditions of thin Te-doped InSb films and first principles studies of their physicochemical properties. Appl Surf Sci 537:147715
Resch-Genger U, Grabolle M, Cavaliere-Jaricot S, Nitschke R, Nann T (2008) Quantum dots versus organic dyes as fluorescent labels. Nat Methods 5(9):763–775
Rizzo MA, Piston DW (2005) High-contrast imaging of fluorescent protein FRET by fluorescence polarization microscopy. Biophys J 88(2):L14–L16
Rosenthal SJ (2001) Bar-coding biomolecules with fluorescent nanocrystals. Nat Biotechnol 19(7):621–622
Saenger AK, Laha TJ, Bremner DE, Sadrzadeh SM (2006) Quantification of serum 25-hydroxyvitamin D2 and D3using HPLC–tandem mass spectrometry and examination of reference intervals for diagnosis of vitamin D deficiency. Am J Clin Pathol 125(6):914–920
Sami R, Li Y, Qi B, Wang S, Zhang Q, Han F, Ma Y, Jing J, Jiang L (2014) HPLC analysis of water-soluble vitamins (B2, B3, B6, B12, and C) and fat-soluble vitamins (E, K, D, A, and β-carotene) of okra (Abelmoschus esculentus). J Chem 2014(2):1–6
Sargazi S, Fatima I, Kiani MH, Mohammadzadeh V, Arshad R, Bilal M, Rahdar A, Díez-Pascual AM, Behzadmehr R (2022) Fluorescent-based nanosensors for selective detection of a wide range of biological macromolecules: a comprehensive review. Int J Biol Macromol 206:115–147
Sekar RB, Periasamy A (2003) Fluorescence resonance energy transfer (FRET) microscopy imaging of live cell protein localizations. J Cell Biol 160(5):629
Shimomura O, Johnson FH, Saiga Y (1962) Extraction, purification and properties of aequorin, a bioluminescent protein from the luminous hydromedusan, Aequorea. J Cell Comp Physiol 59(3):223–239
Steffen V, Otten J, Engelmann S, Radek A, Limberg M, Koenig BW, Noack S, Wiechert W, Pohl M (2016) A toolbox of genetically encoded FRET-based biosensors for rapid L-lysine analysis. Sensors 16(10):1604
Steinegger A, Wolfbeis OS, Borisov SM (2020) Optical sensing and imaging of pH values: spectroscopies, materials, and applications. Chem Rev 120(22):12357–12489
Stryer L (1978) Fluorescence energy transfer as a spectroscopic ruler. Annu Rev Biochem 47(1):819–846
Thaler C, Koushik SV, Blank PS, Vogel SS (2005) Quantitative multiphoton spectral imaging and its use for measuring resonance energy transfer. Biophys J 89(4):2736–2749
Valeur B (2001) Introduction: on the origin of the terms fluorescence, phosphorescence, and luminescence. In: New trends in fluorescence spectroscopy: applications to chemical and life sciences. Springer, Berlin, pp 3–6
Vogel SS, Thaler C, Koushik SV (2006) Fanciful fret. Sci STKE 2006(331):re2
Wang YP, Lei QY (2018) Metabolite sensing and signaling in cell metabolism. Signal Transduct Target Ther 3(1):30
Wang XQ, Ma X, Feng D, Tang J, Wu D, Yang J, Jiao J (2021) Four novel lanthanide (III) metal–organic frameworks: tunable light emission and multiresponsive luminescence sensors for vitamin B6 and pesticides. Cryst Growth Des 21(5):2889–2897
Wu XQ, Feng PQ, Guo Z, Wei X (2020) Water-stable 1D double-chain Cu metal–organic framework-based electrochemical biosensor for detecting l-tyrosine. Langmuir 36(46):14123–14129
Yu Y, Xu X, Su Q, Fu T, Liu W, Chen G (2021) Photo-driven self-powered biosensors for ultrasensitive microRNA detection based on metal–organic framework-controlled release behavior. Analyst 146(3):816–819
Zeug A, Stawarski M, Bieganska K, Korotchenko S, Wlodarczyk J, Dityatev A, Ponimaskin E (2014) Current microscopic methods for the neural ECM analysis. Prog Brain Res 214:287–312
Zimmermann T, Rietdorf J, Pepperkok R (2003) Spectral imaging and its applications in live cell microscopy. FEBS Lett 546(1):87–92

Biomedical Applications of Biosensors

5

Neha Chaurasiya, Imra Mumtaz, and Altaf Ahmad

Abstract

Disease diagnosis is crucial for medical sciences. Good selectivity, lower sensitivity, and minimal interference effects are necessary to execute an effective analysis. Compared to conventional procedures, biosensors have several advantages, including excellent sensitivity, quick analysis time, real-time analysis, affordable instruments, and minimal pretreatment steps. Biosensors enable the quantification of metabolites and degradation products in biological fluids in addition to the active ingredient in pharmaceutical formulations. The main goal of this chapter is about the biomedical applications of biosensors, having a short explanation of the core word biosensor with its general classification. This classification is based on the type of transducing and recognition element involved. This chapter highlights the description of the sensor used in various diseases such as cancer, cardiovascular diseases, neurological disorders, etc. The sensors for disease diagnosis are generally based on the metabolite fluctuations present, which play vital roles and are involved in various physiological mechanisms in the body. Overall this chapter gives the detailed applications of biosensors in the medical field.

Keywords

Biosensor · FRET · Protein kinase · Neurological disorders · Genetically encoded · Cardiovascular diseases

N. Chaurasiya · I. Mumtaz · A. Ahmad (✉)
Department of Botany, Aligarh Muslim University, Aligarh, Uttar Pradesh, India

M. Mohsin, N. Soleja (eds.), *Nanoscale Sensors and their Applications in Biomedical Imaging*, https://doi.org/10.1007/978-981-97-3144-2_5

5.1 Introduction

Biosensors are small devices used in diverse fields and are of great significance. The notion of biosensors was first given by Leyland C. Clark in 1962 (Turner 2013). It is a systematic device that uses a bio-sensing element connected to the transducer, and the linking of the target molecule to the biorecognition molecule generates an electrical signal which is then measured. These biosensors provide real-time and quick results, making them a valuable device for detecting changes in body fluids. Biosensors are tiny and can be used to analyze changes in biological systems. The biosensors have been classified into types based on transducing and biorecognition elements, further divided into sub-classes. Based on transducing elements, biosensors are classified into mass-based, electrochemical, and optical. The biorecognition elements include antibodies, phage, DNA, and enzymes. Biosensors are significant in biomedical applications. These are used to diagnose several diseases, such as cancer, cardiovascular diseases, inflammation, neurological disorders, etc.

Different types of biosensors have been made for the detection of diseases in the biomedical field. Severe illnesses like cancer and cardiovascular diseases can be easily detected using biosensors. The ease in their usage, sensitivity, and rapid detection make them excellent devices. Biosensor in disease diagnosis is a novel and more advanced technique that has made the diagnosis process easy. Biosensors are bringing a change in the medical field. In this chapter, we have briefly explained the types of biosensors and their biomedical applications.

5.2 What Is Biosensor?

There is ample literature available on the internet about the keyword biosensor, which undoubtedly proves that the field is fascinating and interesting. This area is interdisciplinary, and the sharp edges cannot be explained easily. In a very lucid language, a biosensor can be defined as a self-contained analytic device whose response is selective in nature and shows the activity and concentration of chemical entities in a biological sample (Coulet and Blum 2019). In an elite way, the biosensor can also be explained as a sensing device that consists of a biological moiety that serves as a recognition tool and a transducer with physicochemical qualities that detects the signal generated through the bio-sensing element (Baeumner 2003).

5.3 Classification of Biosensor

Based on the type of recognition and transducing element involved, biosensors are classified into many types (Fig. 5.1).

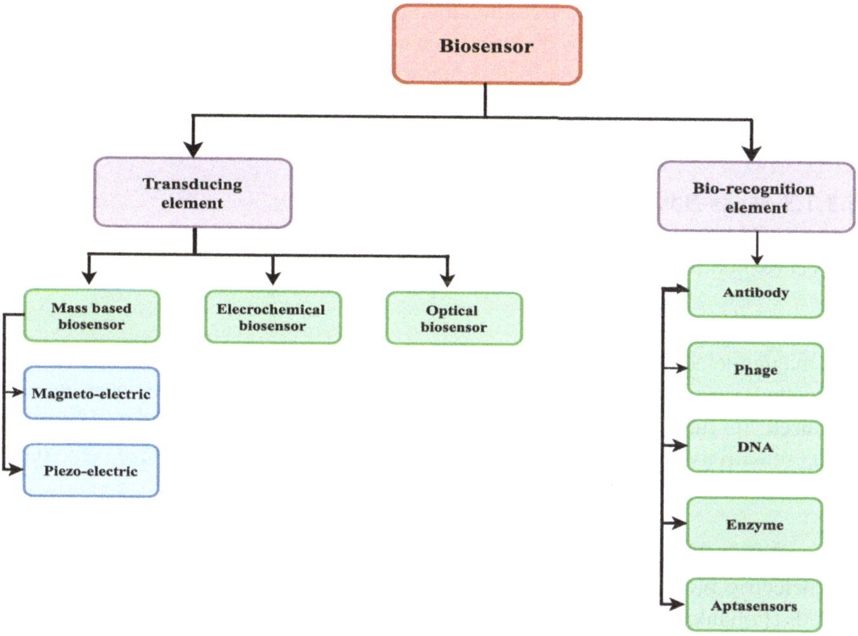

Fig. 5.1 Classification of biosensor

5.3.1 Transducing Element

On the basis of transducing element involved, biosensors are of the following types:

5.3.1.1 Electrochemical Biosensors

An electrochemical biosensor is a device that uses a biorecognition element that acts as a receptor connected with a transducer and provides quantitative and semi-quantitative analytical information (Thevenot et al. 1999). It consists of a working electrode and a reference electrode, and an electrolyte separates them (Hammond et al. 2016). It is a type of biosensor that detects the biomolecule by generating an electrical signal such as voltage, current, impedance, etc. (Singh et al. 2021). These biosensors are composed of biorecognition elements such as proteins, nucleic acids, enzymes, receptors, cells, etc., to which the specific analyte binds and generates an electrical signal which varies depending upon the concentration of the analyte (Ronkainen et al. 2010). Electrochemical biosensors are used in the monitoring of glucose levels in diabetes.

5.3.1.2 Optical Biosensors

Optical biosensors provide qualitative and quantitative information about various biological systems (Myszka 1999). The optical biosensors have been categorized into label-based and label-free types. The label-free type optical biosensors involve the interaction of the analyte with the transducer, generating a signal. While label-based sensors depend upon the label, and then the optical signal is produced by colorimetric, fluorescent, or luminescent method (Damborský et al. 2016). In optical biosensors, transduced light is measured as the output signal (Mohanty and Kougianos 2006). These are used in drug development by determining the binding affinities of different compounds to receptors (Cooper 2002).

5.3.1.3 Mass-Based Biosensors

Mass-based biosensors come under the category of transducing elements and have been classified into two types: magnetoelectric sensors and piezoelectric sensors.

Magnetoelectric Sensors

The magnetoelectric composites have been used in the production of sensors by the magnetoelectric combination, and besides the sensors, several other devices are also produced. Its functioning depends upon different parameters such as the detection limit, sensitivity, the temperature at which it works efficiently, size, etc. (Gao et al. 2021).

Piezoelectric Biosensors

Piezoelectric biosensors are devices used to detect various macromolecular compounds (Pohanka 2018). These biosensors work on the principle of the piezoelectric effect. Jacques Curie and Pierre Curie, the two physicists, discovered the piezoelectric effect and described the anisotropic crystals, which generate electric dipole when squeezed mechanically (Pohanka 2017). Piezoelectric biosensors have also been incorporated into the biomedical field, which involves piezoelectric immunosensors and genosensors (Tombelli 2012). These biosensors rapidly detect pests (Marrazza 2014).

5.3.2 Biorecognition Elements

Biorecognition element provides specificity to the analyte in a biosensor (Morales and Halpern 2018). The biorecognition elements are produced in the laboratory because they do not occur naturally (Chambers et al. 2008). The biorecognition elements used in the biosensor include antibodies, phage, DNA, enzyme, biomimetic recognition element, etc.

5.3.2.1 Antibody Recognition Elements

Antibody recognition elements are used in immunosensors. Immunosensors are of two types: direct or indirect. Hence, the detection of binding depends upon whether antigen and antibody interact directly or whether a label has been used for detection, such as an enzyme or fluorescent molecule (Holford et al. 2012). Due to their strong affinity and specificity to antigens, antibodies are quintessential for use as biorecognition elements (Conroy et al. 2009).

5.3.2.2 Phage Recognition Elements

The phage recognition elements consist of phage virus, which attacks bacteria, and these phage viruses are modified to form a library of phages on which different peptides are present for their specific targets (Nanduri et al. 2007). The phage-based biosensors combine target moieties and fluorescent molecules on the capsid's surface to detect a transduced signal (Lee et al. 2013).

5.3.2.3 DNA Recognition Elements

In DNA-based biosensors, short single-stranded DNA is immobilized on the surface of an electrode, and an electrical signal is produced when the target DNA binds to the immobilized DNA strand. These biosensors are called DNA hybridization-based biosensors (Kavita 2017).

5.3.2.4 Enzyme Recognition Elements

Enzymes are used as recognition elements in biosensors. The enzyme-based biosensor is a device in which a transducer produces a signal when it comes in close contact with the enzyme, and the intensity of the signal depends upon the concentration of the analyte (Kataria et al. 2014).

5.3.2.5 Aptamer Biosensors

Aptamers are short, single-stranded nucleic acid molecules that can undergo three-dimensional conformational change (Sun and Zu 2015). Aptasensors are small in size and highly efficient, which can be reused, and these properties make them versatile (Mairal et al. 2008).

Due to their speed, specificity, field applicability, economies, and simplicity of mass production, biosensor devices are becoming one of the essential applicable diagnostic tools for clinical, food, and environmental monitoring. Their biological binding reaction, which results from a variety of interactions such as receptor/ligand, enzyme/cofactor/substrate, chemical interactions, antigen/antibody, and nucleic acid hybridization in combination with a variety of transducers, gives them their specificity (Thakur and Ragavan 2013). In this chapter, we will discuss the biomedical applications of biosensors.

5.4 Biomedical Applications

Genetically encoded fluorescence biosensors have provided a wealth of information on a cell's biochemical pathways since they were developed in the early 2000s. Researchers in the field of biomedicine have become more interested in genetically encoded fluorescence biosensors over the past 10 years to collect information on pathological processes emerging due to exposure to toxic chemicals or physiological disturbances. To improve drug development and increase the effectiveness of medical care, biosensors provide direct measurements of variations in chemical and physical parameters (such as pH changes or ions flow), the dynamics of macromolecule activities, and the registration of drug potency and its effects on the target pathways. Any scale, from a single cell cultivated in vitro to the complete tissue, can be used to make these measurements (Ovechkina et al. 2021).

5.4.1 Cancer

Cancer is a broad category of diseases distinguished by uncontrollable cell growth, the capacity to spread into many body tissues, and the development of tumors. In addition, the sickness is well recognized to exhibit significant intratumoral and intertumoral heterogeneity. After undergoing a malignant transformation, cancer tumors continue to grow due to the cells' varied exposure, which results in genetic heterogeneity and a range of molecular mechanisms in the cells of a single tumor type. The fact that cancer has different pathogenic processes, even in cells from the same tumor, challenges research and treatment development in the disease (Meacham and Morrison 2013; Prasetyanti and Medema 2017; Dagogo-Jack and Shaw 2018). As a result, it provides exciting new opportunities for using genetically encoded fluorescent biosensors for pathological process research and anti-cancer drug screening. Several intrinsic pathways accompany tumor formation, such as genetic changes in transcription factors or antigens linked to tumors' appearance. Genetically encoded fluorescent biosensors are intended to serve as a suitable toolkit for these molecular approaches.

5.4.1.1 Measurement of Protein Kinase Activity

Living cells accumulate genetic changes through proliferation and aging as the basis for the development of cancer, which can then result in the inactivation of tumor suppressor genes. Thus, one of the critical networks implicated in the development of cancer is protein kinase (PK) signaling. The network controls cellular survival and proliferation; hence, mutations in the genes encoding the proteins that make up this route may impact cellular malignancy. As a result, the study of cancer development is intensely interested in what they observe. For the first time, a group of scientists studied protein kinase with the help of genetically encoded fluorescent biosensors in 2001. The scientists introduced a FRET-based biosensor called AKAR, made of CFP and associated with a phosphoramide acid-binding domain (14-3-3) and YFP, to study cAMP-dependent protein kinase A (PKA). The domain can recognize and bind the phosphorylated peptide selectively—two fluorophores' emission changes due to altered phosphorylation brought on by an elevated amount of cAMP. The intramolecular binding of 14-3-3 to the substrate is increased by phosphorylation, and the effect is reversed by dephosphorylation. The FRET shifts between the combination of CFP and YFP enable this construct to reflect variation in kinase activity because the kinase activity often varies against a relatively steady phosphatase activity (Zhang et al. 2001).

Later versions of biosensors, however, were built on an FHA1 domain rather than the 14-3-3 domain, which has less affinity and does not cause an irreversible stress reaction. Additionally, in 2011, a team of researchers created a bioluminescent sensor for measuring the activity of protein kinase A and protein kinase C

(PKC). In response to kinase activity, the reporter can switch between luminous and FRET reporter units. As a result, the enhanced PKA activity restores the brilliant component's catalytic activity (Herbst et al. 2011). Later, FLIM technology was added to the FRET-based biosensors. By swapping out YFP for mRFP1 and CFP for EGFP, a Picchu-FLIM sensor was created using the CrkII chimeric unit's FRET EGFR activity biosensor phosphorylation indication (Kurokawa et al. 2001; Weitsman et al. 2017). The elaborated data about protein kinase is given in Table 5.1.

Utilizing fluorescent biosensors that are genetically encoded allows for the quantification of kinase activity, as well as the detection of pathogenic dysregulations and the choice of therapeutic medications.

5.4.1.2 Measurement of p53 Activity

Cancer progression and carcinogenesis are linked to decreased p53 activity. In addition, the overexpressed hDM2 protein targets p53 for destruction after it has been ubiquitinylated (Moll and Petrenko 2003). A framework for automated high-content screening for protein–protein interaction disruptors was built to uncover novel inhibitors of the p53–hDM2 interaction. First, an NLS was added to p53 to help localize the construction in the nucleolus. NLS and NES were both carried by hDM2 and were fused to RFP. In the nucleolus, both proteins were found together. hDM2 was exported to the cytoplasm when Nutlin-3, a known inhibitor of the p53–hDM2 interaction, was present. 220,000 small-molecule compounds were screened in a library, and the assay showed remarkable repeatability. Finally, it was established that three chemicals related to methylbenzonaphthyridin-5-amine cause cell cycle arrest and growth inhibition in a p53-dependent manner, increase p53 protein level, and promote apoptosis (Dudgeon et al. 2010).

Table 5.1 The studying of protein kinase using fluorescent-based sensors

Kinase	Name	Type of sensors	Proteins	References
Protein kinase A	AKAR	FRET	ECFP/YFP	Zhang et al. (2001)
	AKAR2	FRET	ECFP/Citrine	Zhang et al. (2005)
	AKAR3	FRET	YPet	Allen and Zhang (2006)
Protein kinase B	*Akt*AR	FRET	CFP/Venus	Gao and Zhang (2008)
Protein kinase C	CKAR	FRET	CFP/YFP	Violin et al. (2003)
Serine/threonine protein kinase	TORCAR	FRET	Cerulean/YPet	Zhou et al. (2015)
LATS kinase	LATS-BS	Bioluminescence	*Photinus pyralis* luciferase	Azad et al. (2018)
P38 MAP kinase	PerKy-38	FRET	YPet/CFP	Tomida et al. (2015)

5.4.1.3 Measurement of Caspase-3 Activity

There is now a reporter for caspase-3 activity based on FRET. It comprises a linker sequence with a caspase-3 recognition site, CFP, and YFP. The sensor protein is cut after caspase-3 activation by apoptotic stimuli, and the FRET ratio falls. In necrotic cell death, the FRET ratio does not drop. The sensor was able to reflect the pro-apoptotic effect for many chemicals with established biological activity, including vincristine, paclitaxel, and hydroxyurea, as well as for several novel plant-derived substances, when expressed in HeLa-C3 cells cultured on microplates (Tian et al. 2007).

5.4.2 Neurological Disorders

Any nervous system condition, ranging from inherited neurodegenerative processes to physical trauma-induced nerve injury, is included in the category of neurological disorders (Fig. 5.2).

Regarding the biological reason for diseases, anomalies in brain tissue development are brought on by particular protein deposits in the matrix or post-translational modification mistakes. The bulk of molecular processes involved in the pathogenic process and the ways and treatments for preventing neurological illnesses are still largely unknown today (Lehn et al. 2016). Different neurological disorders share several essential traits and mechanisms that cause a malfunction in neurons. Surprisingly, the only in vivo biomarkers now accessible for identifying diseases are reverse genetics techniques (Dugger and Dickson 2017). In light of this, real-time analysis of various chemical pathways in biological systems is particularly intriguing. These conditions are addressed by genetically encoded fluorescence biosensors, which can be used to visualize multiple cellular metabolic and signaling pathways and shed light on the causes of pathological disease processes and the effects of medications on defective neurons.

5.4.2.1 Measurement of Lactate Level

Hypoxic tissues release a large quantity of lactate. The overproduction of lactate during hypoxia was often considered a pathogenic process. Still, it is now evident that lactate plays a crucial role in normally oxygenated cells because its level rises dramatically during neuronal activity (Gladden 2004). In addition to being engaged in myelination processes, lactate serves as an intracellular signal for sodium sensing and neurovascular coupling (Shimizu et al. 2007). As a result, the "laconic" previously revealed allows for lactate monitoring in the physiological range and can be transduced into specific cells for in vivo lactate measurements (San Martin et al. 2013).

Some researchers used laconic to image neurons and cortical astrocytes both in vitro and in vivo (Machler et al. 2016). It is understood that lactate is transported across the membrane by monocarboxylate transporters (MCTs) and that lactate dehydrogenase catalyzes the conversion of lactate to pyruvate. Astrocytic lactate synthesis and neuronal lactate consumption are caused by variations in the lactate

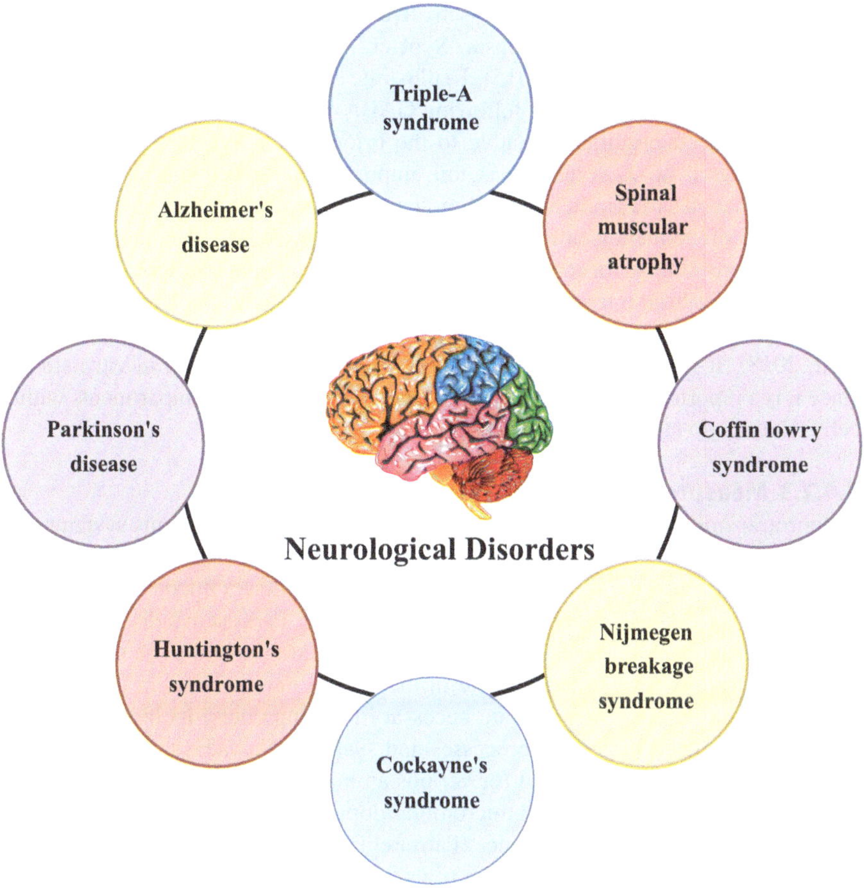

Fig. 5.2 Neurological disorders

affinities of LDH and MCTs in astrocytes and neurons, respectively (Pierre and Pellerin 2005). In the research, these scientists first demonstrated a lactate gradient in mice models in vivo from astrocytes to neurons (Machler et al. 2016). The information gathered may help with medication testing for the treatment of epilepsy because LDH inhibition makes neurons more active during seizures (Sada et al. 2015) and may also help with understanding how Huntington's disease develops (Solis-Maldonado et al. 2018). Additionally, the laconic sensor can be used for drug screening to look for mitochondrial dysfunction caused by toxicant exposure (Contreras-Baeza et al. 2019).

5.4.2.2 Measurement of pH Level

The ability of the genetically encoded reporters to measure the H^+ rate at the cellular and subcellular levels offers considerable promise for elucidating the underlying mechanisms of brain diseases and advancing our understanding of neural regeneration. The well-known H_2O_2-sensitive sensor HyPer2 (Markvicheva et al. 2011) gave rise to the new pH sensor, SypHer2, used in regenerative biology and neuroscience (Matlashov et al. 2015). To illustrate epileptiform activity, the SypHer2 was used to monitor pH activity following GABA-receptor blockage. The SypHer2 showed enhanced brightness relative to the prior sensor SypHer (Poburko et al. 2011). Research on *Xenopus laevis* tail amputations showed that cells near the wound displayed a pH gradient with significant acidification. The information might imply that a pH reduction can affect regeneration. However, SypHer and SypHer2's brightness was subpar for a performance with excellent spatiotemporal resolution. As a result, the brightness and dynamic pH range of a new SypHer-based sensor called SypHer3s with mutations Y145F and D129G were increased (Ermakova et al. 2018). It has considerable promise for neuronal in vivo and in vitro studies since it is a reporter that can identify pH variations in cells and compartments within cells (Raimondo et al. 2016; Chebotarev et al. 2021).

5.4.2.3 Measurement of Dopamine Level

A neurotransmitter called dopamine is crucial for the central nervous systems and cardiovascular systems. High dopamine levels, therefore, signal cardiotoxicity, result in a fast heartbeat, hypertension, and heart failure (Bucolo et al. 2019). Conversely, several neurological conditions, including Parkinson's disease, schizophrenia, Alzheimer's disease, stress, and depression, are thought to be significantly influenced by low levels of dopamine in the central nervous system (Klein et al. 2019). Dopamine measurements are necessary to comprehend its biological roles and those of related biological processes and systems. Nowadays, sensitive dopamine readouts have been created for various animal models. Accordingly, dLight1 sensors can detect changes in submicromolar dopamine concentration at individual dendritic spines and dendrites in mice (Patriarchi et al. 2018). In contrast, GRABDA sensors had optimized brightnesses in trials with zebrafish, flies, and not just mice (Sun et al. 2018).

5.4.2.4 Measurement of Metabolite Level

Multiple neurotransmitters can detect glucose, the brain's energy source, and cell metabolites also play a crucial part in developing neurons. For instance, numerous reporters are able to see glucose, which serves as the brain's energy supply. The intracellular glucose concentration in the brain was estimated to be around 1.4 mM using the data collected by sensors. Additionally, a researcher with his team decided on the function of hydrogen sulfide (H_2S), another cell metabolite, during the onset of neurodegenerative diseases (Youssef et al. 2019). In living things, H_2S is released during sulfur metabolism and functions as a gasotransmitter (Abe and Kimura 1996). Dysregulation of this chemical has been linked to several neurological illnesses (Kamoun et al. 2003). As a result, a brand-new ratiometric biosensor called hsFRET was developed and tested in HEK293T for sensitive real-time monitoring of H_2S dynamics. The study proved that hsFRET can be used for in vitro investigations in mammalian cells (Youssef et al. 2019).

5.4.3 Inflammation

A gut-adapted strain of *E. coli* with *Shewanella* sp. insertions that were sensitive to the hydrogen sulfide (H_2S) metabolites tetrathionate and thiosulfate was presented by Kristina N-M Daeffler's biosensor (Daeffler et al. 2017). Engineered bacteria were fed to the mice, and flow cytometry was used to examine their feces. The H_2S metabolites' level may reflect colitis because this study shows intestinal inflammation triggers the reporter (Fig. 5.3).

Another *E. coli*-based sensor responds to elevated nitric oxide levels (Woo et al. 2020), a biomarker of intestinal inflammation (Lundberg et al. 1994). The device can react to physiological nitrate levels in the stomach and be confirmed in vivo using a mouse model. In addition, the scientists suggest switching out the reporter gene for an anti-inflammatory gene to establish a colitis treatment tool.

Since most biomarkers for intestinal inflammation are still being studied, one interesting synthetic biology strategy uses biosensors to identify many pathological processes and signaling molecules in a single experiment. So, using the lambda cI/Cro system (Kotula et al. 2014), Harvard Medical School researchers developed an *E. coli* strain with a genetic memory component and then modified it with a transcriptional promoter activated in the presence of a specific stimulus (Naydich et al. 2019). Building biosensor starts libraries which can foretell probable signs of intestinal inflammation that are conceivable using data from cell responses. The developed system enables the detection of pathogenic pathways that cannot be rationally built using the current understanding of colitis biomarkers.

Fluorescent sensors based on bacteria may identify not only the whole organ inflammation but also the inflammation of the tissue. For instance, when mice's ears were injected with an *E. coli* strain that expressed GFP, the immunological response was activated, and an increase in fluorescence accompanied the inflammation, followed by an increase in fluorescence (Liu et al. 2019).

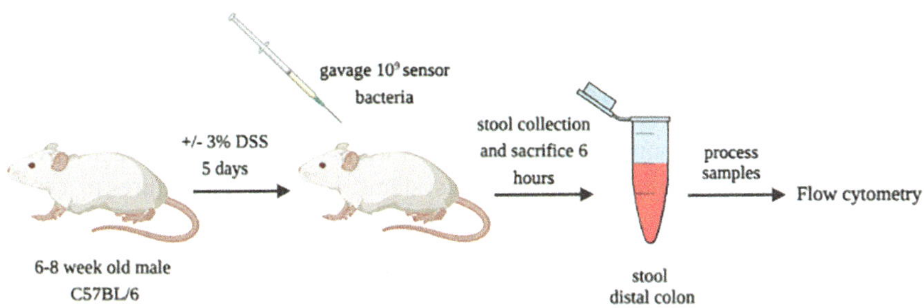

Fig. 5.3 Thiosulfate and tetrathionate levels were measured in vivo in mice with inflammation and health experimental approach. C57BL/6 mice 6–8 weeks old were given water for 5 days with or without 3% DSS before receiving an oral gavage of sensor bacteria. Samples from the mice were taken after 6 h, processed, and examined by flow cytometry to gauge GFP production

5.4.4 Cardiovascular Diseases

Cardiovascular disorders are one of the leading causes of death globally. Voltage-gated ion channels are critical for operating cells with electrical excitability, like cardiomyocytes and many other cellular activities. This makes them one of the main pharmacological targets for cardiovascular diseases, so it is critical to create high-throughput approaches for identifying the inhibitors of voltage-gated ion channels and modulators. There is now potential for all-optical electrophysiology, thanks to the discovery of genetically encoded voltage indicators and optogenetic actuators (Braubach et al. 2015; Joshi et al. 2019).

Action potential imaging of specific cardiomyocyte subtypes was performed using VSFP-CR, a FRET-based voltage sensor comprising a potassium channel's voltage-sensing domain and a GFP/RFP FRET pair (Sakai et al. 2001; Lam et al. 2012). It was possible to detect and measure fluctuations in action potential duration and the frequency of early afterdepolarizations brought on by harmful mutation or brought on by medication in patient-specific human-caused pluripotent stem cell-derived ventricular-, atrial-, or nodal-like cardiomyocytes by placing the sensor under the control of lineage-specific cardiomyocyte promoters (Chen et al. 2017).

Some scientists used ArcLight A242, a variation of Ciona intestinalis voltage-sensing phosphatase-based sensors with fluorescent protein super ecliptic pHluorin (Jin et al. 2012; Shinnawi et al. 2015; Sun et al. 2020). To optically map the electrical activity in a two-dimensional cardiac tissue model under various experimental conditions, including electrically and drug-induced arrhythmias and interventions to prevent arrhythmias, they used cardiac cell sheets derived from human-induced pluripotent stem cells that express the ArcLight protein.

5.5 Conclusion

Over the past 50 years, genetically encoded fluorescence biosensors have become promising replacements for established conventional approaches. Molecular interactions can now be observed and studied in their natural environment because of genetic alteration of the cell genome. New biochemical, cytological, and physiological research instruments are genetically encoded fluorescence biosensors, particularly in biomedicine. Both in vitro and in vivo, they enable real-time monitoring with excellent spatiotemporal precision of various routes within the live cell. Sensors can be included in multiple model systems, from 2D cell cultures to a complete organism, which is a huge advantage. Today, biosensors reveal the underlying causes of numerous disorders, including neurodegeneration, cancer, inflammation, etc. They have also emerged as a viable method for drug screening. Many exciting approaches exist in increasing sensor sensitivity and creating novel model systems. Therefore, genetically encoded fluorescence biosensors appear to have a bright future. Numerous experiments are being conducted to enhance imaging methods and create novel model systems for understanding protein interactions, drug screening, and personalized treatment (Ovechkina et al. 2021).

References

Abe K, Kimura H (1996) The possible role of hydrogen sulfide as an endogenous neuromodulator. J Neurosci 16:1066–1071

Allen MD, Zhang J (2006) Subcellular dynamics of protein kinase A activity visualized by FRET-based reporters. Biochem Biophys Res Commun 348:716–721

Azad T, Janse van Rensburg HJ, Lightbody ED, Neveu B, Champagne A, Ghaffari A, Kay VR, Hao Y, Shen H, Yeung B et al (2018) A LATS biosensor screen identifies VEGFR as a regulator of the Hippo pathway in angiogenesis. Nat Commun 9:1061

Baeumner AJ (2003) Biosensors for environmental pollutants and food contaminants. Anal Bioanal Chem 377:434–445

Braubach O, Cohen LB, Choi Y (2015) Historical overview and general methods of membrane potential imaging. Adv Exp Med Biol 859:3–26

Bucolo C, Leggio GM, Drago F, Salomone S (2019) Dopamine outside the brain: the eye, cardiovascular system and endocrine pancreas. Pharmacol Ther 203:107392

Chambers JP, Arulanandam BP, Matta LL, Weis A, Valdes JJ (2008) Biosensor recognition elements. Curr Issues Mol Biol 10(1–2):1–12

Chebotarev AS, Pochechuev MS, Lanin AA, Kelmanson IV, Kotova DA, Fetisova ES, Panova AS, Bilan DS, Fedotov AB, Belousov VV et al (2021) Enhanced-contrast two-photon optogenetic pH sensing and pH-resolved brain imaging. J Biophotonics 14:e202000301

Chen Z, Xian W, Bellin M, Dorn T, Tian Q, Goedel A, Dreizehnter L, Schneider CM, Ward-van Oostwaard D, Ng JK et al (2017) Subtype-specific promoter-driven action potential imaging for precise disease modelling and drug testing in hiPSC-derived cardiomyocytes. Eur Heart J 38:292–301

Conroy PJ, Hearty S, Leonard P, O'Kennedy RJ (2009) Antibody production, design and use for biosensor-based applications. Semin Cell Dev Biol 20(1):10–26

Contreras-Baeza Y, Ceballo S, Arce-Molina R, Sandoval PY, Alegria K, Barros LF, San Martin A (2019) MitoToxy assay: a novel cell-based method for the assessment of metabolic toxicity in a multiwell plate format using a lactate FRET nanosensor, Laconic. PLoS One 14:e0224527

Cooper MA (2002) Optical biosensors in drug discovery. Nat Rev Drug Discov 1(7):515–528

Coulet PR, Blum LJ (eds) (2019) Biosensor principles and applications. CRC Press, Boca Raton, FL

Daeffler KN, Galley JD, Sheth RU, Ortiz-Velez LC, Bibb CO, Shroyer NF, Britton RA, Tabor JJ (2017) Engineering bacterial thiosulfate and tetrathionate sensors for detecting gut inflammation. Mol Syst Biol 13:923

Dagogo-Jack I, Shaw AT (2018) Tumour heterogeneity and resistance to cancer therapies. Nat Rev Clin Oncol 15:81–94

Damborský P, Švitel J, Katrlík J (2016) Optical biosensors. Essays Biochem 60(1):91–100

Dudgeon DD, Shinde S, Hua Y, Shun TY, Lazo JS, Strock CJ, Giuliano KA, Taylor DL, Johnston PA, Johnston PA (2010) Implementation of a 220,000-compound HCS campaign to identify disruptors of the interaction between p53 and hDM2 and characterization of the confirmed hits. J Biomol Screen 15:766–782

Dugger BN, Dickson DW (2017) Pathology of neurodegenerative diseases. Cold Spring Harb Perspect Biol 9:a028035

Ermakova YG, Pak VV, Bogdanova YA, Kotlobay AA, Yampolsky IV, Shokhina AG, Panova AS, Marygin RA, Staroverov DB, Bilan DS et al (2018) SypHer3s: a genetically encoded fluorescent ratiometric probe with enhanced brightness and an improved dynamic range. Chem Commun 54:2898–2901

Gao X, Zhang J (2008) Spatiotemporal analysis of differential Akt regulation in plasma membrane microdomains. Mol Biol Cell 19:4366–4373

Gao J, Jiang Z, Zhang S, Mao Z, Shen Y, Chu Z (2021) Review of magnetoelectric sensors. Actuators 10(6):109

Gladden LB (2004) Lactate metabolism: a new paradigm for the third millennium. J Physiol 558:5–30

Hammond JL, Formisano N, Estrela P, Carrara S, Tkac J (2016) Electrochemical biosensors and nanobiosensors. Essays Biochem 60(1):69–80

Herbst KJ, Allen MD, Zhang J (2011) Luminescent kinase activity biosensors based on a versatile bimolecular switch. J Am Chem Soc 133:5676–5679

Holford TR, Davis F, Higson SP (2012) Recent trends in antibody based sensors. Biosens Bioelectron 34(1):12–24

Jin L, Han Z, Platisa J, Wooltorton JR, Cohen LB, Pieribone VA (2012) Single action potentials and subthreshold electrical events imaged in neurons with a fluorescent protein voltage probe. Neuron 75:779–785

Joshi J, Rubart M, Zhu W (2019) Optogenetics: background, methodological advances and potential applications for cardiovascular research and medicine. Front Bioeng Biotechnol 7:466

Kamoun P, Belardinelli MC, Chabli A, Lallouchi K, Chadefaux-Vekemans B (2003) Endogenous hydrogen sulfide overproduction in Down syndrome. Am J Med Genet A 116A:310–311

Kataria M, Kaur J, Punia S, Aggarwal D, Kumar K (2014) Immobilized enzymes and its implications. In: Industrial enzymes. Nova, p 73

Kavita VJJBBS (2017) DNA biosensors—a review. J Bioeng Biomed Sci 7(2):222

Klein MO, Battagello DS, Cardoso AR, Hauser DN, Bittencourt JC, Correa RG (2019) Dopamine: functions, signaling, and association with neurological diseases. Cell Mol Neurobiol 39:31–59

Kotula JW, Kerns SJ, Shaket LA, Siraj L, Collins JJ, Way JC, Silver PA (2014) Programmable bacteria detect and record an environmental signal in the mammalian gut. Proc Natl Acad Sci USA 111:4838–4843

Kurokawa K, Mochizuki N, Ohba Y, Mizuno H, Miyawaki A, Matsuda M (2001) A pair of fluorescent resonance energy transfer-based probes for tyrosine phosphorylation of the CrkII adaptor protein in vivo. J Biol Chem 276:31305–31310

Lam AJ, St-Pierre F, Gong Y, Marshall JD, Cranfill PJ, Baird MA, McKeown MR, Wiedenmann J, Davidson MW, Schnitzer MJ et al (2012) Improving FRET dynamic range with bright green and red fluorescent proteins. Nat Methods 9:1005–1012

Lee JW, Song J, Hwang MP, Lee KH (2013) Nanoscale bacteriophage biosensors beyond phage display. Int J Nanomedicine 8:3917–3925

Lehn A, Gelauff J, Hoeritzauer I, Ludwig L, McWhirter L, Williams S, Gardiner P, Carson A, Stone J (2016) Functional neurological disorders: mechanisms and treatment. J Neurol 263:611–620

Liu Y, Fu L, Xu M, Zheng J, Yuan Z (2019) Dual-modal in vivo fluorescence/Photoacoustic microscopy imaging of inflammation induced by GFP-expressing bacteria. Sensors 19:238

Lundberg JO, Hellstrom PM, Lundberg JM, Alving K (1994) Greatly increased luminal nitric oxide in ulcerative colitis. Lancet 344:1673–1674

Machler P, Wyss MT, Elsayed M, Stobart J, Gutierrez R, von Faber-Castell A, Kaelin V, Zuend M, San Martin A, Romero-Gomez I et al (2016) In vivo evidence for a lactate gradient from astrocytes to neurons. Cell Metab 23:94–102

Mairal T, Cengiz Özalp V, Lozano Sánchez P, Mir M, Katakis I, O'Sullivan CK (2008) Aptamers: molecular tools for analytical applications. Anal Bioanal Chem 390:989–1007

Markvicheva KN, Bilan DS, Mishina NM, Gorokhovatsky AY, Vinokurov LM, Lukyanov S, Belousov VV (2011) A genetically encoded sensor for H_2O_2 with expanded dynamic range. Bioorg Med Chem 19:1079–1084

Marrazza G (2014) Piezoelectric biosensors for organophosphate and carbamate pesticides: a review. Biosensors (Basel) 4(3):301–317

Matlashov ME, Bogdanova YA, Ermakova GV, Mishina NM, Ermakova YG, Nikitin ES, Balaban PM, Okabe S, Lukyanov S, Enikolopov G et al (2015) Fluorescent ratiometric pH indicator SypHer2: applications in neuroscience and regenerative biology. Biochim Biophys Acta 1850:2318–2328

Meacham CE, Morrison SJ (2013) Tumour heterogeneity and cancer cell plasticity. Nature 501:328–323

Mohanty SP, Kougianos E (2006) Biosensors: a tutorial review. IEEE Potentials 25(2):35–40

Moll UM, Petrenko O (2003) The MDM2-p53 interaction. Mol Cancer Res 1:1001–1008
Morales MA, Halpern JM (2018) Guide to selecting a biorecognition element for biosensors. Bioconjug Chem 29(10):3231–3239
Myszka DG (1999) Improving biosensor analysis. J Mol Recognit 12(5):279–284
Nanduri V, Sorokulova IB, Samoylov AM, Simonian AL, Petrenko VA, Vodyanoy V (2007) Phage as a molecular recognition element in biosensors immobilized by physical adsorption. Biosens Bioelectron 22(6):986–992
Naydich AD, Nangle SN, Bues JJ, Trivedi D, Nissar N, Inniss MC, Niederhuber MJ, Way JC, Silver PA, Riglar DT (2019) Synthetic gene circuits enable systems-level biosensor trigger discovery at the host-microbe interface. mSystems 4:e00125–e00119
Ovechkina VS, Zakian SM, Medvedev SP, Valetdinova KR (2021) Genetically encoded fluorescent biosensors for biomedical applications. Biomedicines 9(11):1528
Patriarchi T, Cho JR, Merten K, Howe MW, Marley A, Xiong WH, Folk RW, Broussard GJ, Liang R, Jang MJ et al (2018) Ultrafast neuronal imaging of dopamine dynamics with designed genetically encoded sensors. Science 360(6396):eaat4422
Pierre K, Pellerin L (2005) Monocarboxylate transporters in the central nervous system: distribution, regulation and function. J Neurochem 94:1–14
Poburko D, Santo-Domingo J, Demaurex N (2011) Dynamic regulation of the mitochondrial proton gradient during cytosolic calcium elevations. J Biol Chem 286:11672–11684
Pohanka M (2017) The piezoelectric biosensors: principles and applications. Int J Electrochem Sci 12:496–506
Pohanka M (2018) Overview of piezoelectric biosensors, immunosensors and DNA sensors and their applications. Materials 11(3):448
Prasetyanti PR, Medema JP (2017) Intra-tumor heterogeneity from a cancer stem cell perspective. Mol Cancer 16:41
Raimondo JV, Tomes H, Irkle A, Kay L, Kellaway L, Markram H, Millar RP, Akerman CJ (2016) Tight coupling of astrocyte pH dynamics to epileptiform activity revealed by genetically encoded pH sensors. J Neurosci 36:7002–7013
Ronkainen NJ, Halsall HB, Heineman WR (2010) Electrochemical biosensors. Chem Soc Rev 39(5):1747–1763
Sada N, Lee S, Katsu T, Otsuki T, Inoue T (2015) Epilepsy treatment. Targeting LDH enzymes with a stiripentol analog to treat epilepsy. Science 347:1362–1367
Sakai R, Repunte-Canonigo V, Raj CD, Knopfel T (2001) Design and characterization of a DNA-encoded, voltage-sensitive fluorescent protein. Eur J Neurosci 13:2314–2318
San Martin A, Ceballo S, Ruminot I, Lerchundi R, Frommer WB, Barros LF (2013) A genetically encoded FRET lactate sensor and its use to detect the Warburg effect in single cancer cells. PLoS One 8:e57712
Shimizu H, Watanabe E, Hiyama TY, Nagakura A, Fujikawa A, Okado H, Yanagawa Y, Obata K, Noda M (2007) Glial Nax channels control lactate signaling to neurons for brain [Na^+] sensing. Neuron 54:59–72
Shinnawi R, Huber I, Maizels L, Shaheen N, Gepstein A, Arbel G, Tijsen AJ, Gepstein L (2015) Monitoring human induced pluripotent stem cell-derived cardiomyocytes with genetically encoded calcium and voltage fluorescent reporters. Stem Cell Rep 5:582–596
Singh A, Sharma A, Ahmed A, Sundramoorthy AK, Furukawa H, Arya S, Khosla A (2021) Recent advances in electrochemical biosensors: applications, challenges, and future scope. Biosensors 11(9):336
Solis-Maldonado M, Miro MP, Acuna AI, Covarrubias-Pinto A, Loaiza A, Mayorga G, Beltran FA, Cepeda C, Levine MS, Concha II et al (2018) Altered lactate metabolism in Huntington's disease is dependent on GLUT3 expression. CNS Neurosci Ther 24:343–352
Sun H, Zu Y (2015) A highlight of recent advances in aptamer technology and its application. Molecules 20(7):11959–11980
Sun F, Zeng J, Jing M, Zhou J, Feng J, Owen SF, Luo Y, Li F, Wang H, Yamaguchi T et al (2018) A genetically encoded fluorescent sensor enables rapid and specific detection of dopamine in flies, fish, and mice. Cell 174:481–496.e19

Sun YH, Kao HKJ, Chang CW, Merleev A, Overton JL, Pretto D, Yechikov S, Maverakis E, Chiamvimonvat N, Chan JW et al (2020) Human induced pluripotent stem cell line with genetically encoded fluorescent voltage indicator generated via CRISPR for action potential assessment post-cardiogenesis. Stem Cells 38:90–101

Thakur MS, Ragavan KV (2013) Biosensors in food processing. J Food Sci Technol 50:625–641

Thevenot DR, Toth K, Durst RA, Wilson GS (1999) Electrochemical biosensors: recommended definitions and classification. Pure Appl Chem 71(12):2333–2348

Tian H, Ip L, Luo H, Chang DC, Luo KQ (2007) A high throughput drug screen based on fluorescence resonance energy transfer (FRET) for anticancer activity of compounds from herbal medicine. Br J Pharmacol 150:321–334

Tombelli S (2012) Piezoelectric biosensors for medical applications. In: Biosensors for medical applications. Woodhead Publishing, Cambridge, pp 41–64

Tomida T, Takekawa M, Saito H (2015) Oscillation of p38 activity controls efficient pro-inflammatory gene expression. Nat Commun 6:8350

Turner AP (2013) Biosensors: sense and sensibility. Chem Soc Rev 42(8):3184–3196

Violin JD, Zhang J, Tsien RY, Newton AC (2003) A genetically encoded fluorescent reporter reveals oscillatory phosphorylation by protein kinase C. J Cell Biol 161:899–909

Weitsman G, Mitchell NJ, Evans R, Cheung A, Kalber TL, Bofinger R, Fruhwirth GO, Keppler M, Wright ZVF, Barber PR et al (2017) Detecting intratumoral heterogeneity of EGFR activity by liposome-based in vivo transfection of a fluorescent biosensor. Oncogene 36:3618–3628

Woo SG, Moon SJ, Kim SK, Kim TH, Lim HS, Yeon GH, Sung BH, Lee CH, Lee SG, Hwang JH et al (2020) A designed whole-cell biosensor for live diagnosis of gut inflammation through nitrate sensing. Biosens Bioelectron 168:112523

Youssef S, Zhang S, Ai HW (2019) A genetically encoded, ratiometric fluorescent biosensor for hydrogen sulfide. ACS Sens 4:1626–1632

Zhang J, Ma Y, Taylor SS, Tsien RY (2001) Genetically encoded reporters of protein kinase A activity reveal the impact of substrate tethering. Proc Natl Acad Sci U S A 98:14997–15002

Zhang J, Hupfeld CJ, Taylor SS, Olefsky JM, Tsien RY (2005) Insulin disrupts beta-adrenergic signalling to protein kinase A in adipocytes. Nature 437:569–573

Zhou X, Clister TL, Lowry PR, Seldin MM, Wong GW, Zhang J (2015) Dynamic visualization of mTORC1 activity in living cells. Cell Rep 10:1767–1777

Exploring the Potential of Nanosensors in Medicine and Their Characteristic Features

6

Rahila Nazir, Kanchan Birat, and Mohd. Mohsin

Abstract

Across all scientific disciplines, nanotechnology is one of the topics that has drawn a lot of interest. The creation of novel materials, the development of new machinery and molecularly controlled systems, and the application of material properties at nanoscale sizes are all possible, thanks to nanotechnology. As a result, using nanoscale items, such as different biological nanoparticles, as medicine transporters and tumor-detecting agents might be viewed as a significant accomplishment. The same is the cause of the recent rise in the number of publications and articles in this area. The complexity of this area necessitates the formation of multidisciplinary research teams that include specialists in fields like biotechnology, engineering, physics, and medicine. This chapter focuses on how nanotechnology is used in medicine, how functional working nanosensors are, and how they are used to identify various diseases. Different sections discuss different types of nanosensors, such as biosensors and nanochemical sensors. With a focus on sensor technology and its use in biological systems, it then evaluates the potential and progress of nanotechnology from its origin to the present.

Keywords

Nanotechnology · Nanosensors · Medicine · Cancer · Diseases

R. Nazir
Johns Hopkins University, Baltimore, MD, USA

Jamia Millia Islamia, New Delhi, Delhi, India

K. Birat
Jamia Hamdard, New Delhi, Delhi, India

M. Mohsin (✉)
Department of Biosciences, Jamia Millia Islamia, New Delhi, Delhi, India
e-mail: mmohsin1@jmi.ac

M. Mohsin, N. Soleja (eds.), *Nanoscale Sensors and their Applications in Biomedical Imaging*, https://doi.org/10.1007/978-981-97-3144-2_6

6.1 Introduction

The need for nanosensors has increased significantly during the last few decades. Innovative applications of nanosensors in medical diagnostics and other devices have been observed in the market. It can be applied on a large scale and is used in the design of the constructions as well. These aid environmental chemists in creating more efficient, effective, and focused nanomaterials for monitoring a variety of contaminants (Wang 2009; Yang and Duncan 2021). A chemical sensor's ability to expose and detect the target molecule at low concentrations will also be improved by having a larger surface area. The fragile structure of nanomaterials strains everything to its limit. They are more powerful and efficient due to their larger surface area. Some nanoparticles, such as palladium, platinum, and silver nanoparticles, are employed in nanosensors. The characteristics of nanomaterials, which are also readily modifiable in terms of size, structure, and makeup, can all be combined and have an impact on these parameters. Nanosensors have been adopted into our society and are present in many of the everyday items we use since they have undergone extensive testing (Jin et al. 2017). The surface properties of the nanomaterial that the colorimetric test is using have a significant impact on the analytical performance of nanosensor systems with a colorimetric basis. Nanosensors with electromagnetic and chemical properties are two solutions to this problem. While molecular nanosensors transform biological communication systems into coded messages, electromagnetic nanosensors detect changes in electromagnetic waves while accounting for quantum events. The mechanical nanosensors may be used to power the molecular nanosensors, which mechanically collect energy from ambient biological material and nanosensor vibration. To identify the aroma of melanoma with a sensor, scientists have developed wearable sensors that can recognize diabetes when acetone is found in the breath. These state-of-the-art devices can detect more gases than human nostrils can by carefully overlapping multiple gas sensors and a component that rearranges patterns. Biological nanosensors monitor biomolecular activities such as cellular communication, interactions between enzymes, and interactions between antibodies and antigens. These biological recognition mechanisms are frequently enzymes, antibodies, or proteins. In the military and national security fields, nanosensors are used to detect biotoxins like smallpox and anthrax. Nanosensors in the biomedical sector make it simpler to detect diseases like cardiovascular disease or radioactive materials. These are used to keep track of a cell's lipid membrane potential following direct implantation. The devices based on inorganic semiconductor nanoparticles can record both electrical impulses at the nanoscale and the potential actions of numerous neurons. Materials with physicochemical properties on the nanoscale scale are advantageous for biological sensing applications because they provide new insights into changes in critical physiological indicators in real time. Because nanomaterials are not found in bulk materials that form at the nanoscale, a variety of processes, including top-down lithography, bottom-up mounting, and molecular self-assembly, are used to create nanosensors today. Sensors constructed of traditional materials are sensitive and exact in this regard. By interacting with chemical and biological molecules, nanosensors are employed in concert with naturally occurring biological processes to cause observable physical changes. The sources of the increase in sensitivity are high nanomaterial surface-to-volume ratios and novel physical characteristics of nanomaterials, such as nanophotonics, which can be employed for detection (Xie et al. 2015). As nanosensor technology advances, it becomes possible to exponentially connect nanodevices. The data acquired in the healthcare sector is likely the most significant potential for service transformation, and as a result, privacy and security issues are sympathetic. Like the Internet of Things, before widespread use, individuals must have confidence in the data received, gathered, and stored on devices in terms of its privacy and safety components. An extensive examination of the global nanosensors market yields valuable data on the nanosensors' forecasted rates of growth (Cash and Clark 2013; Vo-Dinh et al. 2015). This chapter examines nanosensors, as well as the developments and their potential in their sensing methods.

6.2 Nanotechnology in Medicine

The measuring scale at which molecular nanotechnology functions shapes the field. It denotes hundreds or thousands of nanometers or molecule sizes. The number of atoms in a nanometer is thought to be 3 or 4. Assumedly, viruses are 100 nm in size while discussing nanotechnology. The average human hair measures about 200 μm in diameter (Giljohann and Mirkin 2009). The fundamental building block of molecular nanotechnology is the atom. Groups of atoms enhance the properties of the final product, such as tension, conductivity, or power. A good example of this process is the transformation of graphite into the robust and durable structure of a diamond. Additionally, sand atoms can be improved and reorganized to make a medium silicone plate that can be used to construct semiconductor devices. Consequently, these nanostructures' physicochemical properties and biotic components can help to improve biomedical applications (Akbarian et al. 2018). It is conceivable to make the world a safer place to live if comprehensive efforts are made to develop and complete nanotechnology. As a result, the industry-available sensors are not exempt from this restriction. Nanosensors have been used in research throughout the last 10 years. A sensor constructed on an atomic size and based on nanometer measurements is referred to as a nanosensor. Nanosensor development and assessment can be seen in fields like medicine, native safety, aircraft, integrated circuits, etc. Nanosensors have a wide range of specific applications, and there are several types and production techniques for them as well. The creation of these nanosensors faces difficulties. On the contrary, their benefits grow and their function in daily life is expanded whenever they are developed and refined for widespread usage. Obtaining information on a nuclear scale and transmitting it to information that can be easily inspected might be the primary function of nanosensors. Such sensors can be referred to as "physical or chemical sensors" since they often use nanocomponents at microscopic or submicroscopic dimensions. They have a strong sensitivity, which is one of their qualities. As a result, they can identify viruses at the subatomic level or even very small quantities of naturally dangerous substances. Typically, a nanotechnology-based object may be designed and controlled. However, based on the researchers' viewpoints, it can be claimed that they each have a unique perspective on nanotechnology, and definitions of the term are as varied as the technology's uses. Some people have concentrated on using an electron microscope to analyze materials' microstructures and classifying thin sheets as nanotechnology. Others have investigated the synthesis process and construction materials as nanotechnology. A protein delivery system is a type of nanotechnology that is used to deliver proteins to parts of the body. Although each of the descriptions has a distinct significance in a particular field of study, none of them fully encompasses nanotechnology. Numerous definitions of nanotechnology have been put out because it encompasses a diverse range of research topics and necessitates transdisciplinary and multidisciplinary efforts. It may be deduced from the principles and definitions of nanotechnology that its small size is its single shared characteristic. Nanotechnology is typically thought of as a technique for creating, utilizing, and designing nanoparticles and nanostructures. The fundamental study of the relationship between physical qualities and the size of materials in the nanometer range is included in the field of nanotechnology, which also covers the fundamental understanding of physical properties, nanomaterials, and nanostructure phenomena. A complete definition of nanotechnology would include materials and systems with novel structures and components as well as greatly improved biological, chemical, and physical characteristics. This could be related to nanoparticle size. Although the word "nanotechnology" is new, research on nanoscale scales has existed for a while (Akbarian et al. 2018). The study of biological systems and the engineering of various materials, such as colloidal metal quantum dots and catalysts, are acknowledged to have existed for a long time. Our ability to monitor and work with materials at the nanoscale might be viewed as a component of what is distinctly novel about nanotechnology. New tools for defining, characterizing, and manipulating nanomaterials and nanostructures have been developed through the invention and development of tools including the transmission electron microscope (TEM), scanning tunneling microscope (STM), and scanning probe microscope (SPM). Therefore, it may be said that these technologies allow for the atom-level analysis and manipulation of nanomaterials and nanostructures. One of the most advantageous microorganisms that are found in various industrial and environmental operations is iron-reducing bacteria (IRB). The bacteria are now intriguing in nanobiology technology because they can create iron nanostructures. Recent investigations have shown that these materials have a wide range of applications and can be employed as nanosensors as well (Ravanshad et al. 2018).

6.3 Different Types of Nanosensors

Nanosensors have already had a significant impact on the environment, food, aerospace, and medicine. Materials with properties at the nanoscale differ greatly from those with qualities at greater scales. As an example, consider the melting point of gold. In contrast to bulk gold, which melts at 1064 °C, a 4 nm gold particle does so at roughly 850 °C. There are numerous subcategories of nanosensors. In biological nanosensors, a bio-receptor and a transducer are the key elements. The high sensitivity and relative simplicity of fluorescence make it the leading transduction method now (Park et al. 2006). Real-time measurements, site-directed mutagenesis to produce indicator proteins, attaching active nanoparticles to active proteins inside the cell, making a nanomaterial with bio-receptor attachment sites, or producing indicator proteins can all be used to measure. Electrochemical nanosensors can be used to monitor intracellular properties, however, because they lack bio-receptors like antibodies, DNA, and enzyme specificity, they are frequently less selective for biological measurements. Physical sensors, which operate on the nanoscale, convert signals from physical qualities like force, pressure, flow temperature, and other factors into detectable and measurable signals. In the examination of environmental sample residues, the development of pharmaceuticals, the monitoring of pollution, the assay of organophosphorus compounds, and other fields, chemical nanosensors are frequently used to detect a variety of substances. In diagnostic applications, chemical nanosensors are also used (Parveen and Sahoo 2008).

6.3.1 Biosensors

Because of its superior ability to detect cancer and even serious sickness, biosensors are one of the most significant and practical types of nanosensors. Additionally, it can be argued that specific DNA types can be identified using biosensors. Dendrimers are composite polymer-produced sensors that are built layer by layer onto surfaces with a diameter of less than five nanometers. These sensors are so small that it is possible to organize them via the epidermis. It is possible to say that a nanobiosensor can be used to detect asthma up to several weeks in advance. It can unfold and accumulate, and structural alterations are related to insulin. It should be noted that even minor structural integrity alterations to insulin have the potential to significantly affect its biological efficacy in connection to its physiological and pharmacological actions. A random or deliberate chemical shift must occur to enhance its medicinal capabilities. Gene therapy can be defined conceptually as the introduction of DNA or RNA to cells that can be utilized to treat or prevent genetic uncleanliness. To improve the capacity of genes to transfer, numerous strategies and gene transporters have recently been developed (Zakeri et al. 2018). The use of polyethylenimine (PEI) based transfer material is a wise decision to provide a fresh way to treat illnesses and fuse gene therapy with chemotherapy. It can be said that when ketones are present, techniques for enhancing the chemoselective reduction of aldehydes are very important and have attracted a lot of attention, or even using some additive materials like thermoplastic, nanotube resins, and graphene oxide in polymer composites. to examine the characteristics of nanocomposites, linear low-density polyethylene, ethylene-co-vinyl acetate, and nano-clay particles using electron rays. Nanosensors can be utilized for this as well because of the properties outlined above. An analyte-recognizing biological component (such as a cell, an organelle, an enzyme, an antibody, a nucleic acid, etc.) that may be coupled to a transducer to produce a signal proportional to the analyte's concentration is called a biosensor. Electrochemical, physical, optical, and wearable biosensors are just a few of the many available types of biosensors. The stability, low cost, great sensitivity, and reproducibility of biosensors are their key traits. These biosensors can be used in a variety of settings, including routine medical examinations, the measurement of metabolites, illness screening, the treatment of insulin, therapeutic psychotherapy, and the identification of disease (Akbarian et al. 2018).

6.3.2 Nanosensors and Their Role in Modern Technologies

A nanosensor is a sensing device having at least one dimension smaller than 100 nm that is used to gather data for analysis and convert nanoscale data into information. Nanotechnology is interested in the physical and chemical features of materials that are usually different from their bulk characteristics at the nanoscale. They can be used by nanosensors. Nanosensors have extraordinary properties that can be improved upon in comparison to conventionally built sensors, which have fewer detecting parts. Because of this, nanosensors might not always be little enough to employ the characteristics of nanomaterials to identify and quantify events occurring at the nanoscale but rather might be bigger apparatuses. A new generation of technologies can be developed, thanks to nanosensors. They are simple to interact with at the nanoscale (in the case of proteins, for example), where they can identify processes that are not noticeable at the macroscale. The creation of nanosensors can be accomplished in a variety of ways today, including molecular self-assembly, bottom-up assembly, and top-down lithography (Ravanshad et al. 2018). Unlike the macro world, the nanoworld is an unusual place where things happen frequently. Because of the crucial role that nanosensors play in carrying out several daily tasks, this technology is successfully applied in a variety of industries to meet continuous needs. Nanosensors are employed to monitor the spread of viruses and diseases in public settings. Blood, sickness, and breath testing in-body networks are a few examples of uses. Furthermore, these technologies are easily able to link up with wearable environmental and health trackers. The IoT undoubtedly gives us significantly more control over our lives than adding nanotechnology and biotechnology. The need for an external light source and a device capable of tracking and comparing multiple particle resonance frequencies is one of the main drawbacks of this sensor technology. Nanosensors that cooperate and communicate with one another can help with this. As an illustration, nanosensors may generate the same particles at much lower power levels and pick up the emitted energy at various frequencies. After processing, the results of this transaction can be delivered to a data sink (Nile et al. 2020).

6.3.3 The Efficacy of Nanosensors in Different Applications

It is a component of analytical machinery that combines chemical and physical transducers with physiologically sensitive parts. The expansion of the global market for nanosensors is attributed to the usage of nanosensors in medical diagnostic and imaging processes as well as to technological developments in nanotech equipment. The main obstacles that prevent the market for nanosensors from expanding, however, are issues with the deployment of expensive technology. Market forecasts nanosensors are anticipated to benefit financially from autonomous nanotechnology developments. As miniaturization expands the range of applications, micro- and nanoscale sensors are frequently coupled with biosensors. Some of the most common kinds of nanosensor sensors include optical, mechanical, vibrational, and electromagnetic ones. There are nanosensors on the market and in research for a variety of uses, primarily in the defense, environmental, and healthcare industries. The core workflow and signal creation for these sensors both involve interaction between the nanosensor and the biological element as well as signal processing. Therefore, nanosensors can be developed to understand people's fundamental measures. The current state of portable sports technology is being used to monitor the internal and external workload of athletes. The medical needs of the sports community to learn more about an athlete's internal workload to modify their recuperation schedules are still unmet (Bayda et al. 2019).

6.4 Significant Potentials of Nanosensors in the Modern World

The creation and utilization of physical, chemical, and biological systems and instruments are all included in the market for nanosensors. In many industrial fields, including communications, medicine, transportation, agriculture, energy, materials and production, consumer goods, and residences, nanosensor plays a significant role as a cutting-edge and practical technology. It offers several advantages over its micro-contractors, including lower power consumption, high sensitivity, lower analyte concentration, and a closer proximity between the object and the sensor. Using sensors containing nanotubes for gas sensing applications, it just takes a tiny number of gas molecules to alter the sensor's electrical conductivity. Chemical vapors can thus be found even in small quantities. Table 6.1 discusses the significant potential of nanosensors. In intelligent buildings, nanosensors are used for self-sensing and self-monitoring. Because they can be connected to a wireless network, nanosensors have also been used to monitor traffic and environmental pollutants. In the sphere of medicine, it offers point-of-care diagnostic, multi-agent detection, and in vitro and in vivo medical applications. Nanosensors can be used for mechanical or chemical applications. They can be used to identify chemical substances and gases for both medical diagnosis and pollution monitoring. It can work as an accelerometer or monitor the physical properties of a microelectromechanical system like temperature, displacement, and flow (Cho et al. 2020; Vahedifard and

Table 6.1 Significant potential of biosensors and their applications

Significant potential	Description
• Monitor biological cells	The astonishing ability of nanosensors to track individual biological cells and animals as well as to record exact spatial and temporal measurements, even at the molecule level, has been demonstrated. In order to detect and assess a wide range of infectious diseases, including viruses and dangerous bacteria, they make advantage of the special features of nanomaterials and nanoparticles. Chemicals can be found in this device in as little as one part in a billion. The field of biosensors is currently booming, with expected annual growth rates in the healthcare sector, food quality assessment, and environmental monitoring (Wei et al. 2020)
• Health data	Nanosensors can be used to easily collect private health information. One of the major barriers to widespread adoption is the protection of this data, which guarantees that they are only used for gathering purposes. The frequency of monitoring is increasing, and the Internet of Nano Things makes it more likely to monitor certain individuals. The data could be used to classify socioeconomic groups according to their genetic makeup. Care for chronic patients may be enhanced by the creation and continual observation of minute physiological changes in nanosensors and micro-networks (Zhang et al. 2014)
• Identify tiny particles	Nanosensors can now identify minute components of viruses, malignant tissues, and diseases in plants, animals, and people, thanks to advancements in nanotechnology. These are extremely helpful in the diagnosis of diseases. Using this technology to diagnose illnesses early can reduce medical costs and save many lives, often even before symptoms appear. With real-time cloud computing, digital breath biomarkers analyze exhaled breath to identify chronic health issues and disorders. This medication adherence and early detection in medical care
• Detect human physiological data	Semiconductivity and flexible electronics packaging technologies could be used to enable biomedical sensor systems to continually, unobtrusively, and in real-time detect human physiological data. Sensors incorporate several innovations in neurosensing platforms based on microelectromechanical, biological, chemical, electrocardiograms, and electromyograms. Biological and chemical sensors can potentially replace expensive analytical instruments when the prerequisites for specificity and selectivity are met (Yang et al. 2017)
• Virus detection	Globally, all species are seriously threatened by infectious virus outbreaks such as those brought on by COVID-19, HIV-1, hepatitis, influenza, dengue, and zika. These infections, which frequently result in health issues or problems, occasionally affect many people. The creation of vaccines is a continuing challenge for the entire planet. To develop a vaccination against the illnesses, it is therefore required to experience it successfully and consistently. In this sense, biosensors can make a significant difference (Bayda et al. 2019; Javaid et al. 2020)

(continued)

Table 6.1 (continued)

Significant potential	Description
• Bacteria detection	The natural world and the surroundings are filled with bacteria. An average person has more than 150 different species of germs inside and outside of their body. Bacteria may grow quickly and need nutrients, moisture, and a warm environment to thrive. Optical biosensors can be used to detect germs thanks to advancements in portable, quick, and sensitive biosensor technology. Bacteria like *Salmonella typhimurium* and *Staphylococcus aureus* can be found using optical biosensors
• Wireless medical biosensor patch for COVID-19 monitoring	Under development is a brand-new wireless medical biosensor patch for use in monitoring COVID-19 patients. This novel biosensor technology, known as the biosensor patch 1AX, may be applied to the patient's chest without help and can be disposed away properly after use. This biosensor has the potential to be very beneficial for both patients and medical establishments. It is equipped to capture real-time patient data on body temperature, respiration rate, ECG trace, and heart rate (28)
• Helpful in predicting the future	Future diseases may be predicted by biosensors, which could become commonplace in our daily lives. We can easily keep an eye on our health with this technology at home. Both fitness apps and patient status can be tracked by it. It assists in the detection of specific chemicals and provides useful information about what is occurring inside our bodies. Future applications of this technology will involve accurate patient health monitoring to prevent new diseases

Chakravarthy 2021). One of the most promising methods for understanding the biology and treatment of diseases is the development of nanosensors. This is utilized to determine the "molecular signature" of a physiological disease condition at a specific time and is therefore crucial for an early diagnosis and accurate disease progress. Additionally, it provides details on how diseases develop and finally offers effective methods for diagnosis and therapy. Endoscopy, computed tomography, radiation, positron emission tomography, mammography, and magnetic resonance imaging are the main methods for diagnosing cancer in particular (Li et al. 2020).

6.4.1 Nanodiagnostics and Personalized Cancer Therapy

The development of nanobiotechnology has the promise of enhancing early detection of cancer and enabling personalized cancer treatment. Molecular diagnostics is the most crucial part of personalized medicine, and nanobiotechnology will be crucial in improving it, which is why the term nanodiagnostics has been coined. Nanobiotechnologies will enhance the identification of cancer biomarkers, which will serve as a foundation for developing both diagnostic and treatment strategies.

6.4.2 The Role of Nanotechnology in Personalized Medicines for Cancer

Some of the technical advancements in nanotechnology that are being used to enhance drug development are important to personalized medicine. To hasten the development of novel treatments for numerous diseases, including cancer, miniature devices are being built to examine synthetic cell membranes. Several proteins are found in cell membranes, some of which function as small pumps that swiftly remove chemotherapy medications from tumor cells, lessening the effectiveness of the treatment. Finding medications that turn off the pumps and increase the potency of chemotherapeutic treatments is the goal of research. The chips have the potential to dramatically improve the number of tests that can be carried out with a very small amount of protein (Ullah et al. 2020).

6.4.3 Role of Nanotechnology in Gene Therapy of Cancer

The implantation of genetic material, genetically altered DNA, genes, cells, and/or other substances for the regulation of gene function is commonly referred to as gene therapy. The cell's genetic makeup could be changed so they release drugs. Ex vivo gene therapy is genetically modifying patient cells in vitro, frequently with the use of viral vectors, and then re-injecting those altered cells into the patient's bodily tissues. Some cancer vaccines are made utilizing cells from the patient's tumor. This is an example of individualized treatment. An advanced method of drug delivery, gene therapy for cancer, has been improved with the help of nanotechnology. An injectable nanoparticle formulation of the tumor suppressor gene FUS1 has been explored in test animals, and nanoparticles have been employed for p53 gene therapy for cancer. Other methods include immunolipoplex and integrin-targeted nanoparticles for site-specific delivery (Yao et al. 2021).

6.4.4 Biosensors for COVID-19

The COVID-19 virus in the air can be found and measured with the help of a biosensor. It effectively addresses the difficulties posed by biological constraints and technical constraints. With most of its transmission occurring from person to person, this technology helps determine how long a virus can survive in the air. Additionally, it aids in the diagnosis of illnesses caused by infections. It automatically evaluates numerous laboratory tests as well as the source of several infectious diseases in the atmosphere. The important uses of biosensors for the COVID-19 pandemic are covered in Table 6.1. Healthcare institutions can use the data from biosensors for remote screening of a wide population, such as those in quarantine, patients in nursing homes, and those who are weak and at high risk while at home. The idea is that the current biosensing techniques need to be continuously improved to address the growing difficulties in virus diagnosis. Since the COVID-19 virus is known to move from person to person in the current world, it is imperative to identify this viral type as soon as possible (Kim et al. 2021).

6.5 Conclusion and Future Perspective

The development of nanosensors and nanobiosensors during the previous 10 years has been amazing in terms of both scientific effort and creativity. This chapter tried to outline the most recent innovations and trends in nanosensor and nanobiosensor construction and design, as well as potential medical applications. Nanosensors are now widely used in many industries, including pharmaceuticals, medicine, and industry. Due to their ease of environmental adaptation, nanosensors can be used to treat disease and find solutions to various problems affecting humans. Nanosensors span a wide variety of fields, including biochemistry, physics, biology, materials science, electrical engineering, and more. This can look at many fundamental ideas that underpin nanotechnology and nanomaterials, as well as how crucial a role they play in developing novel applications and sensing properties. With the use of nanosensors, chemical and biological processes can be studied on a small scale. Building gas/vapor sensors, ultrasonic stamp-sized, and lab-on-a-chip biosensors with nanoparticles can be used to monitor water quality and human health. Due to the intrinsic small size of nanosensors and their distinctive optical, magnetic, catalytic, and mechanical properties, tremendous progress has been made in the development and application of optical nanotechnology. It is possible to transmit information from nanoparticles to the macroscopical environment using any physical, biological, or chemical points. They are mostly employed in the development of novel nanoproducts as well as in the biomedical, nanotechnology, and healthcare industries. Tissue creation is facilitated using nanosensors in various biofabrication procedures, and developments in the science and technology of bioprinting may facilitate the production of nanosensors for increasingly challenging and ambitious applications.

References

Akbarian M, Ghasemi Y, Uversky VN, Yousefi R (2018) Chemical modifications of insulin: finding a compromise between stability and pharmaceutical performance. Int J Pharm 547(1–2):450–468. https://doi.org/10.1016/j.ijpharm.2018.06.023

Bayda S, Adeel M, Tuccinardi T, Cordani M, Rizzolio F (2019) The history of nanoscience and nanotechnology: from chemical-physical applications to nanomedicine. Molecules 25(1):112. https://doi.org/10.3390/molecules25010112

Cash KJ, Clark HA (2013) Phosphorescent nanosensors for in vivo tracking of histamine levels. Anal Chem 85(13):6312–6318. https://doi.org/10.1021/ac400575u

Cho I-H, Kim DH, Park S (2020) Electrochemical biosensors: perspective on functional nanomaterials for on-site analysis. Biomater Res 24:6. https://doi.org/10.1186/s40824-019-0181-y

Giljohann DA, Mirkin CA (2009) Drivers of biodiagnostic development. Nature 462(7272):461–464. https://doi.org/10.1038/nature08605

Javaid M, Haleem A, Vaishya R, Bahl S, Suman R, Vaish A (2020) Industry 4.0 technologies and their applications in fighting COVID-19 pandemic. Diabetes Metab Syndr 14(4):419–422. https://doi.org/10.1016/j.dsx.2020.04.032

Jin P, Wiraja C, Zhao J, Zhang J, Zheng L, Xu C (2017) Nitric oxide nanosensors for predicting the development of osteoarthritis in rat model. ACS Appl Mater Interfaces 9(30):25128–25137. https://doi.org/10.1021/acsami.7b06404

Kim C, Raja IS, Lee J-M, Lee JH, Kang MS, Lee SH, Oh J-W, Han D-W (2021) Recent trends in exhaled breath diagnosis using an artificial olfactory system. Biosensors 11(9):337. https://doi.org/10.3390/bios11090337

Li S, Jiang W, Huang J, Liu Y, Ren L, Zhuang L, Zheng Q, Wang M, Yang R, Zeng Y, Wang Y (2020) Highly sensitive and specific diagnosis of COVID-19 by reverse transcription multiple cross-displacement amplification-labelled nanoparticles biosensor. Eur Respir J 56(6):2002060. https://doi.org/10.1183/13993003.02060-2020

Nile SH, Baskar V, Selvaraj D, Nile A, Xiao J, Kai G (2020) Nanotechnologies in food science: applications, recent trends, and future perspectives. Nano Micro Lett 12(1):45. https://doi.org/10.1007/s40820-020-0383-9

Park TG, Jeong JH, Kim SW (2006) Current status of polymeric gene delivery systems. Adv Drug Deliv Rev 58(4):467–486. https://doi.org/10.1016/j.addr.2006.03.007

Parveen S, Sahoo SK (2008) Polymeric nanoparticles for cancer therapy. J Drug Target 16(2):108–123. https://doi.org/10.1080/10611860701794353

Ravanshad R, Karimi Zadeh A, Amani AM, Mousavi SM, Hashemi SA, Savar Dashtaki A, Mirzaei E, Zare B (2018) Application of nanoparticles in cancer detection by Raman scattering based techniques. Nano Rev Exp 9(1):1373551. https://doi.org/10.1080/20022727.2017.1373551

Ullah Z, Witjaksono G, Nawi I, Tansu N, Irfan Khattak M, Junaid M (2020) A review on the development of tunable graphene nanoantennas for terahertz optoelectronic and plasmonic applications. Sensors (Basel) 20(5):1401. https://doi.org/10.3390/s20051401

Vahedifard F, Chakravarthy K (2021) Nanomedicine for COVID-19: the role of nanotechnology in the treatment and diagnosis of COVID-19. Emerg Mater 4(1):75–99. https://doi.org/10.1007/s42247-021-00168-8

Vo-Dinh T, Liu Y, Fales AM, Ngo H, Wang H-N, Register JK, Yuan H, Norton SJ, Griffin GD (2015) SERS nanosensors and nanoreporters: golden opportunities in biomedical applications. Wiley Interdiscip Rev Nanomed Nanobiotechnol 7(1):17–33. https://doi.org/10.1002/wnan.1283

Wang J (2009) Biomolecule-functionalized nanowires: from nanosensors to nanocarriers. Chemphyschem 10(11):1748–1755. https://doi.org/10.1002/cphc.200900377

Wei M, Lin P, Chen Y, Lee JY, Zhang L, Li F, Ling D (2020) Applications of ion level nanosensors for neuroscience research. Nanomedicine (Lond) 15(29):2871–2881. https://doi.org/10.2217/nnm-2020-0320

Xie X, Gutiérrez A, Trofimov V, Szilagyi I, Soldati T, Bakker E (2015) Charged solvatochromic dyes as signal transducers in pH independent fluorescent and colorimetric ion selective nanosensors. Anal Chem 87(19):9954–9959. https://doi.org/10.1021/acs.analchem.5b02566

Yang T, Duncan TV (2021) Challenges and potential solutions for nanosensors intended for use with foods. Nat Nanotechnol 16(3):251–265. https://doi.org/10.1038/s41565-021-00867-7

Yang B, Gong H, Chen C, Chen X, Cai C (2017) A virus resonance light scattering sensor based on mussel-inspired molecularly imprinted polymers for high sensitive and high selective detection of hepatitis A virus. Biosens Bioelectron 87:679–685. https://doi.org/10.1016/j.bios.2016.08.087

Yao X, Zhang Y, Jin W, Hu Y, Cui Y (2021) Carbon nanotube field-effect transistor-based chemical and biological sensors. Sensors (Basel) 21(3):995. https://doi.org/10.3390/s21030995

Zakeri A, Kouhbanani MAJ, Beheshtkhoo N, Beigi V, Mousavi SM, Hashemi SAR, Karimi Zade A, Amani AM, Savardashtaki A, Mirzaei E, Jahandideh S, Movahedpour A (2018) Polyethylenimine-based nanocarriers in co-delivery of drug and gene: a developing horizon. Nano Rev Exp 9(1):1488497. https://doi.org/10.1080/20022727.2018.1488497

Zhang Z, Shi Y, Pan Y, Cheng X, Zhang L, Chen J, Li M-J, Yi C (2014) Quinoline derivative-functionalized carbon dots as a fluorescent nanosensor for sensing and intracellular imaging of Zn^{2+}. J Mater Chem B 2(31):5020–5027. https://doi.org/10.1039/c4tb00677a

Recent Innovations in Sensing Techniques to Detect SARS-CoV-2

7

Reshma Bano, Rahila Nazir, Neha Soleja, and Mohd. Mohsin

Abstract

The emergence of the novel coronavirus, SARS-CoV-2, in 2019 has led to a global pandemic, affecting individuals across all age groups. As a member of the coronaviridae family, SARS-CoV-2 is a positive-sense RNA virus with helical symmetry nucleocapsid and distinctive spike proteins facilitating host cell binding. This virus shares characteristics with other coronaviruses, including Middle East respiratory syndrome (MERS-CoV) and severe acute respiratory syndrome (SARS-CoV), causing a spectrum of illnesses from mild cold-like symptoms to severe respiratory distress and organ failure. Common symptoms of COVID-19 include fever, dry cough, fatigue, and various other manifestations affecting multiple organ systems. The detection and diagnosis of COVID-19 rely on diverse methods, including nucleic acid-based testing, imaging, antigen or antibody testing, and emerging biosensing tools. These techniques enable the rapid identification of the virus in infected individuals, aiding in patient management and public health efforts. Additionally, the angiotensin-converting enzyme 2 (ACE2) receptor affinity of the spike protein contributes to viral cell entry, highlighting potential targets for therapeutic intervention. This chapter reviews current coronavirus detection methods and anticipates future advancements for early-stage virus detection. The development of innovative sensing techniques based on genomic and structural details of the virus holds promise for enhancing diagnostic capabilities and improving patient outcomes in the ongoing battle against COVID-19.

R. Bano · N. Soleja · M. Mohsin (✉)
Department of Biosciences, Jamia Millia Islamia, New Delhi, Delhi, India
e-mail: mmohsin1@jmi.ac.in

R. Nazir
Department of Biosciences, Jamia Millia Islamia, New Delhi, Delhi, India

Johns Hopkins University, Baltimore, MD, USA

M. Mohsin, N. Soleja (eds.), *Nanoscale Sensors and their Applications in Biomedical Imaging*, https://doi.org/10.1007/978-981-97-3144-2_7

Keywords

Coronavirus · Spike protein · Diagnostic methods · Biosensing · Fluorescence-based detection

Abbreviations

ACE2	Angiotensin-converting enzyme 2
FRET	Fluorescence resonance energy transfer
iLACO	Isothermal LAMP-based method for COVID-19
LAMP	Loop-mediated isothermal amplification
LFIA	Lateral flow immunoassays
LSPR	Localized surface plasmon resonance
RPA	Recombinase polymerase amplification
RT-PCR	Reverse transcription-polymerase chain reaction
SARS-CoV-2	Severe acute respiratory syndrome coronavirus-2

7.1 Introduction

The novel coronavirus, identified in 2019 and named SARS-CoV-2, has affected individuals of all age groups on a global scale, causing an infectious disease. Members of the coronaviridae family, coronaviruses (CoVs) have a helical symmetry nucleocapsid and are positive-sense RNA viruses. Various spikes cover the coronavirus surface, which aids in binding host cells (Agostini et al. 2018). Coronaviruses contain several proteins, including the spike (S), envelope (E), membrane (M), and nucleocapsid (N), as shown in Fig. 7.1.

Angiotensin-converting enzyme 2 (ACE2) receptor affinity may be sufficient for the spike protein to allow cell entry (Zhou et al. 2020). The ACE2 mRNA is found in almost all human organs, including the lungs, stomach, small intestine, colon, skin, lymph nodes, liver bile ducts, kidney parietal epithelial cells, and the brain, among others (Hamming et al. 2004). Simple colds to severe illnesses like Middle East respiratory syndrome (MERS-CoV) and severe acute respiratory syndrome are all caused by coronavirus (SARS-CoV) (Ali and Alharbi 2020). Fever, dry cough, and tiredness are the most common COVID-19 symptoms, but some patients may also experience aches and pains, nasal congestion, headache, sore throat, diarrhea, loss of taste or smell, or skin rash or discoloration of fingers or toes. Some people become infected but show no symptoms, and others are found to be asymptomatic. Organ failure, septic shock, pulmonary edema, severe pneumonia, and acute respiratory distress syndrome (ARDS) have all been fatal complications in some patients (Sohrabi et al. 2020). Patients with a suspected infection should have a variety of laboratory tests performed. COVID-19 infection can also be confirmed by full genome sequencing and phylogenetic analysis of bronchoalveolar lavage fluid

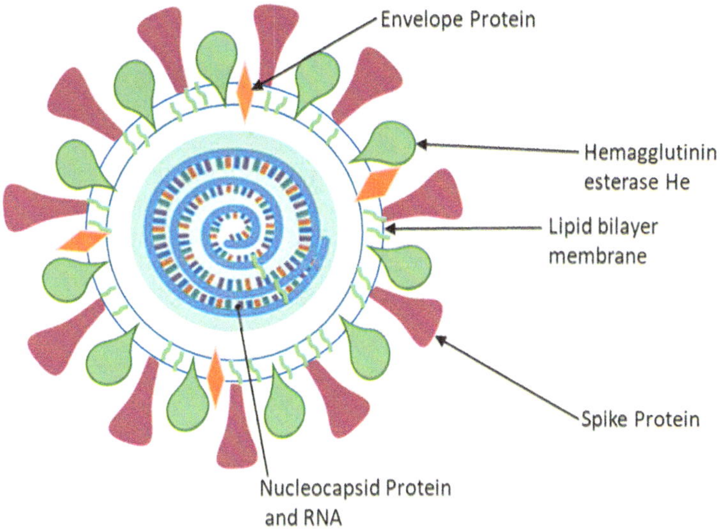

Fig. 7.1 Representation of the SARS-CoV-2 structural viral proteins including the spike, envelope, membrane, and nucleocapsid

(Zhu et al. 2020). Many other markers have been discovered to detect COVID-19, and COVID-19-based kits are also available.

In this chapter, we have taken a look at the current coronavirus detection methods as shown in Fig. 7.2 and speculate on what might be available in the future for detecting the virus at its earliest stages.

7.2 Diagnostic and Prognosis of COVID-19 Infection

Several sensing techniques have been proposed based on the genomic and structural details of coronavirus, including nucleic acid-based testing, serological-based testing, and some biosensing tools. Many other methods should be developed for the rapid identification of the virus in infected patients.

7.2.1 Nucleic Acid-Based Methods of Detection

The main method of SARS-CoV-2 investigation is nucleic acid testing. SARS-CoV-2 has been detected genetically using a variety of reverse transcription-polymerase chain reaction (RT-PCR) kits. For SARS-CoV-2 detection, isothermal amplification has recently been developed. Isothermal amplification methods use nucleic acid tests to target a single temperature and do not require any specialized laboratory equipment to achieve comparable analytical sensitivity to PCR.

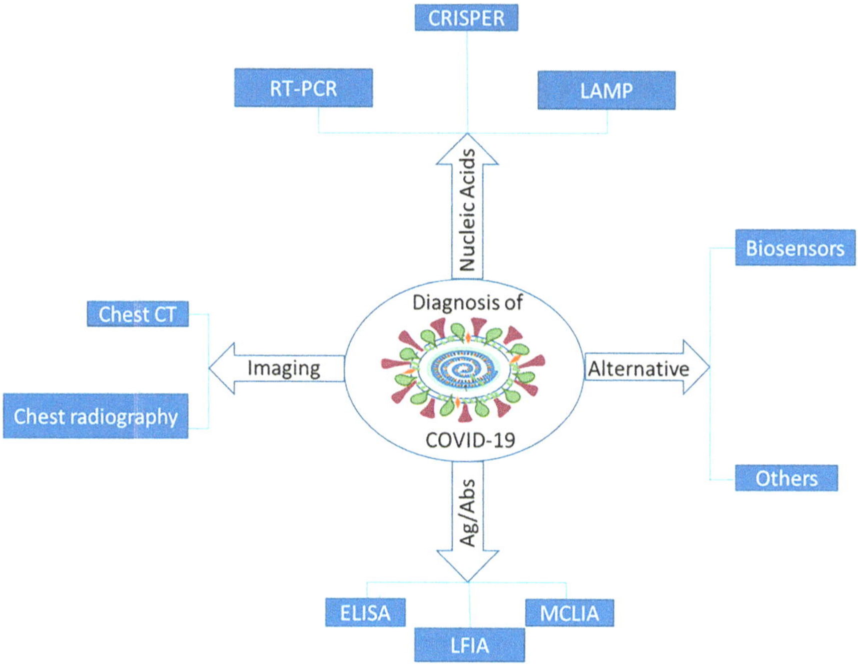

Fig. 7.2 Methods of COVID-19 diagnosis based on nucleic acids, imaging, antigens or antibodies, and others

7.2.1.1 RT-PCR Test

One of the most widely used laboratory methods for detecting COVID-19 infection is RT-PCR. The RT-PCR test is a very effective method for identifying and isolating coronavirus-infectious isolates. The COVID-19 RT-PCR test detects nucleic acid from SARS-CoV-2 in nasal, nasopharyngeal, or oropharyngeal swabs, sputum, bronchoalveolar lavage, and other samples from the lower and upper respiratory tract. The procedure of RT-PCR is described in Fig. 7.3.

The test uses three primers and probe sets to detect three regions in the SARS-CoV-2 nucleocapsid (N) gene and one primer and probe set to detect human RNasP (RP) in a clinical sample in the singleplex format. During the extension phase of the PCR cycle, Taq polymerase's 5′ nuclease activity degrades the bound probe, resulting in a fluorescent signal with the separation of reporter dye from quencher dye. Each PCR cycle can be monitored for fluorescence intensity (Carter et al. 2020).

SARS-CoV-2 has been genetically detected using several RT-PCR kits. As a result, despite the negative outcome of molecular testing, SARS-CoV-2 infection cannot be ruled out. A variety of factors can lead to negative results in an infected person. The results show a trend of fluctuation. Insufficient viral material in the specimen, laboratory error during sampling, or sample transportation restrictions are all possible causes (Henein et al. 2021). The findings of this study also show that specimens used in this study, such as a pharyngeal swab, did not produce

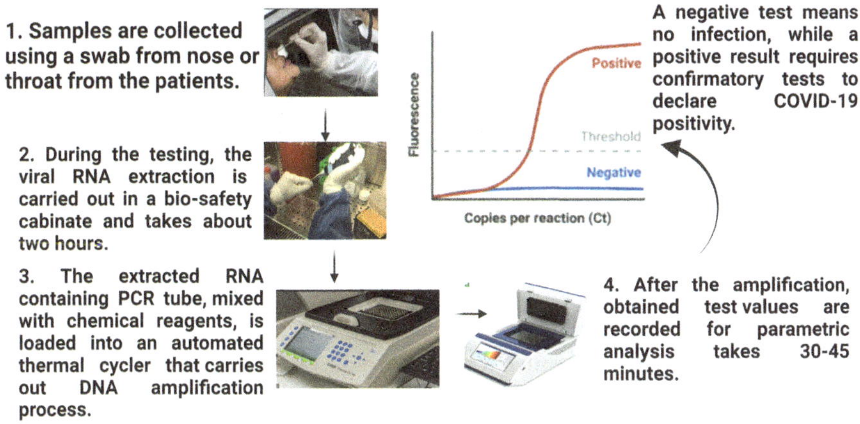

Fig. 7.3 Procedure of RT-PCR to detect the SARS-CoV-2

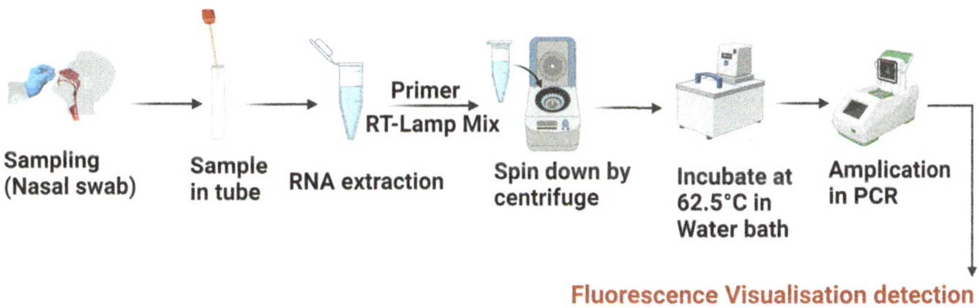

Fig. 7.4 Procedure of RT-LAMP: sample collection, extraction, amplification, and fluorescence-based detection

appropriate results, thus the results were variable and potentially unstable, and should not be used as the sole pointer for diagnosis, treatment, isolation, recovery, and discharge for hospitalized patients clinically diagnosed with COVID-19 (Li et al. 2020).

7.2.1.2 RT-LAMP Test for COVID-19

Loop-mediated isothermal amplification (LAMP) is a technique that uses 4–6 specially designed primers and a DNA polymerase with chain displacement activity (Bst) to amplify nucleic acids in a short amount of time at a constant temperature. The fact that LAMP can be combined with reverse transcription, i.e., RT-LAMP, means that both reverse transcription and amplification can happen at the same time, allowing for direct RNA detection. To diagnose infectious diseases, RT-LAMP, a one-step nucleic acid amplification method based on PCR technology, has been used (Fig. 7.4). RT-LAMP has several advantages, including the ability to be completed in less than an hour, high specificity and sensitivity, and the ability to work at a variety of pH and temperature ranges. Additionally, the reagents are relatively inexpensive and stable at room temperature, which is beneficial for clinical trials

(Sheikhzadeh et al. 2020). Laura E. Lamb et al. discovered that RT-LAMP allows for rapid detection of COVID-19 in the variability of nine common human specimens collected for clinical testing, including serum, urine, saliva, oropharyngeal swabs, and nasopharyngeal swabs, in this viability study.

Lin Yu et al. created the iLACO (isothermal LAMP-based method for COVID-19) method, which uses six primers to amplify a fragment of the ORF1ab gene. They used BLAST to compare the sequences of seven similar coronaviruses, two influenza viruses, and two normal coronaviruses to ensure that iLACO is species-specific. The Taqman-based qPCR detection method detects synthesized RNA equivalent to ten copies of the SARS-CoV-2 in 15–40 min reaction time, depending on virus loading in the collected samples. In the presence of extensive DNA polymerase activity, positive results showed a change in the color phenol red from pink to yellow due to a decrease in pH. Despite this, the negative reaction is kept as pink. The sensitivity of a well-optimized LAMP assay is comparable to that of PCR, and amplification can be observed in real-time. Because the LAMP reaction has a high amplification efficiency, the result can be detected using colorimetric detection methods after observing changes in reaction mixture components. Furthermore, the unpurified sample can be used directly for LAMP. This means that when an unpurified sample is combined with noninstrumental (e.g., colorimetric) detection, high-throughput analysis is promising (Lamb et al. 2020). In their study, Gun-Soo Park et al. designed and screened LAMP primer sets targeting SARS-CoV. The reaction of the specimen was also optimized using specific primer sets. A pair of RT-LAMP assays were designed and evaluated for the detection of SARS-CoV-2 with a LoD of 100 copies per reaction. Cross-reactivity of two SARS-CoV-2 RT-LAMP assays for hCoV-229E, hCoV-OC43, and MERS-CoV that target Nsp3 was also not found. The results of these RT-LAMP assays can be detected within 30 min of starting the amplification reaction (Park et al. 2020; Dao Thi et al. 2020).

7.2.1.3 CRISPR/Cas-Based Detection of COVID-19

For on-site COVID-19 diagnosis, a CRISPR/Cas-based biosensing assay is being developed. CRISPR/Cas is a gene-editing toolkit made up of a guide RNA (CRISPR RNA or crRNA) and the Cas enzyme complex. Cas13a and Cas12a's RNA-guided nuclease activities have focused on the development of novel nucleic acid detection kit as well as tools to manipulate the genomes of cultured cells, animals, and plants. . The target nucleic acid (or activator RNA or DNA) activates general multiple-turnover nuclease activity by correctly base-pairing to the guide RNA/crRNA in both Cas13 and Cas12a. Because the fluorescent protein or enzyme-bound Cas enzyme cleaves the target nucleotides once crRNA binds to the complementary nucleic acid, it generates signals for easy detection. CRISPR-nCoV, a rapid assay for SARS-CoV-2 detection, was developed by a group of scientists who combined investigative method including metagenomic sequencing, RT-PCR, and CRISPR technique. According to the investigators, this is the first report demonstrating that CRISPR-based analysis is a speedy and sensitive method for detecting SARS-CoV-2. It requires minimal instrumentation and is cost-effective based on their evaluation and cross-platform comparison. This CRISPR-nCoV test was mutual with recombinase polymerase amplification (RPA) in the above study. They formed a speedy, highly sensitive, and easy-to-use assay by combining RPA's polymerase-mediated DNA amplification with CRISPR/Cas-initiated enzymatic signal amplification for amplified sensitivity. Multiple sets of RPA primers and CRISPR gRNAs were designed and screened based on the three potential target sequences Orf1ab, N, and E. The set that targeted Orf1ab performed the best in terms of overall sensitivity and specificity, so it was used to develop CRISPR-nCoV in this study for further testing. CRISPR-nCoV validated a sensitivity of 100% after detecting all 52 nCoV positive cases. There were no false positives in any of the 62 negative cases, including the nCoV-infected ones, indicating that CRISPR-nCoV has promising clinical sensitivity and specificity. SHERLOCK (Specific High Sensitivity Enzymatic Reporter UnLOCKing) platform based on CRISPR/Cas for SARS-CoV-2 detection in 60 min was developed in another study (Knott and Doudna 2018). Furthermore, a more advanced version of SHERLOCK, known as CARMEN (combinatorial arrayed reactions for multiplexed evaluation of nucleic acids), was developed with the ability to detect multiple target genes at the same time. A CRISPR-based DETECTOR assay for SARS-CoV-2 detection was also developed, with a 95% positive prediction agreement in 30 min (Hou et al. 2020; Chauhan et al. 2020; Ackerman et al. 2020; Broughton et al. 2020).

7.2.2 Immunoglobulin Detection-Based Test for COVID-19

7.2.2.1 ELISA for Diagnosis of COVID-19

The enzyme-linked immunosorbent assay (ELISA) is a versatile biological assay used to detect a wide range of substances, including antibodies, antigens, proteins, and glycoproteins. In the context of detecting SARS-CoV-2, ELISA plays a crucial role in diagnostic applications. In the development of a human COVID-19 IgG/IgM antibody ELISA, several key components are utilized. Researchers coat a microtiter plate with the SARS-CoV-2 N protein, a specific choice for its ability to bind with antibodies related to the virus. To detect the antibodies of interest, researchers employ goat-anti-human antibodies conjugated with horseradish peroxidase (HRP), a commonly used enzyme in ELISA assays. The ELISA procedure involves coating the wells with the SARS-CoV-2 N protein and then adding the diluted serum samples, allowing antibodies against the virus to bind. After an incubation period, HRP-conjugated goat anti-human IgG antibodies are introduced into the wells, and if SARS-CoV-2 antibodies are present in the sample, they will bind to these secondary antibodies. HRP catalyzes a colorimetric reaction when exposed to a specific substrate, and the intensity of this reaction correlates with the presence and concentration of antibodies in the serum. ELISA serves as an effective tool to detect SARS-CoV-2 antibodies in human serum, indicating past exposure to the virus or an immune response. Moreover, ELISA can be adapted for various purposes, making it a versatile tool in immunology and infectious disease diagnostics (Al-Jighefee et al. 2021; Jääskeläinen et al. 2020; Yong et al. 2020).

7.2.2.2 COVID-19 Serological Diagnosis Using a Peptide-Based Magnetic Chemiluminescence Enzyme Immunoassay (MCLIA)

Synthetic peptide antigens are used as the immunosorbent in a luminescent immunoassay for the diagnosis of SARS-CoV-2 antibodies. As potential antigens, 20 peptides derived from the ORF1a/b, spike, and nucleocapsid proteins of the virus were utilized. The genomic sequence was used to create these peptides. After purifying the peptide-labeled magnetic beads, the corresponding peptides were mixed with serum samples for the IgG and IgM antibody assay. The mixture was incubated at 37 °C for 10 min, washed five times, reacted with an antibody conjugate, washed five more times, and then reacted with the substrate. A luminescence reader was used to determine reactivity (Cai et al. 2020). Real-time PCR detection is used to collect serum samples from confirmed COVID-19 patients. Five confirmed patients' sera and ten normal sera were used to react with these synthetic peptides, which were derived from the amino acid sequence of orf1a/b, S protein, and N protein. Among these peptides, one from the S protein performed the best. Researchers assessed the specificity of this synthetic-peptide-based MCLIA after finding no cross-reactivity, which indicates high specificity. The synthetic peptide as an antigen improves the assay's stability and repeatability, and it is theoretically more specific than using a virus as an antigen. In the assay, this peptide demonstrated excellent specificity, as none of the 167 sera from patients infected with pathogens other than SARS-CoV-2 reacted with it.

7.2.2.3 Lateral Flow Immunoassay (LFIA) Testing to Detect COVID-19

The media has referred to the lateral flow tests as "antibody tests" because they are currently used to detect disease antibodies in a patient's blood. The antigen is also being tested on the skill. Antibodies to the virus can be detected in patient blood, indicating that the patient has COVID-19 or has improved from COVID-19. In some cases, lateral flow experiments can be used to detect the presence of an active virus by directly sensing virus proteins. Antibody lateral flow experiments for SARS-CoV-2 are produced as test kits for professionals and clinicians, not for patients. They will need a droplet of blood from the patient, preferably from a vein or a small finger prick. These experiments differ from RT-PCR and LAMP methods in that they look for the patient's immune antibody response to the virus rather than the virus itself. Antibodies against SARS-CoV-2 that are present in the sample will attach to chemicals in the device as the sample moves through the device, capturing the antibodies on the test and control lines. Depending on the type of antibodies present, this capturing and binding process causes a color change along the test and control lines that can be seen with the naked eye, resulting in one, two, or three lines (IgM or IgG) (Fig. 7.5). Antibody lateral flow immunoassays can be designed to detect either IgM or IgG on their own or both together. The lateral flow immunoassay tests for antibody IgG and IgM show that the patient has been infected with the COVID-19 coronavirus and are very easy to read (Flower et al. 2020; Wu et al. 2020; Yam et al. 2003).

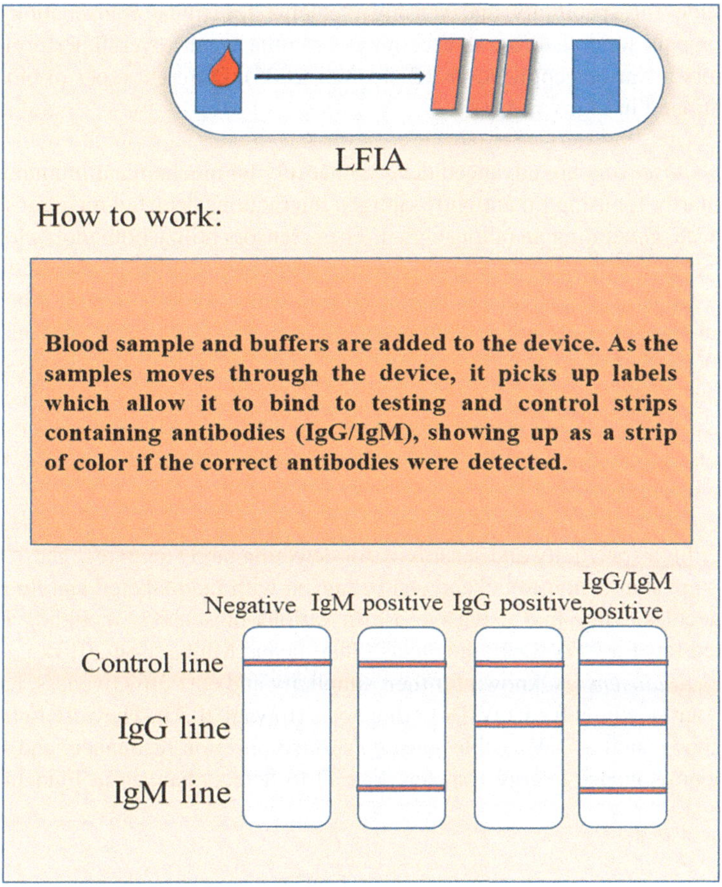

Fig. 7.5 Workflow of LFIA. A strip, displaying three coloured bands, indicates the presence of both IgG and IgM antibodies, signifying a positive COVID-19 result

7.3 Biosensors

Biosensors are sophisticated analytical tools that combine biological and physical components to detect specific molecules of interest. They consist of three essential components: bioreceptors, electrochemical interfaces, and transducers. Bioreceptors interact with target analytes, electrochemical interfaces facilitate reactions, and transducers convert biochemical signals into electrical signals for detection (Sangadevan and Periasamy 2014). To be effective, biosensors must meet the WHO-prescribed assured criteria, ensuring they are affordable, sensitive, specific, user-friendly, rapid and robust, equipment-free, and deliverable to end-users (Land et al. 2019). These devices have become smaller, more cost-effective, sensitive, multi-functional, and powerful analytical tools. Biosensor research now focuses on enhancing their design to overcome limitations. Biosensors consist of bioelements responsible for specificity and sensing elements for signal transduction. These advancements aim to address shortcomings and improve the overall performance of biosensors in various applications (Bisht et al. 2005). Different types of biosensors are mentioned in Fig. 7.6.

(a) *Immunosensors* are advanced devices that rely on biochemical immunoassays, primarily focusing on antibody–antigen interactions detected on a transducer's surface, generating an output signal. These sensors utilize both monoclonal and polyclonal antibodies to bind with complementary antigens for detecting specific pathogens. There are two main categories of immunosensors: labeled and label-free. Several companies have rapidly developed immunosensors for COVID-19 diagnosis, using various detection methods like optical, electrochemical, fluorescence, and others. While labeled immunosensors can enhance sensitivity, they also introduce complexity and cost due to the use of enzyme labels and purification steps, as well as concerns about nonspecific binding (Lassaunière et al. 2020).
(b) *Enzyme-based biosensors*, often using enzymes as biorecognition elements, offer high specificity and sensitivity for detecting target analytes. These biosensors are stable, non-toxic, and can be tagged with radiolabeled and fluorescent molecules. They find extensive use in various biosensors, including ELISA-based kits for COVID-19 diagnosis (Sharma and Mutharasan 2013).
(c) *Optical biosensors*, known for their simplicity and cost-effectiveness, have significant potential for COVID-19 diagnosis (Iravani 2020). They use light-based methods such as UV-visible sensors, surface plasmon resonance, and fluorescence resonance energy transfer (FRET) to detect changes in light intensity,

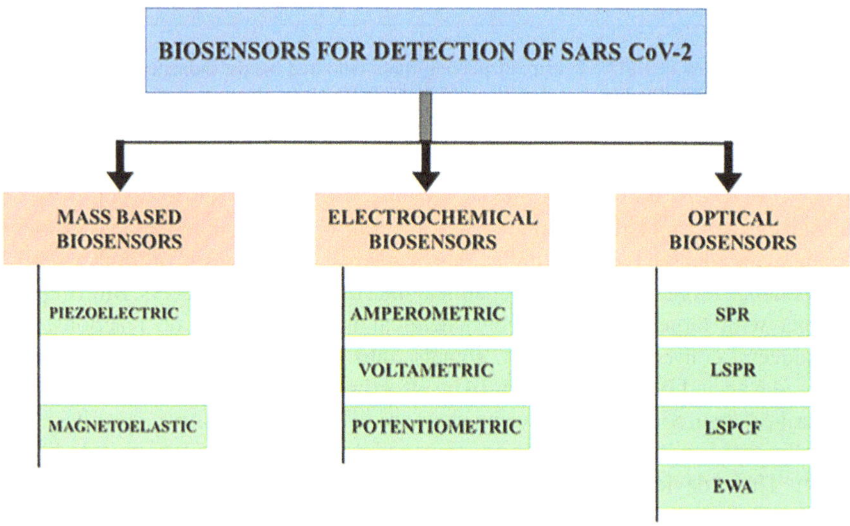

Fig. 7.6 Different types of biosensors for SARS-CoV-2 detection

providing a straightforward means of detection. Localized surface plasmon resonance (LSPR)-based optical biosensors, for example, use gold nanoparticles to detect viral genomes by monitoring changes in heat generation. FRET biosensors, based on fluorescence resonance energy transfer, offer high sensitivity and are used to detect various analytes. These biosensors rely on the energy transfer between donor and acceptor fluorophores. They have been employed for detecting SARS-CoV-2, offering highly sensitive and rapid detection capabilities (Brown et al. 2020; Lukose et al. 2021).

(d) *Colorimetric biosensors* provide a simple and cost-effective means of COVID-19 testing, allowing for qualitative or quantitative results that can be observed with the naked eye. They rely on specific biochemical reactions that lead to a color change when the target analyte is present. Colloidal gold nanoparticles are often used to enhance sensitivity in colorimetric assays (Yuan et al. 2014).

(e) *Electrochemical biosensors*, stemming from the discovery of glucose biosensors, offer a highly sensitive and rapid means of detection. They can be classified into various types, including amperometric, voltammetric, and potentiometric biosensors. They use changes in electrical current or potential to detect analytes, and nanotechnology has played a significant role in enhancing their sensitivity (Mahshid et al. 2021).

(f) *Piezoelectric biosensors* detect changes in oscillation frequency during a biorecognition event. These label-free sensors, based on piezoelectric materials, are highly sensitive and capable of detecting very low masses (Narita et al. 2021).

(g) *Magnetoelastic biosensors* rely on magnetic field changes to detect the target analyte. However, they often require magnetic labels for detection. Field-effect transistor (FET) biosensors are rapid, low-cost devices that have gained prominence in recent years. They are highly sensitive and suitable for point-of-care testing. Graphene-based FET biosensors have been developed for detecting SARS-CoV-2 spike proteins. Microfluidic biosensors integrate biosensing with microfluidics, allowing for precise control of small sample volumes, rapid processing, and portability. They are particularly suitable for point-of-care diagnostics and can be integrated with smartphones and wireless communication for remote testing (Pietschmann et al. 2020).

Overall, these diverse biosensor technologies contribute to the global effort to detect and fight the COVID-19 pandemic by providing various options for sensitive and rapid diagnostic testing. Some recently developed biosensors are described in Table 7.1.

7.4 Conclusion and Future Perspective

The global battle against COVID-19 has impelled remarkable advancements in diagnostic methodologies, transforming our approach to virus detection and containment. Initiating our journey, we recognized the pivotal role played by PCR

Table 7.1 Some recently developed biosensors are used for SARS-CoV-2 detection

Biosensors	Analytes	Materials used	Limit of detection (LOD)	References
Optical biosensor	N protein	Optical fiber and gold nanoparticle modified protein	1 pg/mL	Huang et al. (2009)
Electrochemical biosensor	RNA genome	Gold electrode modified with Fe_3O_4	200 copies/mL	Zhao et al. (2021)
Colorimetric biosensor	RNA genome	Oligonucleotide tagged gold nanoparticle	3.6–3.9 copies/mL	Yakoh et al. (2021)
Magnetosensor	Viral S/N protein	Glossy carbon electrodes and magnetic beads	8 ng/mL for S protein, 19 ng/mL for N protein	Fabiani et al. (2021)
Piezoelectric immunosensor	Helicase protein	Quartz crystal and magnetic particle	3.5 ng/mL	Albano et al. (2018)
Field-effect transistor	S protein	Graphene layer	1 fg/mL	Seo et al. (2020)

assays as the initial fortification against the virus. These assays, by detecting viral genetic material, delivered swift and precise results, significantly aiding early detection and containment efforts. As the pandemic unfolded, rapid serological kits emerged, allowing the detection of antibodies and viral antigens in blood samples. These kits, known for their accessibility and quick turnaround times, have played a crucial role in mass screening initiatives and our understanding of virus transmission dynamics. Yet, the most transformative developments came with the introduction of biosensors. Leveraging microfluidics, aptamers, probes, barcoding, nanotechnology, isothermal amplification, and CRISPR-Cas technologies, these biosensors have democratized diagnostics. They offer user-friendly, rapid, and exceptionally sensitive on-site screening for SARS-CoV-2.

However, the virus's insistent mutability serves as a stark reminder of the ongoing need for advanced diagnostic biosensors. These next-generation tools, fortified by nanotechnology and microfluidics, hold immense promise in enabling even swifter, more sensitive, and specific point-of-care diagnostics. In an era of emerging viral variants, biosensors will be indispensable for early identification and containment.

The pursuit for improved diagnostic tools remains persistent. Exploring the metagenomics frontier, the ability to rapidly obtain whole-genome and full-length genome sequences from samples promises enhanced pathogen detection and characterization. This technology, with its potential to decipher complex infections with diverse symptoms, is poised to become an indispensable clinical testing tool. In summary, the journey of diagnostic advancements in the context of COVID-19 exemplifies the power of global scientific communities when confronted with extraordinary challenges. This narrative, enriched with scientific rigor, illustrates the transformative potential of diagnostic technologies and their central role in the ongoing fight against COVID-19 and emerging infectious diseases.

Acknowledgments The first author (RB) kindly acknowledges the University Grants Commission, Govt. of India, for Senior Research Fellowship in the form of Maulana Azad National Fellowship (MANF).

References

Ackerman CM, Myhrvold C, Thakku SG, Freije CA, Metsky HC, Yang DK, Ye SH, Boehm CK, Kosoko-Thoroddsen TS, Kehe J, Nguyen TG (2020) Massively multiplexed nucleic acid detection with Cas13. Nature 582(7811):277–282

Agostini ML, Andres EL, Sims AC, Graham RL, Sheahan TP, Lu X, Smith EC, Case JB, Feng JY, Jordan R, Ray AS (2018) Coronavirus susceptibility to the antiviral remdesivir (GS-5734) is mediated by the viral polymerase and the proofreading exoribonuclease. MBio 9(2):10–128

Albano DRB, Shum K, Tanner JA, Fung YS (2018) Piezoelectric quartz crystal aptamer biosensor for detection and quantification of SARS CoV helicase protein. 31515. In: Proceedings IMCS AMA Service GmbH, Von-Münchhausen-Str Wunstorf, Germany: Vienna, Austria, vol 49, pp 211–213

Ali I, Alharbi OM (2020) COVID-19: disease, management, treatment, and social impact. Sci Total Environ 728:138861

Al-Jighefee HT, Yassine HM, Al-Nesf MA, Hssain AA, Taleb S, Mohamed AS, Maatoug H, Mohamedali M, Nasrallah GK (2021) Evaluation of antibody response in symptomatic and asymptomatic COVID-19 patients and diagnostic assessment of new IgM/IgG ELISA kits. Pathogens 10(2):161

Bisht V, Takashima W, Kaneto K (2005) An amperometric urea biosensor based on covalent immobilization of urease onto an electrochemically prepared copolymer poly (N-3-aminopropyl pyrrole-co-pyrrole) film. Biomaterials 26(17):3683–3690

Broughton JP, Deng X, Yu G, Fasching CL, Servellita V, Singh J, Miao X, Streithorst JA, Granados A, Sotomayor-Gonzalez A, Zorn K (2020) CRISPR–Cas12-based detection of SARS-CoV-2. Nat Biotechnol 38(7):870–874

Brown AS, Ackerley DF, Calcott MJ (2020) High-throughput screening for inhibitors of the SARS-CoV-2 protease using a FRET-biosensor. Molecules 25(20):4666

Cai XF, Chen J, Hu JL, Long QX, Deng HJ, Liu P, Fan K, Liao P, Liu BZ, Wu GC, Chen YK (2020) A peptide-based magnetic chemiluminescence enzyme immunoassay for serological diagnosis of coronavirus disease 2019. J Infect Dis 222(2):189–193

Carter LJ, Garner LV, Smoot JW, Li Y, Zhou Q, Saveson CJ, Sasso JM, Gregg AC, Soares DJ, Beskid TR, Jervey SR (2020) Assay techniques and test development for COVID-19 diagnosis. ACS Cent Sci 6:591–605

Chauhan N, Soni S, Gupta A, Jain U (2020) New and developing diagnostic platforms for COVID-19: a systematic review. Expert Rev Mol Diagn 20(9):971–983

Dao Thi VL, Herbst K, Boerner K, Meurer M, Kremer LP, Kirrmaier D, Freistaedter A, Papagiannidis D, Galmozzi C, Stanifer ML, Boulant S (2020) A colorimetric RT-LAMP assay and LAMP-sequencing for detecting SARS-CoV-2 RNA in clinical samples. Sci Transl Med 12(556):eabc7075

Fabiani L, Saroglia M, Galata G, De Santis R, Fillo S, Luca V, Faggioni G, D'Amore N, Regalbuto E, Salvatori P, Terova G, Moscone D, Lista F, Arduini F (2021) Magnetic beads combined with carbon black-based screen-printed electrodes for COVID-19: a reliable and miniaturized electrochemical immunosensor for SARS-CoV-2 detection in saliva. Biosens Bioelectron 171:112686

Flower B, Brown JC, Simmons B, Moshe M, Frise R, Penn R, Kugathasan R, Petersen C, Daunt A, Ashby D, Riley S (2020) Clinical and laboratory evaluation of SARS-CoV-2 lateral flow assays for use in a national COVID-19 seroprevalence survey. Thorax 75(12):1082–1088

Hamming I, Timens W, Bulthuis ML, Lely AT, Navis GV, van Goor H (2004) Tissue distribution of ACE2 protein, the functional receptor for SARS coronavirus. A first step in understanding SARS pathogenesis. J Pathol 203(2):631–637
Henein MY, Mandoli GE, Pastore MC, Ghionzoli N, Hasson F, Nisar MK, Islam M, Bandera F, Marrocco-Trischitta MM, Baroni I, Malagoli A (2021) Biomarkers predict in-hospital major adverse cardiac events in COVID-19 patients: a multicenter international study. J Clin Med 10(24):5863
Hou T, Zeng W, Yang M, Chen W, Ren L, Ai J, Wu J, Liao Y, Gou X, Li Y, Wang X (2020) Development and evaluation of a rapid CRISPR-based diagnostic for COVID-19. PLoS Pathog 16(8):e1008705
Huang JC, Chang YF, Chen KH, Su LC, Lee CW, Chen CC, Chen YMA, Chou C (2009) Detection of severe acute respiratory syndrome (SARS) coronavirus nucleocapsid protein in human serum using a localized surface plasmon coupled fluorescence fiber-optic biosensor. Biosens Bioelectron 25(2):320–325
Iravani S (2020) Nano-and biosensors for the detection of SARS-CoV-2: challenges and opportunities. Mater Adv 1(9):3092–3103
Jääskeläinen AJ, Kekäläinen E, Kallio-Kokko H, Mannonen L, Kortela E, Vapalahti O, Kurkela S, Lappalainen M (2020) Evaluation of commercial and automated SARS-CoV-2 IgG and IgA ELISAs using coronavirus disease (COVID-19) patient samples. Euro Surveill 25(18):2000603
Knott GJ, Doudna JA (2018) CRISPR-Cas guides the future of genetic engineering. Science 361(6405):866–869
Lamb LE, Bartolone SN, Ward E, Chancellor MB (2020) Rapid detection of novel coronavirus (COVID-19) by reverse transcription-loop-mediated isothermal amplification. MedRxiv 24:2020-02
Land KJ, Boeras DI, Chen XS, Ramsay AR, Peeling RW (2019) Reassured diagnostics to inform disease control strategies, strengthen health systems and improve patient outcomes. Nat Microbiol 4(1):46–54
Lassaunière R, Frische A, Harboe ZB, Nielsen AC, Fomsgaard A, Krogfelt KA, Jørgensen CS (2020) Evaluation of nine commercial SARS-CoV-2 immunoassays. MedRxiv 10:2020-04
Li Y, Yao L, Li J, Chen L, Song Y, Cai Z, Yang C (2020) Stability issues of RT-PCR testing of SARS-CoV-2 for hospitalized patients clinically diagnosed with COVID-19. J Med Virol 92(7):903–908
Lukose J, Chidangil S, George SD (2021) Optical technologies for the detection of viruses like COVID-19: Progress and prospects. Biosens Bioelectron 178:113004
Mahshid SS, Flynn SE, Mahshid S (2021) The potential application of electrochemical biosensors in the COVID-19 pandemic: a perspective on the rapid diagnostics of SARS-CoV-2. Biosens Bioelectron 176:112905
Narita F, Wang Z, Kurita H, Li Z, Shi Y, Jia Y, Soutis C (2021) A review of piezoelectric and magnetostrictive biosensor materials for detection of COVID-19 and other viruses. Adv Mater 33(1):2005448
Park GS, Ku K, Baek SH, Kim SJ, Kim SI, Kim BT, Maeng JS (2020) Development of reverse transcription loop-mediated isothermal amplification assays targeting severe acute respiratory syndrome coronavirus 2 (SARS-CoV-2). J Mol Diagn 22(6):729–735
Pietschmann J, Vopel N, Spiegel H, Krause HJ, Schroper F (2020) Brief communication: magnetic immuno-detection of SARS-CoV-2 specific antibodies. Mol Biol 1:1–16
Sangadevan S, Periasamy M (2014) Recent trends in biosensors and their application. Revised Adv Mater Sci 36:62–69
Seo G, Lee G, Kim MJ, Baek SH, Choi M, Ku KB, Lee CS, Jun S, Park D, Kim HG, Kim SJ, Lee JO, Kim BT, Park EC, Kim SI (2020) Rapid detection of COVID-19 causative virus (SARS-CoV-2) in human nasopharyngeal swab specimens using field-effect transistor-based biosensor. ACS Nano 14(4):5135–5142
Sharma H, Mutharasan R (2013) Review of biosensors for foodborne pathogens and toxins. Sens Actuators B Chem 183:535–549

Sheikhzadeh E, Eissa S, Ismail A, Zourob M (2020) Diagnostic techniques for COVID-19 and new developments. Talanta 220:121392

Sohrabi C, Alsafi Z, O'Neill N, Khan M, Kerwan A, Al-Jabir A, Iosifidis C, Agha R (2020) Corrigendum to "World Health Organization declares global emergency: a review of the 2019 novel coronavirus (COVID-19)". Int J Surg 77:217

Wu JL, Tseng WP, Lin CH, Lee TF, Chung MY, Huang CH, Chen SY, Hsueh PR, Chen SC (2020) Four point-of-care lateral flow immunoassays for diagnosis of COVID-19 and for assessing dynamics of antibody responses to SARS-CoV-2. J Infect 81(3):435–442

Yakoh A, Pimpitak U, Rengpipat S, Hirankarn N, Chailapakul O, Chaiyo S (2021) Paper-based electrochemical biosensor for diagnosing COVID-19: detection of SARS-CoV-2 antibodies and antigen. Biosens Bioelectron 176:112912

Yam WC, Chan KH, Poon LL, Guan Y, Yuen KY, Seto WH, Peiris JS (2003) Evaluation of reverse transcription-PCR assays for rapid diagnosis of severe acute respiratory syndrome associated with a novel coronavirus. J Clin Microbiol 41(10):4521–4524

Yong SEF, Anderson DE, Wei WE, Pang J, Chia WN, Tan CW, Teoh YL, Rajendram P, Toh MPHS, Poh C, Koh VT (2020) Connecting clusters of COVID-19: an epidemiological and serological investigation. Lancet Infect Dis 20(7):809–815

Yuan J, Wu S, Duan N, Ma X, Xia Y, Chen J, Ding Z, Wang Z (2014) A sensitive gold nanoparticle-based colorimetric aptasensor for Staphylococcus aureus. Talanta 127:163–168

Zhao H, Liu F, Xie W, Zhou TC, OuYang J, Jin L, Li H, Zhao CY, Zhang L, Wei J, Zhang YP, Li CP (2021) Ultrasensitive supersandwich-type electrochemical sensor for SARS-CoV-2 from the infected COVID-19 patients using a smartphone. Sens Actuator B Chem 327:128899

Zhou P, Yang XL, Wang XG, Hu B, Zhang L, Zhang W, Si HR, Zhu Y, Li B, Huang CL, Chen HD (2020) A pneumonia outbreak associated with a new coronavirus of probable bat origin. Nature 579(7798):270–273

Zhu N, Zhang D, Wang W, Li X, Yang B, Song J, Zhao X, Huang B, Shi W, Lu R, Niu P (2020) A novel coronavirus from patients with pneumonia in China, 2019. N Engl J Med 382(8):727–733

Two-Dimensional Transition Metal Dichalcogenides and Conducting Polymer Hybrids for Electrochemical Sensing: Recent Progress and Future Outlook

8

Tarab Fatima and Samina Husain

Abstract

Two-dimensional (2D) layered nanomaterials have validated great potential in the field of sensing. The advancement of electrochemical sensors based on nanomaterials has gained tremendous attention owing to their attractive properties such as high sensitivity, low response time, and enhanced selectivity. Amongst them, 2D transition metal dichalcogenides (TMDs) like WS_2 and MoS_2 are the most commonly used nanostructures for the development of electrochemical sensors due to their facile synthesis method, biocompatibility, high surface area, good charge transfer properties, and peculiar capability of functionalization and attachment of biomolecules. In combination with the general benefits of 2D nanomaterials, the incorporation of conducting polymers like polyaniline (PANI) and polypyrrole (PPy) into 2D-TMD structures is a current approach for improving the electrochemical sensing performance of these materials for environmental pollutants. The rising of the industrialization is a direct consequence of the drastically expanding human population, which has inevitably given rise to new biomedical and environmental concerns. In addition to the quality assurance of pharmaceutical manufacturing, there is an urgent need for a sensitive, selective, and cost-effective platform to monitor contaminated water sources and effluent for water pollutants. This chapter will compile a review to comprehend the electrochemical sensing and synergistic effects of interconnected network of 2D-TMDs and conducting polymers. Parameters such as modification of electrodes, analyte–electrode interactions, interfacial interaction mechanism, redox behaviour, selectivity, sensitivity, and stability for the fabrication of 2D TMD-based hybrids as electrochemical sensors will be comprehensively compiled. Approaches for the development of high-performance electrochemical sensors

T. Fatima · S. Husain (✉)
Centre for Nanoscience and Nanotechnology, Jamia Millia Islamia, New Delhi, India
e-mail: shusain3@jmi.ac.in

M. Mohsin, N. Soleja (eds.), *Nanoscale Sensors and their Applications in Biomedical Imaging*, https://doi.org/10.1007/978-981-97-3144-2_8

and providing a comprehensive toolkit for the detection of various pharmaceutical drugs, viz. (a) strategies for the modification of the existing electrodes with the characteristics and functionalities of existing nanomaterials and (b) opportunities obtained by the advancement of the electrochemical sensing of pharmaceutical compounds for the protection of the ecosystem will also be discussed.

Keywords

Two-Dimensional Transition Metal Dichalcogenides · Conducting Polymer · Nanocomposites · Electrochemical sensing

8.1 Introduction

A chemical sensor is a device that measures chemical or physical changes and converts the information into measurable signal. Electrochemical sensing is an analytical technique that employs the properties of electrochemistry to detect and quantify the concentration of a specific analyte in a sample. This method employs an electrode, a conductor of electricity placed in contact with the sample (Murthy et al. 2022). When an electrical potential is applied to the electrode, a signal proportional to the analyte concentration in the sample is generated. Electrochemical sensors are gaining tremendous attention due to its importance in several fields such as food monitoring, clinical diagnosis, environmental remediation, food analysis, and chemical process control (Canevari 2022). This technique is highly sensitive, selective with low response time making it an effective method for detecting and quantifying trace quantities of analytes in complex matrices (Simões and Xavier 2017a).

In the last few years, several strategies have been employed using various nanomaterials such as conducting polymer, 2D nanomaterials, metals, metal oxides, metal organic framework, and carbon-based nanomaterials. The high surface to volume ratio and tunable electronic properties are the most prominent properties of nanoscience and nanotechnology (N&N), the inclusion of these methods and ideas connected to N&N has a considerable influence on the rapid development of nanosensors. In 1950s, the beginning of the use of electrochemical sensors started with the monitoring of industrial oxygen. Leland C. Clark describes the concept of the oxygen electrochemical sensor by using two electrodes of a cell with an oxygen-permeable membrane separating the electrodes from the electrolyte solution. The reduction of the oxygen was appeared at the surface of the electrode by the permeation of the oxygen through the membrane. The oxygen sensor developed by Clark achieved widespread use in clinical, industrial, and environmental monitoring (Islam and Haider 2010).

Nanomaterials are gaining significant attention in the development of electrochemical sensors because of their peculiar properties, which include a large surface area, tunable electronic properties, enhanced surface reactions, improved accuracy, reliability, and ease of miniaturization (Ping et al. 2017). These sensors employ electrochemical processes to detect analytes in a sample, such as chemicals or biomolecules. Nanomaterials such as metal and metal oxide nanoparticles, graphene,

CNT, and other carbon-based materials have been commonly used as an electrode sensing element owing to their high electrical conductivity and surface area (Aragay and Merkoçi 2012). These nanomaterials have the ability to bind to specific molecules, resulting in measurable change in their electrical properties. Additionally, nanomaterials also provide enhanced stability to the electrochemical sensors, as they have the ability to get immobilized on the surface of electrodes and prevent the degradation/agglomeration of the matrix by the encapsulation of the analyte (Scozzari 2008).

Recently, transition metal dichalcogenides (TMDs) have gained significant attention in the area of electrochemical sensing. TMDs are comprised of a transition metal (such as molybdenum (Mo), tungsten (W), or titanium (Ti)) and a chalcogen (such as sulfur (S), selenium (Se), or tellurium (Te)) (Su et al. 2015). TMDs such as semiconducting (MoS_2 and WS_2), semi-metallic ($TiSe_2$ and WTe_2), and metallic (VSe_2 and NbS_2), and insulating (HfS_2) offer peculiar electronic and optical properties such as high surface area, which can increase the sensitivity of sensors, and can be functionalized with chemical groups to enhance their selectivity. This makes them promising candidates for use in electrochemical sensing (Zhang et al. 2017). In addition, TMDs have a high electron mobility and are effective catalysts for electrochemical reactions. Electrochemical sensors based on TMDs have been used for several applications such as sensing of drugs, biomolecules, organic compounds, and heavy metals (Lee et al. 2019; Sun et al. 2017). In order to overcome the existing limitations of TMDs and improve the sensing ability, researchers have fabricated several composites using noble metals (Au, Ag and Pt), metal oxide (ZnO, TiO_2, and SnO_2), and conducting polymers (polyaniline (PANI), polythiophene (PTh), and polypyrrole (PPy)). Conducting polymers (CPs) are the organic materials with conjugated double bonds that increase the conductivity of the polymers (Namsheer and Rout 2021). CPs are widely used in the field of electrochemical sensing due to their unique properties including electronic and optical properties. In 1862, Letheby proposed the first electrochemical synthesis of PANI using oxidative polymerization in the presence of aniline and sulfuric acid. CPs have the ability to functionalize and become more selective and interactive towards target analyte. Nanocomposites containing CPs have enhanced surface area, porosity, stability, conductivity, and flexibility (Ramanavicius et al. 2022; El Rhazi et al. 2018).

This chapter focuses on providing the deep knowledge of the TMDs, conducting polymers and their hybrid material for the electrochemical sensing of various pollutants. In order to improve the sensing ability, TMDs–conducting polymers hybrid materials are being studied extensively to improve the sensing ability of the fabricated sensors. The synergistic association of TMDs and CPs enhances the sensitivity, selectivity, and response time of the hybrid material. The optimization of CP plays a crucial role as different kinds of polymers provide varying degrees of selectivity and sensitivities towards the specific analyte. The conducting polymer may be deposited onto the TMD surface using a different methods, including solution mixing, layer-by-layer assembly, and electrochemical deposition. In conclusion, TMDs and CPs may work together to provide enhanced electrochemical sensing materials.

The resultant hybrid materials can act as a promising candidate for several applications including healthcare, environmental monitoring, and food safety.

8.2 Fundamentals of Electrochemical Sensing

Electrochemical sensors are devices that detect and measure chemical substances in a sample by converting chemical information into an electrical signal as shown in Fig. 8.1. They are widely used in various fields such as environmental monitoring, medical diagnosis, food safety, and industrial process control. The basic mechanism of an electrochemical sensor involves an electrode, an electrolyte, and an interfacial electrode–electrolyte interaction. The electrode is a conducting material (usually gold or platinum) that supports as a platform for the electrochemical processes (Kimmel et al. 2012). The analyte of interest is dispersed in an electrolyte solution, which also acts as a medium for ion transfer between the electrode and the solution at the interface. The detecting mechanism is based on the conversion of the chemical reaction into a detectable electrical signal (Yan et al. 2017).

8.2.1 Characteristic Parameters

The basic components of an electrochemical sensor comprises of an analyte, recognition element, a transducer, amplifier, processor, and a display. The characteristic parameters of an electrochemical sensors are:

Sensitivity: Sensitivity is the degree by which a change in analyte concentration leads to a corresponding change in the measured signal. The lower the detected concentration of the analyte, the greater the sensitivity.

Selectivity: Selectivity refers to a capability of the sensor to differentiate between the target analyte and the interferants species present in a sample. A highly selective sensor can detect the target analyte even in the presence of interfering chemicals.

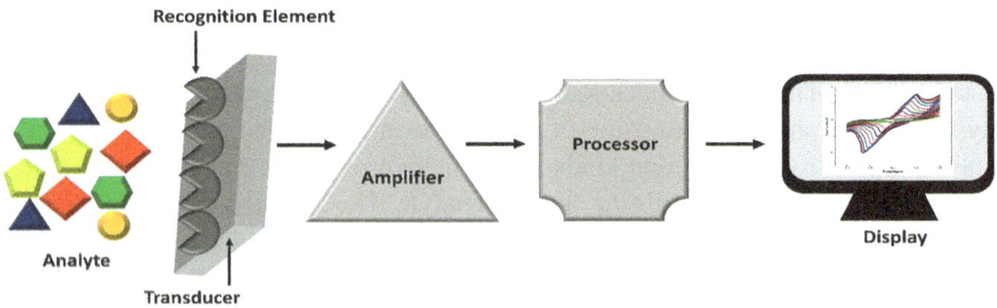

Fig. 8.1 Basic components of electrochemical sensors

Response time: The sensitivity of a sensor to changes in analyte concentration is measured in terms of its response time.

Stability: Stability of the sensor is the capability to retain its sensitivity and selectivity over time. A sensor is said to be stable when it provides consistent and reliable measurements over a specific interval of time.

Detection limit: The detection limit may be defined as the lowest analyte concentration that can be detected by the sensor. For applications requiring a high degree of sensitivity, a lower detection limit is desirable.

Range: The range of the sensor is the concentration range across which it provides reliable measurements of the analyte.

Reproducibility: Reproducibility is the capability to provide consistent measurements when analysing the same reaction multiple times. For accurate measurements and data analysis, high reproducibility is required.

8.2.2 Types of Electrochemical Sensor

Electrochemical sensors are divided into three wide classes including amperometric, potentiometric, and conduct-metric sensors as shown in Fig. 8.2.

Amperometric sensors are the most commonly used electrochemical sensors which measure the current generated by a change in the chemical reaction that occurs between the target analyte and the electrode at the interface. The mechanism is based on the electro-oxidation/reduction of the analyte at the modified electrodes. Amperometric technique was used for the sensing of hydrogen peroxide using MoS_2 nanosheets/Au-modified electrodes. This provides the direct charge transfer between the electrodes and enzyme with high selectivity. Figure 8.3 represents the amperometric response at different concentrations of H_2O_2 on MoS_2/Au-modified glassy carbon electrode (Shu et al. 2017). In another report, WS_2 nanosheets were used for the non-enzymatic amperometric detection of the imidacloprid (Haritha et al. 2023).

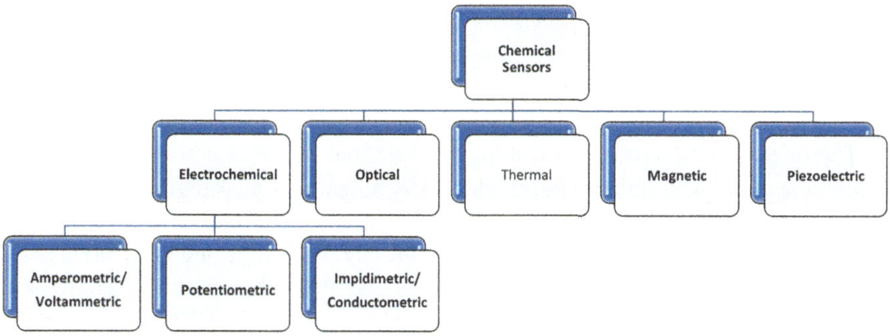

Fig. 8.2 Types of electrochemical sensors

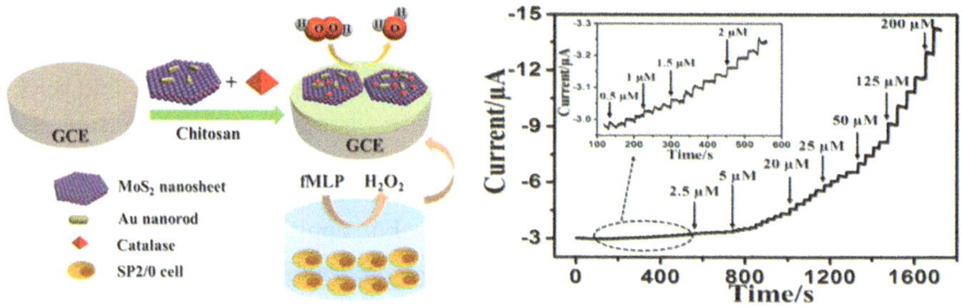

Fig. 8.3 Amperometric response of CAT/MoS_2–Au/chitosan/GCE for the detection of H_2O_2 at various concentrations (Shu et al. 2017)

Potentiometric sensor is based on the use of the ion-selective membrane separating two solutions to generate the potential difference at a constant rate of current. This is a transducer-based mechanism based on reference electrode and an ion-selective electrode (ISE). These membranes interact with the produced charged ion that leads to the change in the relative potential difference at the reference electrode. The change in the concentration of target analyte gives the linear change in the potentiometric response which is governed by the Nernst equation. Potentiometric biosensor was used for the detection of acrylamide based on the transduction mechanism using an ammonium ion selective electrode (Silva et al. 2011).

Conductometric or impedimetric sensors are based on the mechanism of changes in the conductivity or the charge transfer resistance at the interface of electrode–electrolyte solution. The WS_2 nanostructure was used for the impedimetric sensing of Listeria monocytogenes. The paper-based two-electrode system was modified with WS_2 nanostructure and aptamer to achieve high selectivity. The charge transfer resistance (R_{ct}) was observed to be increased with increasing concentration of aptamer as shown in Nyquist plot in Fig. 8.4 (Mishra et al. 2022).

The electrodes can be two- or three-electrode system. The two-electrode system consists of a working and a counter electrode, while the three-electrode system has a working, counter, and a reference electrode system as shown in Fig. 8.5.

The working electrode is the part of the sensor that is in contact with the sample being analysed. It is typically made of a conductive material, such as gold or platinum, and is coated with a sensing layer that interacts with the target analyte. When the analyte comes into contact with the sensing layer, it undergoes a chemical reaction that generates an electrical signal. The sensing layer can be designed to selectively react with a specific analyte or a group of analytes (Simões and Xavier 2017b).

The reference electrode is a stable electrode that provides a constant electrical potential against which the working electrode potential is measured. It is responsible to provide drift-free potential. It typically consists of a metal wire coated with a salt bridge and a reference electrolyte solution. The salt bridge and electrolyte solution maintain a constant concentration of ions that allow the reference electrode to maintain a constant potential. Standard hydrogen electrode (SHE) was commonly

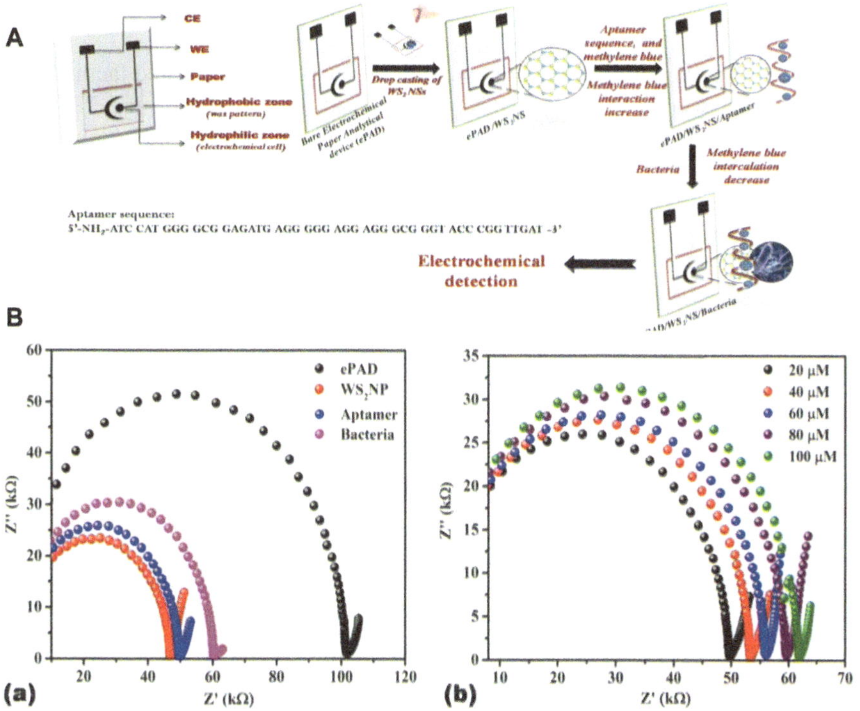

Fig. 8.4 (**A**) Schematic representation of paper-based two-electrode system. (**B**) Nyquist plot. (**a**) Control experiment of WS_2/aptamer-modified electrodes. (**b**) Optimization of aptamer for impedimetric sensing of Listeria monocytogenes (Mishra et al. 2022)

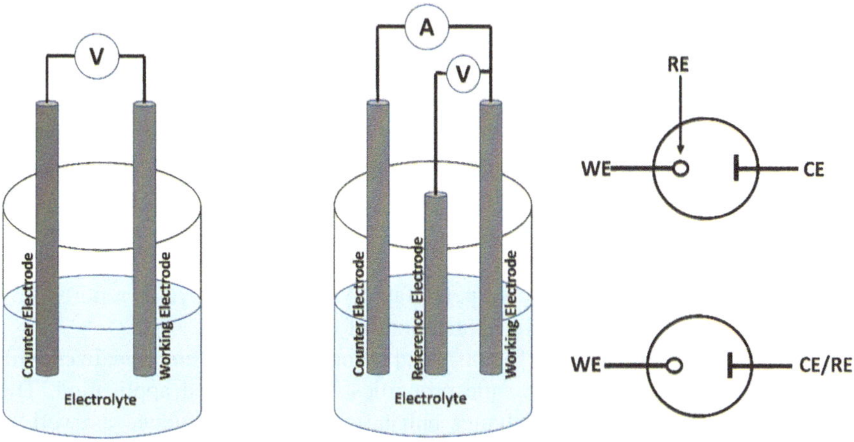

Fig. 8.5 Schematic representation of (**a**) two-electrode system and (**b**) three-electrode system

used to provide standard electrode potentials in terms of reduction/oxidation potential.

The counter electrode is used to complete the electrical circuit and facilitate the reaction at the working electrode. It is typically made of a conductive material, such as silver or platinum, and is placed in close proximity to the working electrode. When an electrical potential is applied between the working and counter electrodes, the target analyte is oxidized or reduced at the working electrode, and the counter electrode is responsible to maintain the stability. When the electrode system is exposed to the solution of the analyte, a current flows between the working and counter electrodes. The observed current is proportional to the concentration of the target analyte in the solution.

8.3 Properties

8.3.1 Properties of Conducting Polymers

CPs are considered as an emerging class of functional polymers that have gained considerable attention because of their desirable characteristics, including their low cost, flexibility, good electrical, mechanical, and optical properties. The first material to exhibit conductivity was polyacetylene, whose conductivity increased by a factor of 10 upon oxidation with iodine vapour. Many conducting polymers such as polyaniline (PANI), polypyrrole (PPy), polythiophene, polyfluorene, and polyterthiophene have been developed to overcome the limitations of the existing polymers (Terán-Alcocer et al. 2021). Generally conducting polymers are semiconducting in their pure form and can be modified by doping to enhance their conductivity. CPs can be termed as p and n doped by the oxidation/reduction of the existing conjugated system by doping. The conducting polymers can be modified by using chemical and electrochemical techniques. CPs may exist as cation radicals, dications, anion radicals, or dianions in their oxidized or reduced forms, respectively. Therefore, it is crucial to improve their conductivity, particularly for the use in electrochemical sensing (Naveen et al. 2017). CPs have relatively high redox reactivity and electron affinity, unlike other conventional organic polymers. The characteristic properties of CPs are directly dependent on the size, length, and molecular weight CPs. Generally, CPs are synthesized by oxidizing the monomer with oxidising compounds and initiator or by applying potential or current through electrodes. In situ polymerization of polyaniline was shown in Fig. 8.6 (Fatima et al. 2022a) in the presence of an oxidizer Ammonium persulfate (APS). In the previous reports, several researchers have explored the applications of CPs.

These conducting polymers (Fig. 8.7) and monomers can be functionalized with various nanomaterials to achieve the properties of the required application. The modification of the existing polymer enhances functional moieties as well as improves the chemical, electrical, and mechanical properties. These conductive polymers can also act as co-catalyst with many existing nanomaterials to improve their performance.

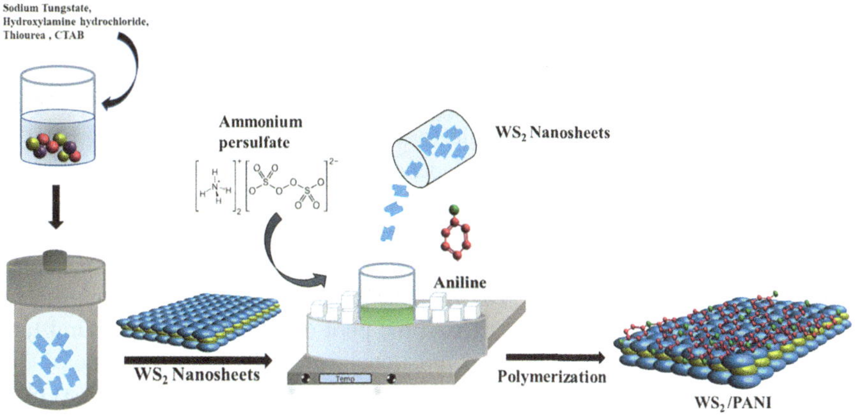

Fig. 8.6 In situ polymerization of PANI for the synthesis of WS_2/PANI nanocomposite (Fatima et al. 2022a)

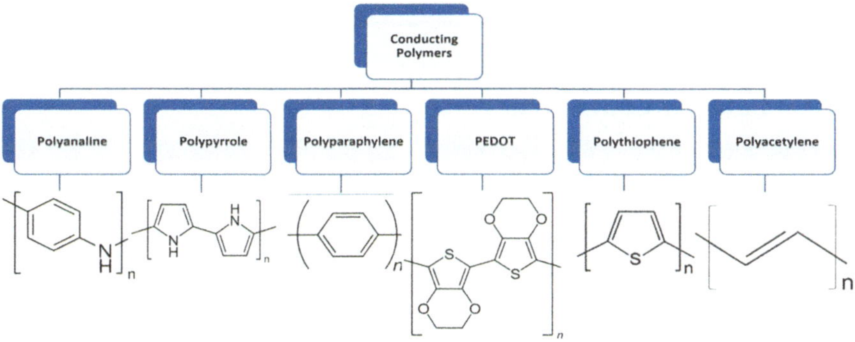

Fig. 8.7 Classification of conducting polymers

8.3.2 Properties of TMDs

In recent years, transition metal dichalcogenides (where M is a transition metal element and X is a chalcogen (Se, S, or Te)) have gained a huge attention. TMDs have several advantages, including their semiconducting (WS_2 and MoS_2), insulating (HfS_2), semi-metallic ($TiSe_2$ and WTe_2), and pure metallic (VSe_2 and NbS_2) properties (Han et al. 2015).

TMDs are layered material that exhibits the MX_2 type structure, formed by strong covalent bonding between X–M–X with Van der Waals inter-layer interaction. However, TMDs in groups 8–10 exhibit non-layered nature (Zhang et al. 2017). The TMDs exist in several structural phases including trigonal, octahedral (1 T), prismatic (2H), or rhombohedral (3R) coordination. The d-electron present in the TMDs is responsible for the electrical conductivity and the coordination condition of metal atoms. TMDs exhibit a wide range of electronic features, from insulating to metallic band structure characteristics to topological and correlated phases, depending on their structural phases and chemical composition. The phases 1H and 1 T, non-bonding d bands of TMDs are located between the antibonding (*) and bonding bands of the M–X bonds (Choi et al. 2017). The properties of TMDs, such as the quantum distribution of electron states, the continuous energy band to discrete level transition, and the size-dependent tunable band gap, are expected to change as a result of this size-induced quantum mechanism, leading to enhanced surface area and optical and electronic properties. These unique properties of TMDs can provide a new avenue for developing various applications for a low-dimensional material (Chhowalla et al. 2013). The simple combination of X and M can produce novel materials and exponentially expand the pool of possibilities, structural tunability, control, and required chemical functionalization. TMDs at the scale of 0D, 1D, and 2D nanostructures can be used for the variety of applications such as photocatalysis, electrochemical sensing batteries, supercapacitors, photovoltaics, and transistors. They can be used as a next-generation electrode and electrochemical sensor due to their impressive physical, chemical, optical, and electrical properties (Choi et al. 2017).

8.4 Electrochemical Sensing Using TMDs–Conducting Polymer Composite

In a report, highly sensitive 3D MoS_2–PANI/rGO was synthesized via hydrothermal method to detect ascorbic acid (AA), dopamine (DA), and uric acid (UA). The nanocomposite displayed improved functionality, good bioaffinity, and enhanced electrochemical activity. Differential pulse voltammetry measurements revealed wide linear range of concentrations with low limit of detection of 50–8.0 mM: 22.20 mM, 5.0–500 mM: 0.70, and 1.0–500 mM: 0.36 mM for AA, DA, and UA (Li et al. 2019). Chloramphenicol (CAP) has been considered as an effective antibacterial drug with a wide range of bactericidal activity. However, it has deleterious effects on the haematopoietic system, including aplastic anaemia. Furthermore, chronic exposure to trace amounts of CAP can cause an imbalance in the human body. Another group of researchers has synthesized MoS_2/PANI nanocomposite using in situ polymerization method by adding previously prepared MoS_2 nanosheets for the electrochemical detection of chloramphenicol (CAP). This sensor exhibited high selectivity with LOD of 6.9×10^{-8} mol/L and linear range of 1×10^{-7} to 1×10^{-4} mol/L (Chen et al. 2016).

Bisphenols (BPs) have been identified as significant environmental pollutants and endocrine-disrupting compounds that have detrimental effects on human health. In a report, MoS_2 nanosheets were incorporated into self-doped PANI (SPAN) for the electrochemical sensing of bisphenols. The fabricated MoS_2/SPAN/GCE exhibited enhanced sensitivity with a limit of detection of 0.6 nM. The enhanced sensing ability was assigned to high electroactivity, high electroactive surface area, and high anionic exchange ability (Yang et al. 2015). The interference experiment was also performed to confirm the high selectivity of the developed sensor. A novel molecularly imprinted based nanozyme sensor was developed using conducting polymer poly(3,4-ethylenedioxythiophene) (PEDOT) with molybdenum disulfide (MoS_2) for the detection of luteolin in Gnaphalium affine (Lu et al. 2021). The sensor was fabricated by electrochemical polymerization in the presence of MoS_2 and luteolin as a template molecule. This modification enhances the surface area, electronic conductivity, and selectivity of the interface of the electrode. The electrochemical performance revealed the limit of detection of 0.04 μM with linear range of 3×10^{-7} to 3×10^{-5} μM. Another conducting polymer polypyrrole (PPy) was used for the detection of glucose by synthesizing the nanocomposite with gold nanoparticle modified 2D MoS_2 nanosheets on glass carbon electrode (GCE). Ultrathin MoS_2 nanosheets were synthesized by liquid phase exfoliation (LPE) using DMF. The electro-polymerisation of PPy was done on the electrochemically deposited gold-modified MoS_2 on GCE using amperometry. At an oxidation potential of +0.45 V, glucose has been converted into gluconolactone using a Cu(II)/Cu(III) redox pair as a catalytic centre. The developed electrode exhibited high reproducibility, selectivity with LOD of 0.08 nM and the LOQ was 0.26 nM (Ma et al. 2019). A β-lactam antibiotic ampicillin (AMP) has been widely used for the treatment of several infections such as urinary tract infections (UTI), stomach gastrointestinal infections, pneumonia, and bronchitis. This is a broad-spectrum antibiotic that can be used by human and animals. The widespread use of these antibiotics can lead to anti-microbial resistance (AMR) which can risk human health. MoS_2 nanosheets/PPY-based electrochemical sensor was developed for the detection of ampicillin (AMP) with a limit of detection (LOD) of 10 pg/L (0.28 pM) as shown in Fig. 8.8 (Hamami et al. 2021).

A psychoactive drug clonazepam (5-(2-chlorophenyl)-7-nitro-2,3-dihydro-1,4-benzodiazepin-2-one) was detected using mesoporous WS_2 nanorods. The mesoporous WS_2 nanorods modified screen printed electrodes exhibited enhanced

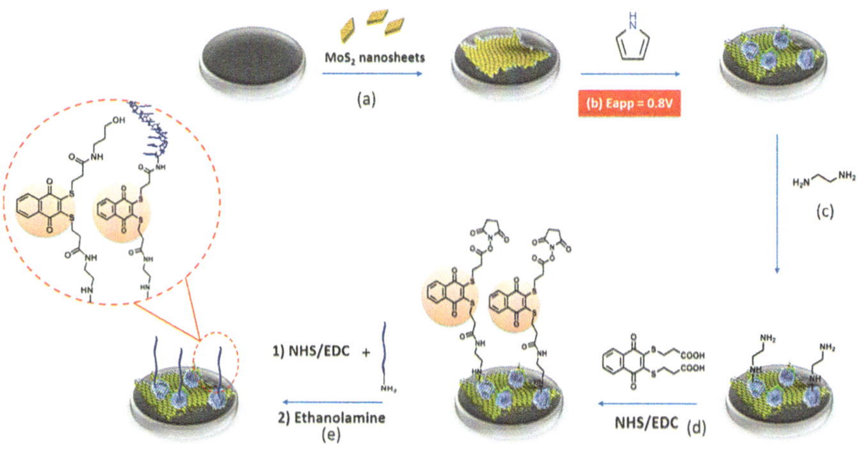

Fig. 8.8 Schematic representation of MoS_2/PPy-based biosensor (**a–e**) for the detection of ampicillin (Hamami et al. 2021)

reduction peak with lower detection limit of 2.37 nM. This also displayed excellent selectivity in the presence of interfering compounds with high sensitivity (24.32 $\mu A \mu M^{-1}$ cm^{-2}) (Chen et al. 2019). Clonazepam is a sedative drug affecting the imbalanced neurotransmitter in the brain neurons, primarily used to treat panic disorder, chronic fatigue syndrome, seizures, and akathisia in humans. Nitrofurantoin is a widely used antibiotic and belongs to the family of nitrofuran. Generally, it is used for the treatment of urinary tract infections and pulmonary diseases. A group of researchers used WS_2 and WS_2–PANI nanocomposite modified paper-based two-electrode system for the electrochemical sensing of pulmonary toxic drug nitrofurantoin. A conducting polymer, polyaniline (PANI) enhances the charge transfer property at the interface of the electrode and improves the electrocatalytic activity as shown in Fig. 8.9. As compared to WS_2, nanocomposite with PANI shows synergistic association and improves the limit of detection from 0.05 to 0.01 μM, respectively (Fatima et al. 2022a).

A molecularly imprinted PEDOT-based electrochemical sensor was developed using WS_2 nanosheets for the detection of vitamin B2. The WS_2-PEDOT film was developed by co-polymerization of monomer and VB2 template molecule in the presence of WS_2 nanosheets. WS_2 provides affinity and stability in the fabricated sensor with the LOD of 0.7 nM (Zhang et al. 2018).

The long-term use of rifamycin antibiotics shows adverse effects on human beings (Table 8.1). The persistence of these kinds of antibiotics can pose threats to aquatic as well as human life. The $MoSe_2$ microsphere-based electrochemical sensor modified with the polymer β-cyclodextrin for the trace level detection of rifampicin was developed. The obtained results revealed the excellent performance of the sensor with high sensitivity of 11.64 μA μM^{-1} cm^{-2} and limit of detection of 0.019–374.5 μM (Ganguly et al. 2022). The nanocomposite of WSe_2 nanosheets was developed for the detection of ascorbic acid (AA), dopamine (DA), and uric acid (UA) with LOD of 3.2, 0.18, and 0.42 μM, respectively (Bhethanabotla 2022).

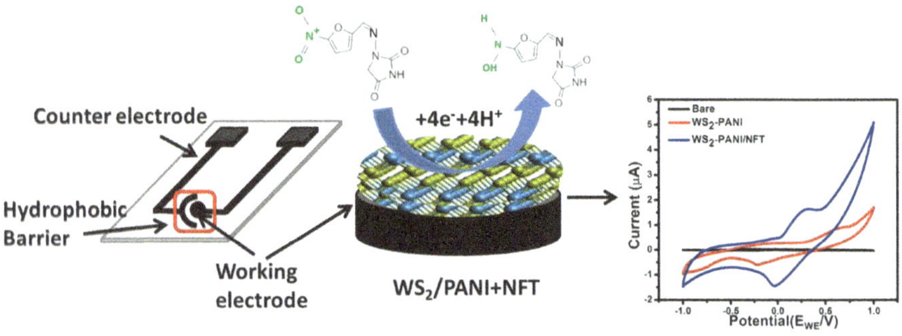

Fig. 8.9 Schematic illustration of WS_2/PANI modified paper electrode for the electrochemical sensing of nitrofurantoin (Fatima et al. 2022a)

Table. 8.1 TMDs–conducting polymers composite for the electrochemical detection of various analyte

S. no.	Material	Electrode	Analyte	Linear range	Limit of detection	References
1.	MoS_2-PANI/rGO	Glassy carbon	Ascorbic acid (AA), Dopamine (DA), and Uric acid (UA)	50 μM to 8.0 mM, 5.0 to 500 μM, and 1.0 to 500 μM	22.20 μM 0.70 μM 0.36 μM	Li et al. (2019)
2.	MoS_2/PANI	Carbon paste electrode	Chloramphenicol	1×10^{-7} mol/L to 1×10^{-4} mol/L	6.9×10^{-8} mol/L	Chen et al. (2016)
3.	MoS_2/PEDOT	Glassy carbon	Luteolin	3×10^{-7} to 3×10^{-5} M	0.04 μM	Lu et al. (2021)
4.	MoS_2/Au/PPy	Glassy carbon	Glucose	0.1–80 nM	0.08 nM	Ma et al. (2019)
5.	MoS_2/PPy	Screen printed carbon	Ampicillin	50-250 pg/L	0.28 pM	Hamami et al. (2021)
6.	Mesoporous WS_2	Screen printed carbon	Clonazepam	10–551 μM	2.37 nM	Chen et al. (2019)
7.	WS_2 Nanosheets	Paper electrodes	Nitrofurantoin	100–0.01 μM	0.05 μM	Fatima et al. (2022b)
8.	WS_2-PANI	Paper electrodes	Nitrofurantoin	100–0.01 μM	0.01 μM	Fatima et al. (2022a)
9.	PANI/C-dots	Glassy carbon	Estradiol	0.001–100 μmol L^{-1}	43 nmol L^{-1}	Supchocksoonthorn et al. (2021)
10.	WS_2/PEDOT	Glassy carbon	VB2	0.002 to 0.9 μM	0.7 nM	Zhang et al. (2018)
11.	$MoSe_2$/Beta-Cyclodextrin	Glassy carbon	Rifamycin	0.019 to 374.5 μM	28 nM	Ganguly et al. (2022)
12.	WSe_2 nanosheets	Glassy carbon	Ascorbic acid (AA), Dopamine (DA), and Uric acid (UA)	8.0 to 2456 μM for, 72.8–381.4 μM 36.9–185.2 μM	3.2 μM 0.18 μM 0.42 μM	Bhethanabotla (2022)

8.5 Conclusion

In conclusion, the peculiar properties of 2D-TMDs and their potential for composite formation have drawn the considerable attention of the researchers for the use in sensing applications. The development of TMDs–conducting polymer hybrid has a potential with an infinite possibilities of material combinations and functionalization for the development of highly selective and sensitive electrochemical sensors. Therefore, the development of novel nanocomposites or functional materials can be used as a platform for the various applications of electrochemical sensing. In spite of the previous reported literature, the development of TMDs and their composite-based sensors are still in their early stages. Several characteristic factors such as chemical stability, biocompatibility, tunability, and interfacial behaviour are still need to be addressed. The majority of the reported studies focussed on MoS_2 and WS_2 with polyaniline (PANI) and polypyrrole (PPY) for the electrochemical sensing with unique features have been discussed. The versatile properties of their nanocomposite with other conducting polymers make them pleasing to explore in other areas. The synergistic association of conducting polymers provides large surface area, high charge transfer efficiency, enhanced interfacial diffusion which improves the overall performance of the fabricated sensor with enhanced stability, sensitivity, and selectivity.

References

Aragay G, Merkoçi A (2012) Nanomaterials application in electrochemical detection of heavy metals. Electrochim Acta 84:49–61. https://doi.org/10.1016/j.electacta.2012.04.044

Bhethanabotla SCVR (2022) Electrochemical sensor based on carbon-incorporated WSe_2 nanosheets for simultaneous detection of ascorbic acid, dopamine, and uric acid. IEEE Sensors J 22(15):14952–14958

Canevari TC (2022) Application of hybrid nanomaterials for development of electrochemical sensors. Elsevier Ltd. https://doi.org/10.1016/B978-0-12-823788-5.00002-8

Chen HY, Wang J, Meng L, Yang T, Jiao K (2016) Thin-layered MoS_2/polyaniline nanocomposite for highly sensitive electrochemical detection of chloramphenicol. Chin Chem Lett 27(2):231–234. https://doi.org/10.1016/j.cclet.2015.09.018

Chen TW, Rajaji U, Chen SM, Jothi Ramalingam R (2019) A relative study on sonochemically synthesized mesoporous WS_2 nanorods & hydrothermally synthesized WS2 nanoballs towards electrochemical sensing of psychoactive drug (clonazepam). Ultrason Sonochem 54(January):79–89. https://doi.org/10.1016/j.ultsonch.2019.02.012

Chhowalla M, Shin HS, Eda G, Li LJ, Loh KP, Zhang H (2013) The chemistry of two-dimensional layered transition metal dichalcogenide nanosheets. Nat Chem 5(4):263–275. https://doi.org/10.1038/nchem.1589

Choi W, Choudhary N, Han GH, Park J, Akinwande D, Lee YH (2017) Recent development of two-dimensional transition metal dichalcogenides and their applications. Mater Today 20(3):116–130. https://doi.org/10.1016/j.mattod.2016.10.002

El Rhazi M, Majid S, Elbasri M, Salih FE, Oularbi L, Lafdi K (2018) Recent progress in nanocomposites based on conducting polymer: application as electrochemical sensors. Int Nano Lett 8(2):79–99. https://doi.org/10.1007/s40089-018-0238-2

Fatima T, Husain S, Khanuja M (2022a) Superior photocatalytic and electrochemical activity of novel WS2/PANI nanocomposite for the degradation and detection of pollutants: antibiotic, heavy metal ions, and dyes. Chem Eng J Adv 12(July):100373. https://doi.org/10.1016/j.ceja.2022.100373

Fatima T, Husain S, Narang J, Khanuja M, Shetti NP, Reddy KR (2022b) Novel tungsten disulfide (WS_2) nanosheets for photocatalytic degradation and electrochemical detection of pharmaceutical pollutants. J Water Process Eng 47(March):102717. https://doi.org/10.1016/j.jwpe.2022.102717

Ganguly A, Hwa K-Y, Santhan A, Tata Sanjay KS (2022) Strategic orchestration of $MoSe_2$ microspheres on β-cd functionalized rGO: a sustainable electrocatalyst for detection of rifampicin in real samples. Chemosphere 307:135373

Hamami M, Bouaziz M, Raouafi N, Bendounan A, Korri-Youssoufi H (2021) MoS_2/ppy nanocomposite as a transducer for electrochemical aptasensor of ampicillin in river water. Biosensors 11:9. https://doi.org/10.3390/bios11090311

Han SA, Bhatia R, Woo KS (2015) Synthesis, properties and potential applications of two-dimensional transition metal dichalcogenides. Nano Converg 2:17. https://doi.org/10.1186/s40580-015-0048-4

Haritha VS, Sarath Kumar SR, Rakhi RB (2023) WS_2-nanosheet-modified electrodes as an efficient electrochemical sensing platform for the nonenzymatic detection of the insecticide Imidacloprid. ACS Omega 8(9):8695–8702. https://doi.org/10.1021/acsomega.2c08077

Islam SK, Haider SMH (2010) Sensors and low power signal processing. Springer Science+Business Media, Springer, US

Kimmel DW, Leblanc G, Meschievitz ME, Cliffel DE (2012) Electrochemical sensors and biosensors. Anal Chem 84(2):685–707. https://doi.org/10.1021/ac202878q

Lee CW, Suh JM, Jang HW (2019) Chemical sensors based on two-dimensional (2D) materials for selective detection of ions and molecules in liquid. Front Chem 7(November):1–21. https://doi.org/10.3389/fchem.2019.00708

Li S, Ma Y, Liu Y et al (2019) Electrochemical sensor based on a three dimensional nanostructured MoS_2 nanosphere-PANI/reduced graphene oxide composite for simultaneous detection of ascorbic acid, dopamine, and uric acid. RSC Adv 9(6):2997–3003

Lu X, Li Y, Duan X et al (2021) A novel nanozyme comprised of electro-synthesized molecularly imprinted conducting PEDOT nanocomposite with graphene-like MoS_2 for electrochemical sensing of luteolin. Microchem J 168:106418. https://doi.org/10.1016/j.microc.2021.106418

Ma K, Sinha A, Dang X, Zhao H (2019) Electrochemical preparation of gold nanoparticles-polypyrrole co-decorated 2D MoS_2 nanocomposite sensor for sensitive detection of glucose. J Electrochem Soc 166(2):B147–B154. https://doi.org/10.1149/2.1231902jes

Mishra A, Pilloton R, Jain S et al (2022) Paper-based electrodes conjugated with tungsten disulfide nanostructure and aptamer for impedimetric detection of listeria monocytogenes. Biosensors 12:2. https://doi.org/10.3390/bios12020088

Murthy HCA, Wagassa AN, Ravikumar CR, Nagaswarupa HP (2022) Functionalized metal and metal oxide nanomaterial-based electrochemical sensors. Elsevier Ltd. https://doi.org/10.1016/B978-0-12-823788-5.00001-6

Namsheer K, Rout CS (2021) Conducting polymers: a comprehensive review on recent advances in synthesis, properties and applications. RSC Adv 11(10):5659–5697. https://doi.org/10.1039/d0ra07800j

Naveen MH, Gurudatt NG, Shim YB (2017) Applications of conducting polymer composites to electrochemical sensors: a review. Appl Mater Today 9:419–433. https://doi.org/10.1016/j.apmt.2017.09.001

Ping J, Fan Z, Sindoro M, Ying Y, Zhang H (2017) Recent advances in sensing applications of two-dimensional transition metal dichalcogenide nanosheets and their composites. Adv Funct Mater 27(19):1–18. https://doi.org/10.1002/adfm.201605817

Ramanavicius S, Samukaite-Bubniene U, Ratautaite V, Bechelany M, Ramanavicius A (2022) Electrochemical molecularly imprinted polymer based sensors for pharmaceutical and biomedical applications (review). J Pharm Biomed Anal:215. https://doi.org/10.1016/j.jpba.2022.114739

Scozzari A (2008) Electrochemical sensing methods: a brief review Andrea Scozzari*. *Methods* Published online :335–351

Shu Y, Chen J, Xu Q et al (2017) MoS_2 nanosheet-Au nanorod hybrids for highly sensitive amperometric detection of H_2O_2 in living cells. J Mater Chem B 5(7):1446–1453. https://doi.org/10.1039/c6tb02886a

Silva NAF, Matos MJ, Karmali A, Rocha MM (2011) An electrochemical biosensor for acrylamide determination: merits and limitations. Port Electrochim Acta 29(5):361–373. https://doi.org/10.4152/pea.201105361

Simões FR, Xavier MG (2017a) Electrochemical sensors. Elsevier Inc. https://doi.org/10.1016/B978-0-323-49780-0.00006-5

Simões FR, Xavier MG (2017b) Electrochemical sensors. Nanoscience and its Applications 1:155–178

Su S, Chao J, Pan D, Wang L, Fan C (2015) Electrochemical sensors using two-dimensional layered nanomaterials. Electroanalysis 27(5):1062–1072. https://doi.org/10.1002/elan.201400655

Sun X, Fan J, Fu C et al (2017) WS_2 and MoS_2 biosensing platforms using peptides as probe biomolecules. Sci Rep 7(1):1–8. https://doi.org/10.1038/s41598-017-10221-4

Supchocksoonthorn P, Alvior Sinoy MC, de Luna MDG, Paoprasert P (2021) Facile fabrication of 17β-estradiol electrochemical sensor using polyaniline/carbon dot-coated glassy carbon electrode with synergistically enhanced electrochemical stability. Talanta 235(August):122782. https://doi.org/10.1016/j.talanta.2021.122782

Terán-Alcocer Á, Bravo-Plascencia F, Cevallos-Morillo C, Palma-Cando A (2021) Electrochemical sensors based on conducting polymers for the aqueous detection of biologically relevant molecules. Nano 11(1):1–62. https://doi.org/10.3390/nano11010252

Yan L, Shi H, Sui X, Deng Z, Gao L (2017) MoS2-DNA and MoS2 based sensors. RSC Adv 7(38):23573–23582. https://doi.org/10.1039/c7ra02649h

Yang T, Chen H, Yang R, Jiang Y, Li W, Jiao K (2015) A glassy carbon electrode modified with a nanocomposite consisting of molybdenum disulfide intercalated into self-doped polyaniline for the detection of bisphenol A. Microchim Acta 182:2623–2628

Zhang C, Gong C, Nie Y et al (2017) Systematic study of electronic structure and band alignment of monolayer transition metal dichalcogenides in Van der Waals heterostructures. 2D Mater 4(1):48. https://doi.org/10.1088/2053-1583/4/1/015026

Zhang Z, Xu J, Wen Y, Wang T (2018) A highly-sensitive VB2 electrochemical sensor based on one-step co-electrodeposited molecularly imprinted WS_2-PEDOT film supported on graphene oxide-SWCNTs nanocomposite. Mater Sci Eng C 92(June):77–87. https://doi.org/10.1016/j.msec.2018.06.029

9 Fluorescence Nanosensors for Monitoring Blood Clotting Disorders

Sana Masroor and Mohd. Mohsin

Abstract

Fluorescence nanosensors are nanoscale sensors that offer high sensitivity, specificity, and rapid detection of biomolecules within living cells. Fluorescence nanosensors have undergone significant development in the detection of blood clotting factors, enhancing their potential for clinical diagnosis. Various fluorescence nanosensors for blood clotting detection have been developed till now. Fluorescent labeling techniques have drastically improved thrombosis monitoring by using fluorophores to track clot molecules and cells, enabling the quantitative measurement of multiple targets simultaneously. Thrombin is the principal clotting factor that plays a central role in the regulation of blood clotting process, thus serving as the key biomarker for the detection of blood clotting. In this chapter, we have discussed different types of fluorescence nanosensors designed to detect thrombin, which directly monitor blood clotting disorders.

Keywords

Fluorescence · Blood clotting · Thrombin · Nanosensors

Abbreviations

CHA	Catalytic hairpin assembly
DVT	Deep vein thrombosis
DSN	Duplex-specific nuclease
EGFP	Enhanced green fluorescence protein
FRET	Fluorescence resonance energy transfer

S. Masroor · M. Mohsin (✉)
Department of Biosciences, Jamia Millia Islamia, New Delhi, Delhi, India
e-mail: mmohsin1@jmi.ac.in

M. Mohsin, N. Soleja (eds.), *Nanoscale Sensors and their Applications in Biomedical Imaging*, https://doi.org/10.1007/978-981-97-3144-2_9

MNPs	Magnetic nanoparticles
MSN	Mesoporous silica nanoparticles
PiCT	Prothrombinase-induced clotting time
ROTEM	Rotational thromboelastometry
TEG	Thromboelastogram
TIC	Traumatic-induced coagulopathy
VTE	Venous thromboembolic

9.1 Introduction

Blood serves as a highly specialized form of connective tissue with a crucial responsibility of circulating throughout the human body. It fulfills a variety of vital functions, including the transportation of essential nutrients, oxygen (O_2), and carbon dioxide (CO_2) to the cells within our bodies. This intricate fluid is primarily comprised of four main components that collectively constitute its solid portion, while the fluid part is known as plasma. Among these components, red blood cells (RBCs) carry oxygen, white blood cells (WBCs) have a defensive action mechanism against infections, and platelets are responsible for blood clotting. Dysfunction and improper production of any of two or more components of blood lead to the development of blood disorders, which may be acute or chronic. This circulating blood serves as a valuable source of information for coagulation or clotting parameters, hypercoagulability and fibrinolysis changes.

At the site of injury, whether it occurs internally or externally, the process of blood clotting plays a vital role in halting bleeding. Abnormalities in blood clotting, such as hypercoagulability, can lead to the formation of excessive blood clots and subsequent obstruction of veins, ultimately resulting in a stroke. Once blood clots have formed, they have the potential to migrate to various regions within the body, leading to adverse effects.

There are various serious conditions that are associated with blood clotting such as deep vein thrombosis (DVT), hemorrhage, heart arrhythmia, heart failure, renal vein occlusion, peripheral artery disease, pulmonary embolism, cancer, and stroke. Therefore, timely and accurate diagnosis of these disorders is crucial for the effective management and prevention of life-threatening complications.

While blood-thinning medications are typically administered to patients to manage blood clotting, accurately determining appropriate dosing, timing, and the necessity for other aggressive strategies are critical in preventing severe complications. Therefore, the development of new methods for quickly assessing the severity of clotting could be extremely beneficial to clinicians.

The thromboelastogram (TEG) and rotational thromboelastometry (ROTEM) are frequently utilized as rheometers in clinical settings to monitor blood coagulation. Several point-of-care devices are available for investigating patients with blood coagulation and for drug monitoring. These devices include the viscoelastic coagulation test, prothrombin time (PT), activated partial thromboplastin time (aPTT), thrombin time (TT), and prothrombinase-induced clotting time (PiCT). One of the primary drawbacks of these tests is that it requires regular calibration which can be challenging. Moreover, the entire testing process requires operator training to ensure adherence to analytical conditions (Mohammadi Aria et al. 2019). This is where nanodiagnostics come into play. Nanodiagnostics have garnered interest due to their small size and expansive surface area attributes inherent to nanostructures and nanodevices. They are extremely stable and biocompatible. Real-time diagnosis is now feasible due to nanotechnology, which saves time and money. Because of the compact nature of nanodevices, they can be designed to be portable and used for point-of-care diagnosis.

Fluorescence nanosensors are nanoscale devices that utilize nanotechnology and fluorescent molecules to achieve high specificity and real-time monitoring of biomolecules. Fluorescence-based nanosensors have been emerging as powerful tools in biomedical research, as they allow for the non-invasive measurement of key biochemical parameters with high spatial and temporal resolution. Fluorescence sensing technology first appeared in the 1970s and 1980s, with the development of voltage-sensitive and calcium-sensitive dyes, respectively. Fluorescent-based molecular probes have been used to create highly sensitive nano-biosensors for detecting various biological and chemical agents. Fluorescence nanosensors have further evolved into nanomedicines that can be applied both in vitro and on organisms. This fluorescence nanosensing technology is an emerging field with enormous potential for identifying diseases and disorders. In this chapter, we will describe available fluorescence-based nanosensors for the detection of blood clotting disorders.

9.2 Blood Clotting Disorders

Humans undergo the homeostasis process to counter excessive bleeding upon injury. During an injury, the ruptured epithelium stimulates a protein response, such as tissue factor, to initiate changes in blood platelet and plasma protein fibrinogen. This process leads to the formation of a plug at the site of injury, occluding the vascular lesion (Ramaswamy et al. 2013). These sequential stages in the formation of the plug, which stops the hemorrhage and repairs the damaged vessel, are referred to as blood clotting (Gopinath 2008). Clotting factors in blood plasma respond in a cascade mechanism to polymerize fibrin and activate platelets. At the site of injury, a series of cofactors modulating the protease cascade control fibrin formation (Furie and Furie 1988).

9.2.1 Mechanism of Blood Clotting

The process of blood clotting involves two distinct pathways, namely the intrinsic and extrinsic pathways. These pathways operate independently but eventually converge to form a unified pathway. This unified pathway facilitates the conversion of factor IX to factor IXa, which ultimately leads to the formation of thrombin as the final product (Lowe 2001) (Matejovic et al. 2007; Gopinath et al. 2006, 2007). When tissue factor in the tissue adventitia is exposed to circulating blood after an injury, a blood clotting cascade is initiated. This exposure activates the extrinsic pathway by binding to factor VII or activating the serine protease factor VIIa. The complex formed by the activated serine factor VIIa is responsible for converting factor IX and factor X into factor IXa and factor Xa, respectively. Factor Xa is responsible for the conversion of prothrombin, which is the inactive form, into thrombin. This conversion occurs in the presence of calcium ions Ca^{2+}, as well as the non-enzymatic cofactor factor Va and anionic phospholipids.

Thrombin serves both as a procoagulant and an anticoagulant, controlling its production by activating coagulation factors V, VIII, and XI. Additionally, it induces platelet activation and, by activating factors XI and XIII, inhibits fibrinolysis of fibrin clots, influencing the development of a stable fibrin clot. This process ultimately leads to the formation of blood clots, preventing bleeding. Anionic phospholipids are exposed on the surface of activated platelets or other cells. The production of thrombin triggers a feedback amplification loop through limited proteolysis, resulting in the generation of factor Va and VIIIa. The intrinsic pathway involves the activation of zymogen factor XII to factor XIIa upon contact with anionic surfaces (Zhu and Diamond 2014). Factor XIIa is generated through a process known as limited proteolysis, in which a zymogen is activated by both factor XIIa and thrombin (Gopinath et al. 2007). Factor IXa is produced through the activation of factor XIIa via a comparable mechanism.

In the common pathway, thrombin functions as an enzyme that catalyzes the cleavage of fibrinogen, an inactive plasma protein, resulting in the formation of fibrin monomers. Blood clot formation initiates when the thrombin concentration in the bloodstream reaches approximately 5–20 nM, and its concentration after clot formation increases to several hundred nM. These individual fibrin monomers will clump together to form an insoluble fibrin polymer, which in turn forms a clot (Fig. 9.1).

9.2.2 Disorders Associated with Blood Clotting

Blood clotting is a physiological response of the body to injury. However, certain thrombotic disorders necessitate the use of anticoagulant therapy for both prevention and treatment purposes. One example is hypercoagulability, which refers to a condition characterized by an increased activation of the coagulation system. This elevated activation is primarily caused by increased platelet activation and the loss of thrombo-resistant properties of the vascular endothelium. There has been a significant surge of interest in perioperative monitoring of blood coagulation. This monitoring aims to predict the likelihood of bleeding and diagnose the underlying causes of hemorrhage during surgical procedures.

Oral anticoagulants such as heparin are used to prevent thromboembolic disorders postoperatively. They offer advantages in terms of being cost-effective, safer, and easier to manage. Downsides of their use encompass a prolonged half-life, the

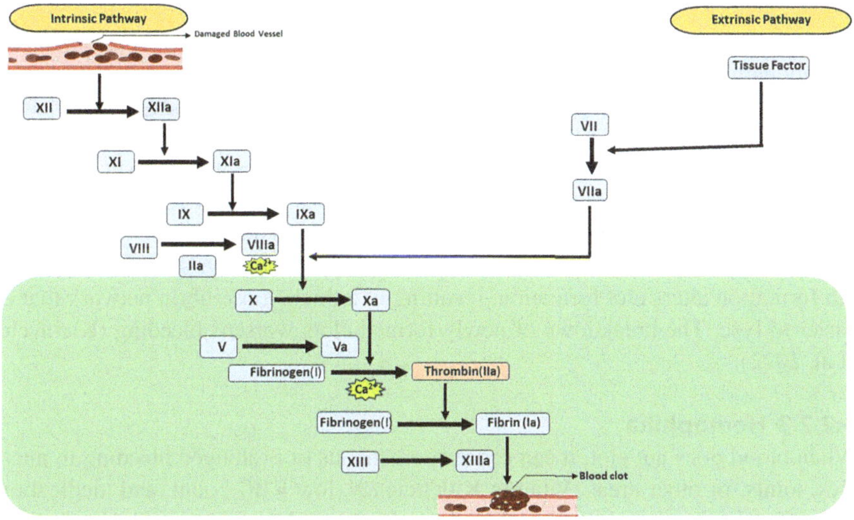

Fig. 9.1 The blood clotting cascade in which several clotting factors are involved and they are sequentially activating one another causing a fibrin clot to form

need for protamine as an antidote, thrombocytopenia, and non-specific plasma binding. Heparin can cause platelet aggregation and dysfunction (Ahmed et al. 2007).

9.2.2.1 Deep Vein Thrombosis

Deep vein thrombosis (DVT) is a pathological condition characterized by the obstruction of blood flow in the deep veins, resulting from hindered venous reflux mechanism. DVT typically affects the venous system in the lower limb. Clot formation typically begins in a deep calf vein and then spreads toward the proximal direction. The incidence of this particular venous thromboembolic (VTE) disorder is 1.6 per 1000 individuals on an annual basis (Waheed et al. 2018). Multiple inherited hypercoagulable disorders have been identified as potential causes of deep vein thrombosis (DVT). Among these, the clotting factor V Leiden mutation and prothrombin gene mutation are the most prevalent, collectively accounting for over 50% of all inherited thrombophilic disorders (McLendon et al. 2017).

9.2.2.2 Trauma-Induced Coagulopathy

Trauma significantly contributes to global mortality rates, with bleeding being identified as the primary preventable factor leading to death. Around 25% of patients experiencing severe trauma are diagnosed with trauma-induced coagulopathy (TIC), a clotting disorder with a fatality rate ranging from 30% to 50%. TIC is a condition that arises from the combination of severe tissue damage and hemorrhagic shock. This condition impedes the body's capacity to form and sustain blood clots, leading to excessive bleeding and further deterioration of the shock state. Consequently, a detrimental cycle is established, exacerbating the severity of TIC. In the context of thrombocytopenia with TIC, dysregulation in the cell-mediated homeostasis occurs. The depletion of coagulation factors in traumatic-induced coagulopathy (TIC) typically occurs as a result of systemic consumption, and degradation caused by the overproduction of various proteins from the anticoagulation and fibrinolytic systems. These proteins, namely activated protein C (APC) and plasmin, play a crucial role in this process. The activated protein C (APC) enzyme contributes to the degradation of coagulation factors FVa and FVIIIa while also facilitating the conversion of plasminogen to plasmin. The increased plasmin production degrades fibrin and fibrinogen, causing hyperfibrinolysis. Reduced thrombin formation alters clot formation, resulting in a thick, loose fibrin network that is easier to lyse. The breakdown of newly formed clots worsens bleeding (Kleinveld et al. 2022).

9.2.2.3 Hemophilia

When blood does not clot, it can cause spontaneous or prolonged bleeding in muscles, joints, or other areas. Vitamin K deficiency, low RBC count, and medication side effects can cause bleeding disorders. Anticoagulants prevent blood clotting. Some bleeding disorders cause severe bleeding after an accident or injury, while others cause unexpected heavy bleeding. Hemophilia is a rare genetic disorder that hinders the blood's capacity to coagulate effectively. Hemophilia A and B are medical conditions characterized by insufficient levels of clotting factors in the bloodstream, resulting in abnormal and excessive bleeding, particularly within the joints. Hemophilia A occurs due to a deficiency in factor VIII, and hemophilia B is caused by a deficiency in factor IX. Although hemophilia is not prevalent, it can give rise to severe complications and adverse reactions that pose a threat to one's life (Zhang et al. 2019; Meenambiga et al. 2022).

9.3 Fluorescence Nanosensors

Fluorescence nanosensors represent an emerging technological frontier facilitating the measurement of crucial biochemical parameters with exceptional spatial and temporal resolution. Nanosensors have the capability to be produced in various sizes, ranging from 10 to 1000 nm in diameter (Elsutohy et al. 2018). This process involves the use of inert biocompatible materials that could bind to or encapsulate fluorophores. A distinguishable feature of fluorescence nanosensor is their composition, often comprising multiple fluorophores, each serving as either a reference or analyte-sensitive component. The reference fluorophore exhibits a consistent fluorescence emission when stimulated, regardless of the presence or absence of the target analyte. The fluorescence emission of an analyte-sensitive fluorophore is dynamically influenced by the quantity of the target (Chauhan et al. 2016).

Fluorescence nanosensors exhibit a high surface-to-volume ratio in relation to the overall volume of pulled optical fibers owing to their small size (Clark et al. 1998). Due to their size, inner matrix, intensity of signals, and ratiometric measurements, they can be utilized to detect analytes in sub-cellular compartments and make real-time measurements. Fluorescence nanosensors provide high sensitivity, selectivity, versatile optical properties, and enable non-invasive monitoring of biomolecules. These nanosensors can be efficiently delivered in large quantities with minimal disruption to cellular function, enabling high-resolution imaging when visualized through fluorescence or confocal microscopy techniques (Clark et al. 1999).

9.3.1 Currently Available Fluorescence Nanosensors for Monitoring Blood Clotting Disorders

Thrombin is typically found at negligible levels in the blood of individuals who are in good health. However, during the coagulation process, thrombin can be produced at concentrations of low micromolar levels. In the early stages of hemostatic progress, thrombin levels can even reach concentrations as low as nanomolar levels. Furthermore, it has been noticed that patients who experience coagulation disorders unrelated to the hemostatic process have higher amounts of thrombin in their blood that can reach high picomolar concentrations. Hence, the detection of thrombin with high sensitivity and selectivity holds immense importance. Various detection techniques have been created so far to identify the recognition interaction between the detection probes and thrombin (Eivazzadeh-Keihan, et al., 2022) (Fig. 9.2).

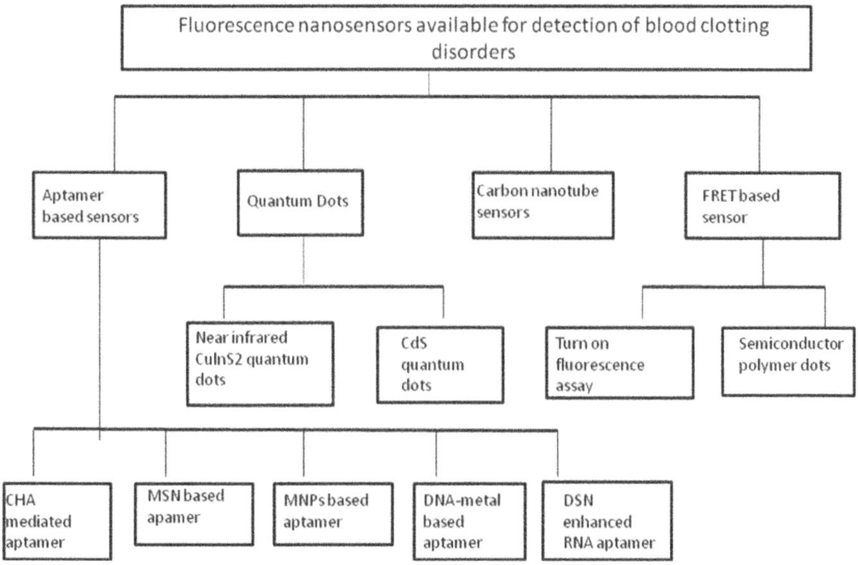

Fig. 9.2 Schematic representation of fluorescence nanosensors available for monitoring of blood clotting disorders

9.3.1.1 Aptamer-Based Sensor

Aptamers are single-stranded DNA or RNA molecules that exhibit high affinity and specificity for a wide range of molecules. Aptamers behave like antibodies and are therefore also known as chemical antibodies. They possess inherent capabilities for molecular recognition and specific binding to their corresponding targets, facilitated by their ability to fold into well-defined three-dimensional structures (Park et al. 2014). Aptamers can be chemically synthesized, ensuring consistent production without batch-to-batch variation. This not only reduces production costs but also shortens the time required. The stability of aptamers is demonstrated by their ability to endure repeated denaturation and renaturation processes, where the temperature is returned to its optimal level. Aptamers, being nucleic acids, possess the advantage of being readily labeled and modified. This characteristic enables them to offer straightforward methods of detection (Toh et al. 2015).

Catalytic Hairpin Assembly Mediated Aptamer

Fluorescent methodology has become a predominant way to translate aptamer recognition events into observable signals. Aptamer recognition events can be signaled by fluorophore-labeled aptamers through the measurement of changes in fluorescence intensity, fluorescence anisotropy, or fluorescence energy transfer. A sensitive fluorescent approach for detecting thrombin employs a proximity recognition-dependent strand translocation strategy with catalytic hairpin assembly (CHA). This sensor involves binding two probes to distinct sites on thrombin. This method increases local concentrations, allowing for ligation or assembly of DNA strands for amplified detection of different target molecules. Specifically, two aptamers, Apt15 and Apt29, simultaneously bind to fibrinogen and heparin sites, inducing strand translocation and CHA-mediated signal amplification. Consequently, the cyclic unfolding of hairpins that are fluorescently quenched, induced by CHA results in significantly amplified fluorescence recovery for highly sensitive detection of thrombin (Li et al. 2018).

MNP-Based Aptamer

Magnetic nanoparticles (MNPs) have been utilized to concentrate target proteins, aiding in the elimination of non-specific absorption of coexisting proteins and reduction of background fluorescence. One thrombin aptamer (TBA1) and one lysozyme aptamer were initially modified on the surface of MNPs to capture the respective target proteins, thrombin and lysozyme. Thrombin was bound with another thrombin aptamer (TBA2) labeled with rhodamine B, and lysozyme was bound with another lysozyme aptamer labeled with fluorescein. An anti-thrombin aptamer was developed by labeling with rhodamine B to target thrombin protein with enhanced specificity and sensitivity.

This converted target-identifying events into fluorescence signals for detection. Utilizing the magnetic field not only eliminated interferences of coexistence proteins but also concentrated trace target proteins enhancing fluorescence signals.

MSN-Based Aptamer

Researchers developed a fluorescence biosensor using aptamer-based stimulus response mesoporous silica nanoparticles (MSN) in which the target aptamer, thrombin was combined with the complementary DNA to form a double-strand DNA (dsDNA) and then it was subsequently modified on N3-modified MSN. The guest molecules, fluorescein, were blocked efficiently in the MSN pores, preventing leakage. The enhanced fluorescence intensity showed a linear correlation with thrombin concentration and reached a detection limit of 28.46 ng mL^{-1} (Wang et al. 2009).

DNA–Metal Based Nanoclusters

A thrombin aptasensor with high sensitivity was developed by utilizing the change in aptamer conformation caused by target recognition and the resulting increase in fluorescence due to the close proximity of two DNA–template copper/silver nanoclusters (DNA–Cu/Ag NCs). Two functional DNA templates were designed, each composed of a Cu/Ag NC-nucleation segment located at either two termini or one terminus, and an aptamer segment positioned in the middle of the DNA template. The fluorescence signal readout increased as a result of the conformational alteration of the aptamer structure, causing two DNA–Cu/Ag NCs with a darkish appearance to come into proximity when the aptamer bound to thrombin. Thrombin was accurately quantified at a minimum concentration of 1.6 nM within the range of 1.6–8.0 nM, demonstrating excellent selectivity. The sensor successfully detected thrombin in a real fetal bovine serum (Chen et al. 2018).

DSN-Enhanced RNA Aptasensor

The duplex-specific nuclease (DSN) is a highly regarded enzyme that has attracted significant interest due to its ability to accurately cleave specific DNA segments within double-stranded DNA or DNA–RNA hybrids. RNA aptamer coupling extends DSN sensing application. Specially designed RNA ligand sequences are employed for the purpose of capturing thrombin protein and concurrently supplying complementary DNA sequences to facilitate DSN-assisted fluorescent signal enhancement. The generation of fluorescent signals can be achieved by combining a specific RNA aptamer with the target molecule. The assay under consideration has successfully attained a detection limit of 0.039 pM, which is the lowest recorded value. The assay has been utilized for the real-time detection of thrombin release from live cells and other biotic media, enabling early disease diagnosis (Zhang and Wei 2020).

MoS_2 Aptamer Nanosheets

Molybdenum disulfide (MoS_2) exhibits great potential as a sensing platform. However, it is important to note that its interaction with DNA through physisorption can lead to unintended probe displacement and the generation of false-positive signals. A novel biosensor utilizing MoS_2–aptamer nanosheets based on the FRET phenomenon was developed for the purpose of detecting thrombin. A solution containing 10% concentrated Tween 80 was used in order to mitigate nonspecific binding and facilitate the formation of gold nanoparticle (AuNP) conjugates. This strategy was developed to enhance the detection limits of thrombin by utilizing a combination of MOS_2 and exonuclease for signal amplification. The detection limit of this method is 6 fM, which shows great potential for use in the field of medical diagnostics (Ashraf et al. 2022).

9.3.1.2 Quantum Dots (QDs)

Quantum dots are nanocrystals composed of a semiconductor material that possess distinct and noteworthy fluorescent characteristics. These nanoparticles have the capability to undergo functionalization with specific biomolecules, enabling them to effectively detect blood clotting factors or fibrin formation. QDs are composed of a semiconductor core, typically comprising CdSe, CdTe, CdS. These QDs are commonly enclosed within an outer shell. QDs possess several advantageous characteristics, including high brightness, photostability, and the ability to tune their emission wavelengths. These properties make QDs well-suited for applications involving multiplexed detection.

Since thrombin is a biomarker for blood clotting disorders, there is considerable interest in development of sensors capable of rapidly detecting thrombin in blood at picomolar concentrations. Luminophores in the near-infrared (NIR) spectrum are quite useful for biological imaging and detection, but only a limited number of organic dyes are suitable in this range. Semiconductor QDs offer several advantages over traditional organic fluorophores, such as elevated photoluminescence efficiency, emission wavelengths that depend on size, narrow emission spectra, and broad absorption. Researchers have reported thrombin activity monitoring through the utilization of water-soluble near-infrared copper indium sulfide (CuInS2) QDs that are capped by mercaptopropionic acid. This assay utilizes the fluorescence quenching phenomenon of the fibrinogen–CuInS2 QDs complex, which is induced by thrombin. The observed decrease in the readout signal can be attributed to the quenching effect of fibrin, which is correlated with the increase in thrombin concentration (Gao et al. 2017).

CdS Quantum Dots

The assay is dependent on the specific interaction between thrombin and fibrinogen, along with the enzymatic amplification step involving the in situ growth of CdS QDs. The process involves incubating biotinylated fibrinogen and thrombin with microplate-immobilized fibrinogen. Insoluble biotinylated fibrin is obtained as a result, which is subsequently subjected to incubation with avidin-labeled alkaline phosphatase (ALP). The growth of CdS QDs is initiated by an enzyme, resulting in the stabilization of phosphate ions. This, in turn, stabilizes the CdS QDs that are grown in situ using cadmium (II) and sulfide. The process of fibrin generation is closely linked to the activity of thrombin. As the concentration of thrombin increases, there is a corresponding increase in fluorescence. The utilization of this enzyme-induced amplification technique enables the quantification of thrombin within the picomolar concentration range (Zhang et al. 2013).

Recent nanomedicine studies on QDs show their potential for medical research, diagnostics, and novel drug delivery. To ensure in vivo efficiency and safety, QD-based systems need detailed physico-chemical and functional characterization. To increase detection sensitivity, traditional fluorescent aptamer sensors must be combined with nucleic acid signal amplification strategies.

9.3.1.3 Carbon Nanotubes (CNTs)

Since the blood clotting process involves the participation of numerous clotting factors, the conversion of fibrinogen to insoluble fibrin, which is facilitated by thrombin, is a crucial step in the coagulation pathway. The significance of developing new techniques to monitor the dynamics of the coagulation cascade arises from the potentially life-threatening nature of coagulation disorders. The near-infrared (NIR)-fluorescent single-walled carbon nanotubes (SWCNTs) were employed for the purpose of imaging and real-time monitoring of fibrin clotting. After the successful binding of fibrinogen to a specifically designed single-walled carbon nanotube (SWCNT) platform, thrombin initiates the conversion of fibrinogen into fibrin monomers. These fibrin monomers then proceed to undergo polymerization. The single-walled carbon nanotubes (SWCNTs) are integrated into the clot and can be easily observed in the near-infrared (NIR) fluorescent channel. This channel provides a better signal-to-noise ratio compared to bright-field imaging in the visible range, which makes clear visualization of the SWCNTs. In addition, it has been observed that the movement of individual single-walled carbon nanotubes (SWCNTs) within the fibrin clot decreases gradually following the introduction of thrombin. This phenomenon indicates that the rate of coagulation is influenced by the concentrations of both fibrinogen and thrombin. This method has the capability to facilitate coagulation disorder diagnostics and enable real-time monitoring of the coagulation cascade using a near-infrared (NIR) optical signal output within the biological transparency window (Gerstman et al. 2023).

9.3.1.4 Fluorescence Resonance Energy Transfer (FRET) Sensor

FRET is a non-radiative energy transfer phenomenon between two close chromophores with spectral overlap. In this phenomenon, the emission spectrum of a donor fluorophore overlaps with the emission spectrum of an acceptor fluorophore. When a donor fluorophore absorbs excitation light, its energy is transferred to a neighboring acceptor fluorophore no more than 10 nm away, resulting in acceptor fluorophore emission and the FRET ON state. The quantification of the FRET signal allows detection of the cellular concentration of target biomolecules in the living cells. FRET involves the use of specific molecular groups, namely donor and acceptor fluorophores. These fluorophores have the ability to emit fluorescence when exposed to an external photon source, specifically in the presence of a particular analyte (Lindenburg and Merkx 2014).

Turn on Fluorescence Assay

Gold nanoparticles (AuNPs) and an enhanced green fluorescence protein (EGFP) probe were used to create a label-free, sensitive fluorescent assay for measuring thrombin activity. A thrombin recognition site was added to the N-terminal of the EGFP probe together with a hexahistidine sequence (His-tag). The His-tag facilitates the adsorption of the probe onto the surface of AuNPs by utilizing the strong affinity of the His–Au bond. As a result, FRET occurs, with EGFP acting as the fluorescence donor and AuNPs serving as the acceptor. The high absorption coefficient of AuNPs causes the fluorescence quenching of EGFP. In the presence of thrombin, the recognition site is cleaved, leading to the detachment of the His-tag from EGFP. The efficient adhesion of EGFP without a His-tag onto AuNPs is not possible, resulting in the preservation of its fluorescence. The fluorescence signal can be influenced by the activity of thrombin. The turn-on fluorescence assay was employed to achieve highly precise and sensitive detection of thrombin activity. Under optimal experimental conditions, the assay achieved a limit of detection of 0.0025 U/mL, with a linear range covering concentrations from 0.01 to 0.25 U/mL. This designed assay was capable of quantifying the inhibition of thrombin activity with IC50 value obtained was 1.38 nM, indicating the concentration required to inhibit 50% of thrombin activity. Furthermore, the assay was successfully conducted using real samples (Grant et al. 2004).

Semiconductor Polymer Dots

In recent years, there has been a growing interest in semiconducting polymer dots (P dots). P dots exhibit exceptional optical characteristics, including a substantial optical absorption cross-section and a high fluorescence quantum yield. Moreover, they also offer high biocompatibility and colloidal stability. P dots are ideal for nanofluorescent probes due to their ability to exhibit both high fluorescent brightness and high stability. A FRET-based sensor was synthesized by using carboxyl-functionalized semiconducting P dots as donor and black hole quenching dye (BHQ) thrombin aptamers as acceptor. The FRET between the aptamers and P dots quenched their fluorescence and thrombin increased fluorescence. Under optimized experimental conditions, the system's fluorescence was restored to its maximum at 130 nM thrombin and a detection limit of 0.33 nM. This sensor had good thrombin specificity and signal response in real samples and was less affected by impurities (Jin et al. 2016).

9.4 Conclusion and Future Perspectives

Various fluorescence nanosensors for blood clotting detection have been developed till now. Fluorescent labeling techniques drastically enhance monitoring of blood clotting by employing fluorophores to track clot molecules, allowing for the simultaneous quantitative measurement of multiple targets. In conclusion, fluorescence nanosensors have emerged as a promising and innovative tool in the detection and monitoring of blood clotting disorders. Through their ability to provide real-time and sensitive measurements, fluorescence nanosensors offer several advantages over traditional methods, such as enhanced accuracy, rapid response times, and reduced invasiveness. The application of fluorescence nanosensors in diagnosing blood clotting disorders, such as deep vein thrombosis, pulmonary embolism, and hemophilia, has shown significant potential in improving patient outcomes. Their ability to detect specific biomarkers associated with these disorders allows for early diagnosis, leading to timely and targeted treatment interventions.

Furthermore, fluorescence nanosensors can be engineered to be biocompatible and can be designed to target specific sites or cells within the circulatory system, enabling precise and localized detection of clotting events. This not only reduces the risk of false positives but also minimizes the need for invasive procedures, which can improve patient comfort and compliance. However, further research and development are needed to optimize the sensitivity specificity and stability of fluorescence nanosensors. Additionally, addressing potential challenges, such as long-term biocompatibility, potential cytotoxicity, and large-scale production will be essential for their successful translation into clinical practice.

Despite these challenges, the progress made in fluorescent nanosensing technology for blood clotting disorders is highly promising. As research in this field continues to advance, fluorescence nanosensors could be further evolved and they have the potential to revolutionize the diagnosis and management of blood clotting disorders.

Acknowledgement The first author is thankful to University Grants Commission for the Non-NET Fellowship.

References

Ahmed I, Majeed A, Powell R (2007) Heparin induced thrombocytopenia: diagnosis and management update. Postgrad Med J 83(983):575–582

Ashraf G, Zhong ZT, Asif M, Aziz A, Song L, Zhang S et al (2022) Extension of duplex specific nuclease sensing application with RNA aptamer. Talanta 242:123314

Chauhan VM, Giuntini F, Aylott JW (2016) Quadruple labelled dual oxygen and pH-sensitive ratiometric nanosensors. Sens BioSensing Res 8:36–42

Chen X, Li T, Tu X, Luo L (2018) Label-free fluorescent aptasensor for thrombin detection based on exonuclease I assisted target recycling and SYBR green I aided signal amplification. Sensors Actuators B Chem 265:98–103

Clark HA, Barker SL, Brasuel M, Miller MT, Monson E, Parus S et al (1998) Subcellular optochemical nanobiosensors: probes encapsulated by biologically localised embedding (PEBBLEs). Sensors Actuators B Chem 51(1–3):12–16

Clark HA, Hoyer M, Philbert MA, Kopelman R (1999) Optical nanosensors for chemical analysis inside single living cells. 1. Fabrication, characterization, and methods for intracellular delivery of PEBBLE sensors. Anal Chem 71(21):4831–4836

Eivazzadeh-Keihan R, Saadatidizaji Z, Maleki A, Guardia MDL, Mahdavi M, Barzegar S, Ahadian S (2022) Recent progresses in development of biosensors for thrombin detection. Biosensors 12(9):767

Elsutohy MM, Selo A, Chauhan VM, Tendler SJ, Aylott JW (2018) Enhanced distance-dependent fluorescence quenching using size tuneable core shell silica nanoparticles. RSC Adv 8(62):35840–35848

Furie B, Furie BC (1988) The molecular basis of blood coagulation. Cell 53(4):505–518

Gao L, Li Q, Deng Z, Brady B, Xia N, Zhou Y, Shi H (2017) Highly sensitive protein detection via covalently linked aptamer to MoS2 and exonuclease-assisted amplification strategy. Int J Nanomedicine:7847–7853

Gerstman E, Hendler-Neumark A, Wulf V, Bisker G (2023) Monitoring the formation of fibrin clots as part of the coagulation cascade using fluorescent single-walled carbon nanotubes. ACS Appl Mater Interfaces 15(18):21866–21876

Gopinath SCB (2008) Anti-coagulant aptamers. Thromb Res 122(6):838–847

Gopinath SC, Shikamoto Y, Mizuno H, Kumar PK (2006) A potent anti-coagulant RNA aptamer inhibits blood coagulation by specifically blocking the extrinsic clotting pathway. Thromb Haemost 95(05):767–771

Gopinath SC, Shikamoto Y, Mizuno H, Kumar PK (2007) Snake-venom-derived factor IX-binding protein specifically blocks the γ-carboxyglutamic acid-rich-domain-mediated membrane binding of human factors IX and X. Biochem J 405(2):351–357

Grant SA, Lichlyter DJ, Lever S, Gallazzi F, Soykan O (2004) A novel sensing technique to detect thrombin. Sens Lett 2(3–4):164–170

Jin X, Liu X, Zhu X, Li H, Li W, Huang Y, Yao S (2016) A label-free fluorescence assay for thrombin activity analysis based on fluorescent protein and gold nanoparticles. Anal Methods 8(18):3691–3697

Kleinveld DJ, Hamada SR, Sandroni C (2022) Trauma-induced coagulopathy. Intensive Care Med 48(11):1642–1645

Li J, Zhou W, Yuan R, Xiang Y (2018) Aptamer proximity recognition-dependent strand translocation for enzyme-free and amplified fluorescent detection of thrombin via catalytic hairpin assembly. Anal Chim Acta 1038:126–131

Lindenburg L, Merkx M (2014) Engineering genetically encoded FRET sensors. Sensors 14(7):11691–11713

Lowe GD (2001) Factor IX and thrombosis. Br J Haematol 115(3):507–513

Matejovic M, Krouzecky A, Radej J, Kralova H, Radermacher P, Novak I (2007) Coagulation and endothelial dysfunction during long term hyperdynamic porcine bacteremia–effects of selective inducible nitric oxide synthase inhibition. Thromb Haemost 97(02):304–309

McLendon K, Goyal A, Attia M (2017) Deep venous thrombosis risk factors

Meenambiga SS, Sakthiselvan P, Hari S, Umai D (2022) Nanotechnology for blood test to predict the blood diseases/blood disorders. In: Nanotechnology for hematology blood transfusion, and artificial blood. Elsevier, pp 285–311

Mohammadi Aria M, Erten A, Yalcin O (2019) Technology advancements in blood coagulation measurements for point-of-care diagnostic testing. Front Bioeng Biotechnol 7:395

Park JH, Cho YS, Kang S, Lee EJ, Lee GH, Hah SS (2014) A colorimetric sandwich-type assay for sensitive thrombin detection based on enzyme-linked aptamer assay. Anal Biochem 462:10–12

Ramaswamy B, Yeh YTT, Zheng SY (2013) Microfluidic device and system for point-of-care blood coagulation measurement based on electrical impedance sensing. Sensors Actuators B Chem 180:21–27

Toh SY, Citartan M, Gopinath SC, Tang TH (2015) Aptamers as a replacement for antibodies in enzyme-linked immunosorbent assay. Biosens Bioelectron 64:392–403

Waheed SM, Kudaravalli P, Hotwagner DT (2018) Deep vein thrombosis

Wang L, Li L, Xu Y, Cheng G, He P, Fang Y (2009) Simultaneously fluorescence detecting thrombin and lysozyme based on magnetic nanoparticle condensation. Talanta 79(3):557–561

Zhang B, Wei C (2020) An aptasensor for the label-free detection of thrombin based on turn-on fluorescent DNA-templated Cu/Ag nanoclusters. RSC Adv 10(58):35374–35380

Zhang X, Hu R, Shao N (2013) Label-free sensing of thrombin based on quantum dots and thrombin binding aptamer. Talanta 107:140–145

Zhang N, Wei MY, Ma Q (2019) Nanomedicines: a potential treatment for blood disorder diseases. Front Bioeng Biotechnol 7:369

Zhu S, Diamond SL (2014) Contact activation of blood coagulation on a defined kaolin/collagen surface in a microfluidic assay. Thromb Res 134(6):1335–1343

Fluorescent Probes for Gasotransmitters (H_2S, CO, and NO) Detection

10

Rashid Ali, Aayaz Ali, and Shakeel Alvi

Abstract

Gasotransmitters (GTs), e.g., carbon monoxide (CO), hydrogen sulfide (H_2S), and nitric oxide (NO), are endogenously fabricated gases, inducing inimitable signaling responses in the "cellular systems," and the generation of these gaseous molecules is regulated in mammals enzymatically. It has been reported in the scientific literature that GTs are allied with a myriad of biological functions. Therefore, it is of high demand to foster the sensitive and distinctive techniques to detect such vital biological messengers. In this particular chapter, we shade light onto the detection of GTs by means of fluorescent probes. Hopefully, the readers will enjoy this particular chapter and will get opportunities to further advance this emerging research topic to advanced level in many years to come.

Keywords

Gasotransmitters · Fluorescent probes · Fluorophore displacement approaches (FDAs) · Nitric oxide synthase · Reactive oxygen species

10.1 Introduction

The gasotransmitters (GTs) or "gaseous transmitters" coined in 2002 are either endogenously engendered or exogenously originated gaseous molecules assisting the transmittance of chemical signals which usually produces an explicit biochemical or physiological signaling responses in both the intracellular and extracellular systems (Wang 2002). Notably, GTs are reactive, toxic, and antimicrobial small

R. Ali (✉) · A. Ali · S. Alvi
Department of Chemistry, Jamia Millia Islamia, Okhla, New Delhi, India
e-mail: rali1@jmi.ac.in

M. Mohsin, N. Soleja (eds.), *Nanoscale Sensors and their Applications in Biomedical Imaging*, https://doi.org/10.1007/978-981-97-3144-2_10

molecules with molar mass in the range of 28–34 amu, e.g., “carbon monoxide (CO), hydrogen sulfide (H_2S), and nitric oxide (NO).” (Coburn 2022). Importantly, their endogenous generation in solution is enzymatically catalyzed in a greatly controlled manner, and they are freely permeable across the membranes (Hosogi et al. 2021). Moreover, these vital molecules have atypical functions at low concentrations, and their roles could be mimicked through the exogenous complements. Noticeably, they are better than the deep-rooted transmitters, for instance, neurotransmitters, hormones, drugs, etc., and can chemically alter their intracellular objects to initiate individual chemical-signaling mechanisms—contrary to the quintessential signaling molecules. The vital functions of NO and CO are to transmit the signals in the brain and to promote anti-tumor activity in the immune system, while H_2S gas acts as a physiological vasodilator.

Basically, NO was determined as the chemical agent for biological function—vasodilation, while creation of the CO in mammalian physiology was acknowledged long-ago for its biological functions, and its existence in biology was solely considered as a waste-product of heme decomposition—catalyzed by the oxygenase (HMOX) (Gallio et al. 2021). Later, CO was recognized as a leading biological messenger in mammalians (Maines 1997; Motterlini and Foresti 2017). Moreover, CO also mediates a wide assortment of the signaling procedures such as cell survival, generation of inflammatory mediators, apoptosis, bacterial infection, signaling in the central nervous system (CNS), etc. (Ryter and Choi 2013) Conversely, though H_2S in tissues was found in late 1990s, its GT status arrangement on the endogenous metabolism as well as physiological functions was revealed even at much later stage. Noticeably, both enzymatic and non-enzymatic routes are responsible to make H_2S gas endogenously. For example, in mammalian tissues, cystathionine-synthase catalyzed the enzymatic manufacture of the H_2S from *L*-cysteine (Shibuya et al. 2009; Rose et al. 2017). Similarly, at low concentrations, H_2S affects the human physiology at cellular, molecular, and tissue levels. Remarkably, the NO gas has numerous acquaintances through the physiological means with prominence in the immune, cardiovascular, and nervous systems. Noticeably, the nitric oxide synthase (NOS) is a vital bio-catalyst to endogenously generate NO whose three isoforms available in the mammals are: (1) neuronal NOS1 (nNOS)—found in both neurons and other nerve cells; (2) inducible NOS2 (iNOS)—found in some tissues; (3) endothelial NOS3 (eNOS)—found in cells as well as tissues, other than the endothelium.

Markedly, apart from their physiological as well as biological functions, they are harmful to the environment, and any perturbation in the homeostasis of these gases results in the pathological abnormalities like neurodegradation and carcinogenesis (Brown and Bal-Price 2003). It is to be pointed out that the excessive concentrations of the NO gas can increase the unessential fabrication of reactive nitrogen species (RNS) which could destroy proteins, DNA, and/or lipids with or without the involvement of reactive oxygen species (ROS). On the other front, convincing affinity of CO gas for the hemoglobin (Hb) creates an enormously stable yet unessential carboxyhemoglobin (COHb) that poses a threat to functioning of the O_2 transport to the tissues (Carrola et al. 2023). Whereas Alzheimer’s disease leading cause has been noticed because of the unwarranted production of H_2S in the human body.

Gratifyingly, electron paramagnetic resonance (EPR)/electron spin resonance (ESR), colorimetry, electrochemistry, fluorometry, metal-induced sulfide precipitation, and gas chromatography (GC) are the most usual instrumental techniques to nicely detect the GTs (Pénicaud et al. 2012). Among them, fluorescence spectroscopy is one of the most effective methods for the detection of GTs because of its simplicity, high versatility, and great sensitivity. Moreover, fluorescence imaging is a powerful technique for studying the intracellular particles without annihilating the cells/tissues. Importantly, utility of the fluorescence probes for the investigation of CO, NO, and H_2S gases is not only indispensable to untangle their toxicological and/or pathological effects but also to apprehend their physiological consequences. In this particular chapter, the authors have focused on the advancements in the fluorescence-based chemosensors for the recognition of CO, NO, and H_2S. Hopefully, this write-up will benefit the readers.

10.2 Gasotransmitters (NO, CO, and H_2S) as Intracellular Signaling Agents

Since endogenous fabrication of the gaseous molecules (e.g., NO, CO, and H_2S) was discovered, their function as extracellular signaling agents in distinctive organisms has fruitfully been unraveled. As the GTs are released, they are not stored in "*vesicular reservoirs*" and their release is administrated through the secretory cell by means of the regulatory biosynthesis mechanisms. Notably, as the GTs are neutral molecules, they can infuse via cytoplasmic membrane and hence eagerly diffuse toward the targeted cell, partaking largely in the adjustment of local physiological activities. And inside the targeted cell, they access their metabolism by virtue of the chemical adjustment of the cytoplasmic protein. Remarkably, such adjustments might encompass redox reactions amid explicit amino acid (AA) residues and metal centers existing in the prosthetic groups and/or non-covalent bonding in pivotal modulatory parts of the targeted proteins. Gratifyingly, hemoproteins because of their unique reactivity pattern prominently contribute to this particular targeted protein. It is to be pointed out that though they show affinity for all the GTs, their responses relating to each are controlled differently. For instance, as can be inspected from Scheme 10.1, inflammatory responses toward the pathogens are being controlled by NO, CO, and H_2S acting as the extracellular messengers. Besides, it is to be pointed out that the regulatory pathways are interlinked for some of these molecules (Fukuto and Collins 2007; Kabil et al. 2011).

10.2.1 Nitric Oxide (NO)

Remarkably, L. Ignarro's research group around 1980 reported endogenous generation of the NO gas in the *endothelial cells*. Moreover, the authors have also revealed

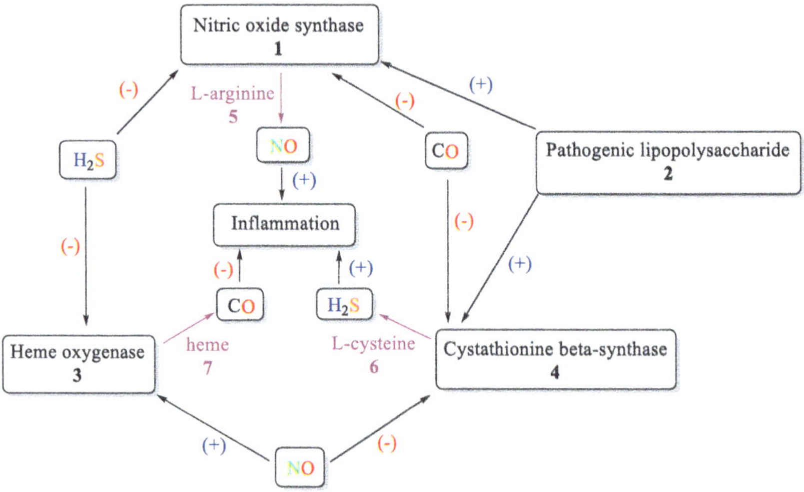

Scheme 10.1 The extracellular signaling in inflammatory responses by the NO, CO, and H_2S

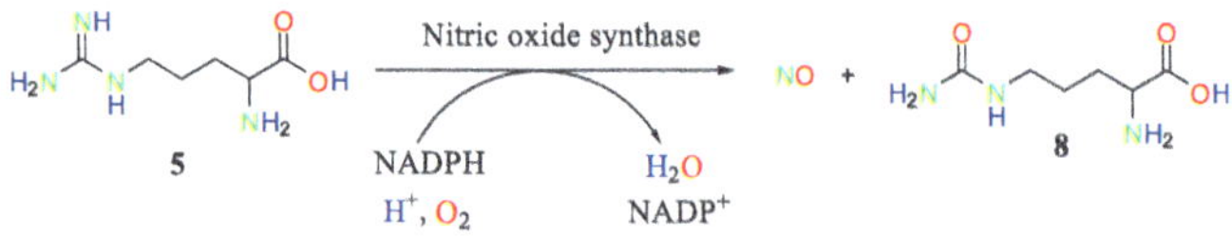

Scheme 10.2 Transformation of "*L*-arginine into *L*-citrulline" using nitric oxide synthase

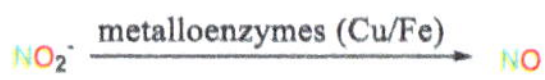

Scheme 10.3 Nitrite reduction to NO using Cu/Fe metalloenzymes

its role in the *arterial vasodilation*. Notably, as the NO gas plays a tremendous role in diverse biological processes, a myriad of comprehensive studies on its chemical as well as biochemical developments have fruitfully been exposed during the succeeding years. As can be seen from Scheme 10.2, the production of NO gas in biological systems is chiefly because of the activity of NOS, which generally uses arginine (Arg) **5**, O_2, and co-factor NADPH to generate NO and *L*-citrulline **8** (Ignarro et al. 1980). Conversely, nitrite reduction utilizing the Cu/Fe metalloenzymes under hypoxic conditions can also produce NO (Scheme 10.3) (Machha and Schechter 2012).

10.2.2 Carbon Monoxide (CO)

Noticeably, CO begin as the biologically intriguing molecule in 1857, when for the first time, CO was found to cause the tissue hypoxia through the formation of complex with the hemoglobin (Hb). It has been pointed that Hb's affinity for CO is approximately 220 times than O_2, and it was also shown that the Hb-CO union influences the

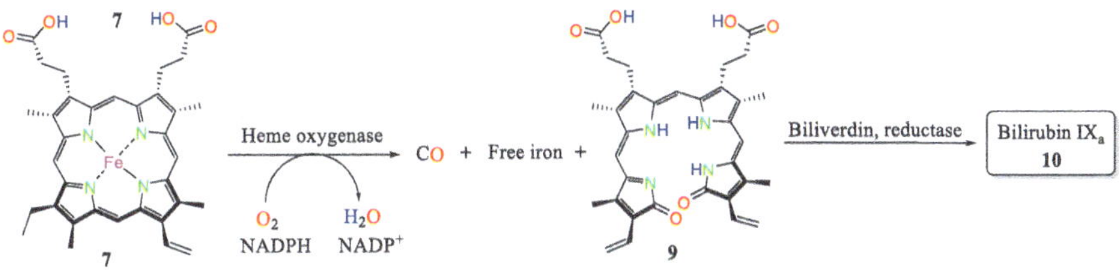

Scheme 10.4 Generation of CO via oxidation of heme-group with the usage of heme oxygenase

Scheme 10.5 The H_2S production from the cysteine through the CBS

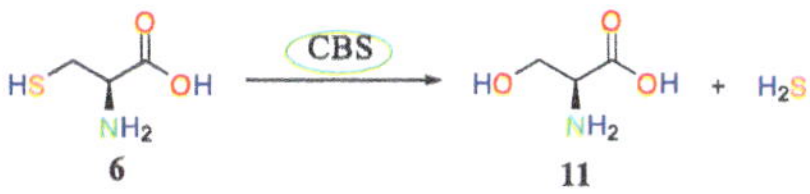

Scheme 10.6 Preparation of H_2S from homocysteine via "*cystathionine gamma-lyase*" (CSE)

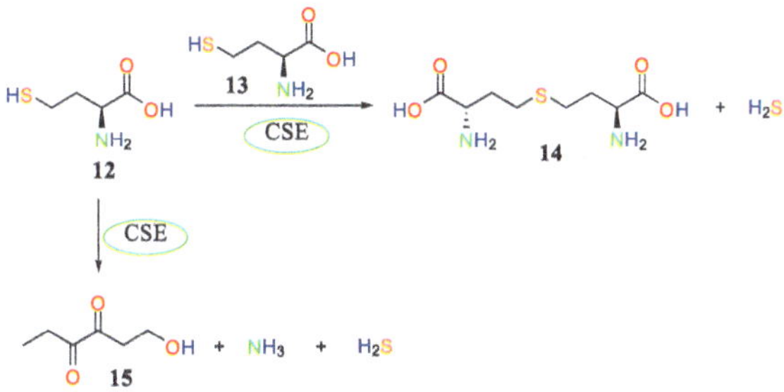

binding kinetics (Roughton and Darling 1944). Notwithstanding, its well-known toxic activity in "*living organisms*," endogenous generation of the CO via the decomposition of Hb was detailed in 1952 (Scheme 10.4) (Platt and Nath 1998).

10.2.3 Hydrogen Sulfide (H_2S)

Importantly, it is to be pointed out that H_2S gas molecule can act as a ligand for the active sites present in the metalloproteins. More importantly, its robust affinity for cytochrome C is trustworthy for inhibiting the mitochondrial respiration procedures and outstanding toxicity at the high concentrations. Endogenous production of H_2S molecule is mainly ascribed to the straight activity of two explicit enzymes: (a) cystathionine *beta*-synthase (CBS) that acts on the cysteine **6** (Scheme 10.5) and (b) cystathionine gamma-lyase (CSE) which acts on the homocysteine **12** (Scheme 10.6).

10.3 Fluorescent Probes for NO Detection

NO, a neutral odd electron molecule which is considered as the first GT molecule, is stowed from the endothelial cells and acts as an endothelium-derived relaxing factor, besides regulating the smooth muscle relaxation as well as vasodilation. Nonetheless, uneven NO secretion is being allied with the creation of

RNS—responsible for the pathological circumstances including neurodegenerative disease, carcinogenesis, and inflammation. Herein, we have categorized the fluorescence probes for the detection of NO into two main different parts.

10.3.1 Fluorescent Probes Based on the Transition Metals

10.3.1.1 Copper Reduction Technique

This vital practice encompasses reduction of the Cu^{II} into Cu^{I} by means of NO for the selective identification of NO molecule (Scheme 10.7).

Generally, fluorophore coordination to the Cu^{II} results in the quenching of the fluorophore emission, whereas NO-induced Cu^{II} (paramagnetic) to Cu^{I} (diamagnetic) favors the emission of fluorophore. To this regard, Tsuge et al. revealed a $Cu\text{-}(DAC)^{2+}$ complex (**19**), the authors noticed that the free DAC (a *1-bis(9-anthracylmethyl) offshoot of tetraamine macrocyclic cyclam*), because of the presence of anthracene moieties is strongly luminescent, whereas $Cu\text{-}(DAC)^{2+}$ complex (**19**) display no luminescence as Cu^{II}, results in intramolecular quenching (Fig. 10.1) (Tsuge et al. 2004). Conversely, Lippard's group reported the copper complexes **20–22** for the detection of NO through the reduction of Cu^{II} to Cu^{I} by NO (Fig. 10.1) (Lim and Lippard 2005).

In a separate report, Hu et al. revealed a Cu(II)-complex (**23**) involving rhodamine as a fluorophore for the selective detection of NO (Hu et al. 2011). Remarkably, their sensing system showed 700-fold fluorescent improvement with the NO molecule, and it may be involved for monitoring the intracellular NO (Fig. 10.1).

10.3.1.2 Fluorophore Displacement-Based Tactic

In recent years, fluorophore displacement approaches (FDAs) have drawn tremendous curiosity of the scientific community worldwide (Rather and Ali 2021, 2023). To this context, transition metal complexes coordinated with fluorophore have also

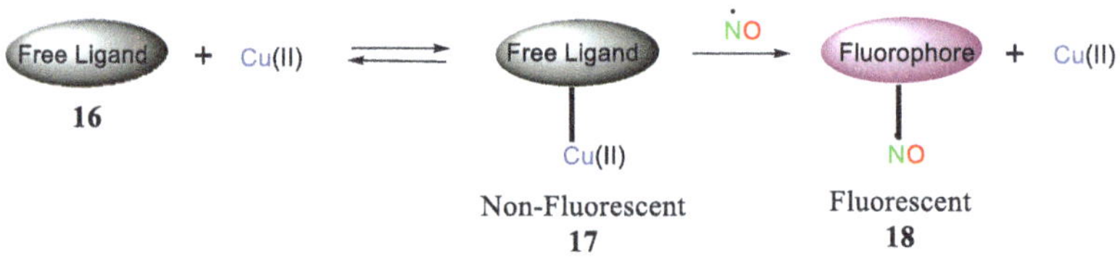

Scheme 10.7 The NO-induced reduction of Cu^{II} to induce fluorescence turn-ON

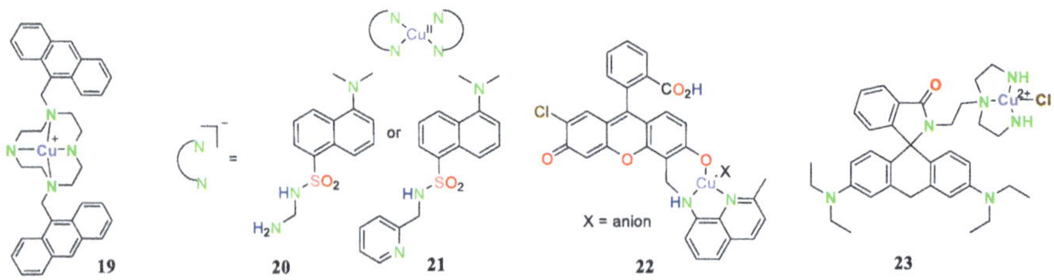

Fig. 10.1 Representation of the Cu-complexes for NO detection through fluorescent probes

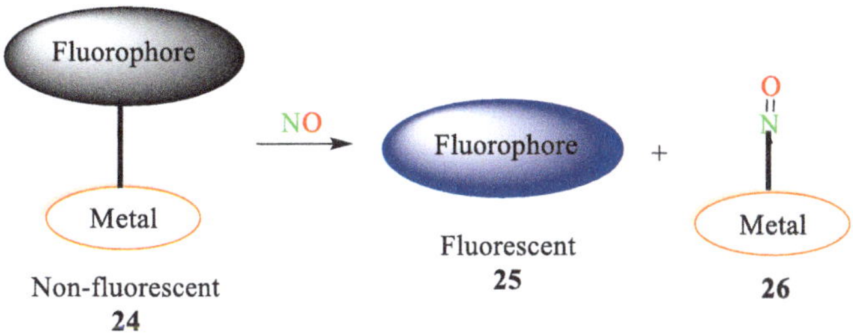

Scheme 10.8 General representation of NO-mediated fluorophore displacement technique

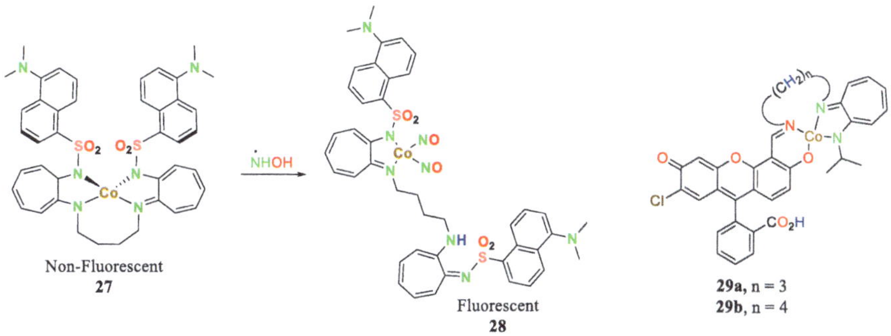

Scheme 10.9 NO-induced formation of cobalt-dinitrosyl complex in **28** and complexes **29a-b**

been noticed to displace fluorophore by nitric oxide, thereby NO detection via the fluorophore displacement strategy (Scheme 10.8). Herein, as can been seen from Scheme 10.8, we have depicted a common NO-induced fluorophore displacement approach. It can be inspected from the same scheme that when fluorophore is coordinated with the metal center, fluorescence is quenched (turn-OFF), while after fluorophore displacement with NO molecule fluorescence turn-ON. The mechanism for this particular process is based upon the fluorescence quenching characteristic of transition metal (TM) ions, having a partly filled *d*-shell.

Lippard and co-workers have reported many fluorescent metal complexes for the detection of NO using the displacement tactic (Hilderbrand et al. 2004). For instance, the Co-complex **27**, where each amino-troponiminate ring is fine-tuned with the dansyl fluorophore, was designed, synthesized, and reported for the NO detection via fluorescent displacement method (Scheme 10.9). From the fluorescence studies, these authors observed that the spectrum of the Co(II) complex under NO atmosphere displayed four-fold fluorescence augmentation with the limit of detection (LOD) of 50–100 μM. Along similar lines, the Co(II) complexes **29a-b** were also employed for the identification of NO.

10.3.2 The Fluorescent Probes Based on Organic Molecules

Herein, fluorescent probes containing *o*-diamino aromatic moiety have been exposed. As shown in the general Scheme 10.12, the probes with free amino groups are non-emissive (non-fluorescent) because of the photoinduced electron transfer

Scheme 10.10 The NO-induced diamine cyclization process

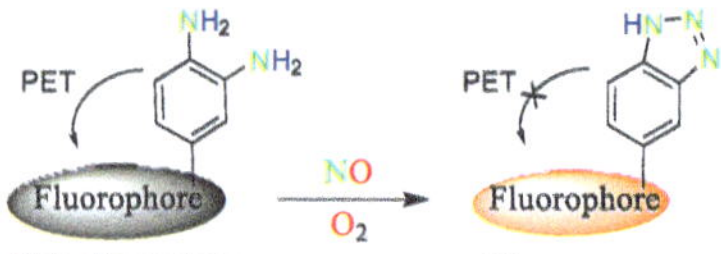

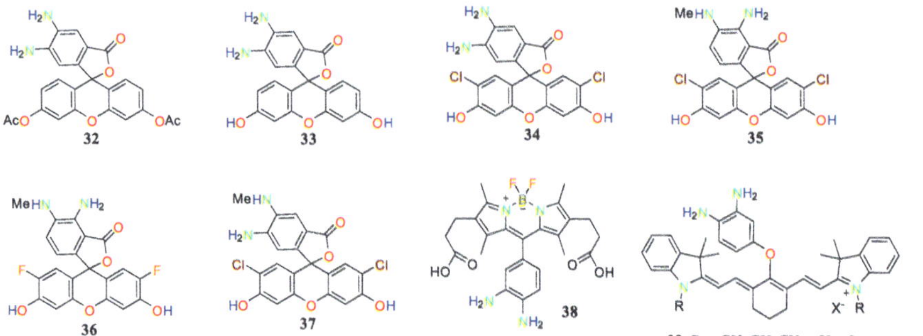

Fig. 10.2 Schematic representation of structures of the compounds from **32** to **40**

(PET), whereas after the reaction with the NO yield the triazole moiety represses the PET phenomenon and the fluorescence is recovered (Scheme 10.10).

Remarkably, for the detection of NO molecule, the spirolactam ring-opening based reactions have successfully been observed, whereby the *o*-diamine groups play the crucial role as a NO-reactive functionality in spirolactam structure. To this regard, diverse diaminofluoresceins (**32–40**) as the fluorescent indicators for NO detection have successfully been reported by Kojima et al. (Fig. 10.2) (Kojima et al. 1998, 1999). Interestingly, it has been noticed by these workers that the fluorescence quantum efficiency for **32** and **33** enhances around 100 times with a LOD of 5 nM. Instead, due to instability in the intensity of fluorescence around neutral pH, they further modified the systems (**35–37**) and find some better results. Pleasingly, because of the electron-donating influence by the methyl group, they observed stable fluorescence. Moreover, in the case of boron dipyrromethene (BODIPY) system (**38**), it was observed by these workers that pH-dependency of the fluorescence intensity was escaped by means of controlling the PET activity, regulating the spectroscopic characteristics of the BODIPY chromophore by following the *Rehm–Weller equation*. Besides these systems, they have also prepared the hydrophilic functionality incorporated systems (**39**, **40**) so as to prevent the fluorescence quenching from the stacking, thereby promoting the biological applications. Importantly, with this probe while detecting the NO molecule near-infrared (NIR)-emission was noticed by these authors. More importantly, by incorporating sulfonate groups to this probe enhances the water solubility, and these NIR probes were found less susceptible toward the interference of biological species and facilitate both in vitro and in vivo imaging of NO.

Nonetheless, Wong and co-workers revealed a spirobifluorene system (**41**) for the fluorescence turn-ON detection of NO utilizing the NO-mediated diamine

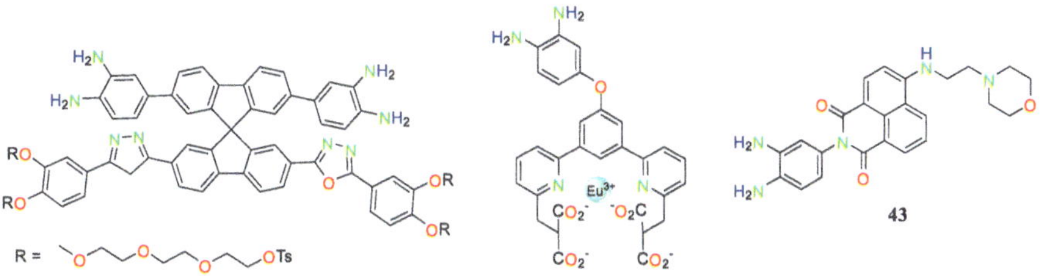

41 42 43

Fig. 10.3 Schematic representation of the structures of fluorescent probes **41–43**

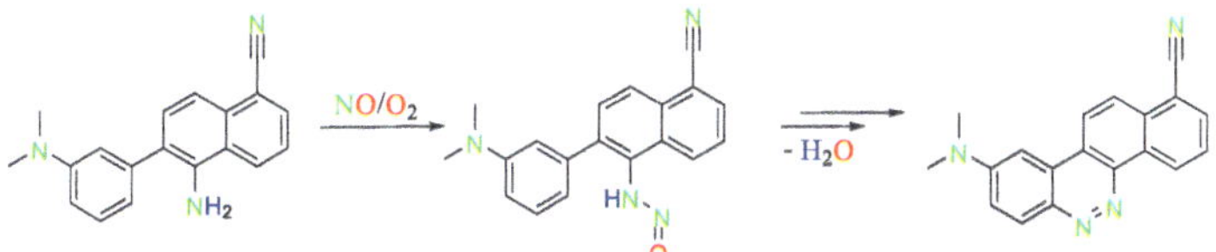

44 45 46

Scheme 10.11 Schematic sensing mechanism of the probe **44** for NO detection

cyclization strategy (Fig. 10.3) (Lin et al. 2011). In their investigation, these authors observed the fluorescence turn-OFF conversion into the turn-ON fluorescence, when NO reacts with the probe **41**. In a separate study, as can be seen from Fig. 10.3, Chen et al. have reported a Eu^{3+} coordinated luminescent probe (**42**) for the selective NO detection (Chen et al. 2011). They noticed that the probe **42** because of PET from *o*-aminophenyl to the Eu^{3+}-chelated luminophore was initially found to be a non-luminescent which upon reacting with the NO in aq. buffer (pH 7.4) under aerobic conditions converted into the luminescent one with the enhancement of the quantum yield around 47-fold and 1 lifetime from 1.08 to 1.30 ms. Similarly, Xiao and his teammates have reported a naphthalimide-based lysosome-targeting two-photon probe (**43**) displaying fluorescence turn-ON with great selectivity for NO as compared to the other ROS with a LOD in nanomolar-scale (Fig. 10.3) (Yu et al. 2012a).

Noticeably, a fluorescent NO-selective probe (**46**) was designed and synthesized by the research group of Anslyn based upon the consolidation of dimethylamino and cyano-groups, allowing an intramolecular charge transfer (CT) upon the photo-excitation (Yang et al. 2010). Gratifyingly, they observed 1500-fold fluorescence enhancement upon the reaction with NO molecule, at 550 nm (Scheme 10.11). Furthermore, this system besides showing high selectivity was also found to be equally effective in a wide pH-window range and finds use for NO bioimaging.

10.4 Fluorescent Probes for CO Detection

We are aware that CO is second GT found in our body which is obtained from collapse of the heme by means of "*enzyme heme oxygenase*" which regulates vascular smooth muscle cell growth, vasorelaxation, and tissue injury. It is to be addressed that for monitoring and detection of the CO, very rare fluorescent probes are available in the scientific literature. To this context, recently, a selective and sensitive

Scheme 10.12 (**a**) CO-mediated fluorescence turn-ON; (**b**) CO-mediated carbonylation

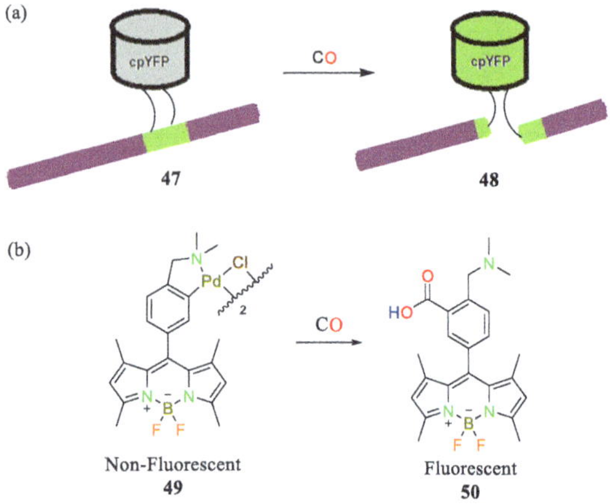

probe for the "CO imaging" in living cells was reported. Genetically, "ciphered fluorescent probes entail circularly altered variant of *yellow fluorescent proteins* (caYFP and caVenus) and a dimeric CO-sensing heme protein (CooA)," and the mechanism is based upon the exchange of Pro2 by CO, activating conformational adjustments in CooA, thereby boosting the fluorescence consequent to the yellow fluorescent proteins (Scheme 10.12a). Interestingly, after breeding for 10 minutes with CO (10 μM) around two-fold enhancement was observed besides the probe showed worthy selectivity for CO over O_2, CN^-, NO, and imidazole.

Chang's group has developed a BODIPY-based cyclopalladated probe (**49**) in which palladium stops the fluorescence of BODIPY subunit through heavy-atom electronic effect, and carbonylation reaction with CO releases the Pd(0) to furnish the fluorescent system **50** (Scheme 10.12) (Michel et al. 2012). Noticeably, the probe **49** was found to be feebly fluorescent in aq. buffered solution at the physiological pH (λ_{ex} = 499 nm). The fluorescence turn-ON response at ca. 507 nm was attributed to a ten-fold hike by the addition of [Ru(CO)$_3$Cl(glycinate)] (50 μM) as a CO source within 60 minutes. Furthermore, the dose-dependent feedback of the probe **49** demonstrated the ppb level sensitivity to the CO and displayed better results as compared to other biologically pertinent reactive oxygen, sulfur, and nitrogen, in living cells for the imaging of CO.

10.5 Fluorescent Probes for H_2S Gas

H_2S is a flammable and odorous gas which is around five times more toxic than CO, and it suppresses the performing of cytochrome C oxidase enzyme. Noticeably, volcanic emissions, natural gas, and bacterial decomposition of the organic matter are some of the environmental sources for the H_2S gas. Currently, H_2S is regarded as a cardiovascular GT that plays a prominent role in physiological phenomena, for instance, inflammation, blood vessel dilation, and vasodilation. Nonetheless, uncontrolled level of the H_2S is allied with the diabetes mellitus and Alzheimer's disease.

Although for the detection of H_2S, there are a plethora of methods, for example, fluorescence-based technique, gas chromatography (GC), colorimetric assays (CAs), polarographic sensors, etc. (Doeller et al. 2005). But because of the sensitive as well as non-destructive nature of the fluorescence probe-based tactic, it seems to be fascinating for differential analysis of the H_2S. In this particular section, we have covered various probes based on the fluorescence technique for the H_2S gas detection.

10.5.1 Nucleophilic Addition Technique

This vital technique embraces a nucleophilic addition of the H_2S gas molecule (nucleophile) onto the electrophilic portion of the probe, leading toward the cyclization to furnish a fluorescent molecule. As can be inspected from the general Scheme 10.13, the H_2S nucleophile first attacks the electrophilic center present in the probe (**51**) to generate a thiol intermediate (**52**). Later, the thiol functionality attacks the carbonyl carbon of the ester functionality to produce cyclic system along with the fluorescent molecule (**53**).

Based upon the above-mentioned strategy, He's research team has reported the fluorescent probes (**55**) and (**57**) consisting of the 1,3,5-triaryl-2-pyrazoline (**54**) and BODIPY (**56**) subunits, respectively, in their structures (Scheme 10.14) (Qian et al. 2011). Similarly, Zhao and co-workers have revealed a BODIPY-based probe

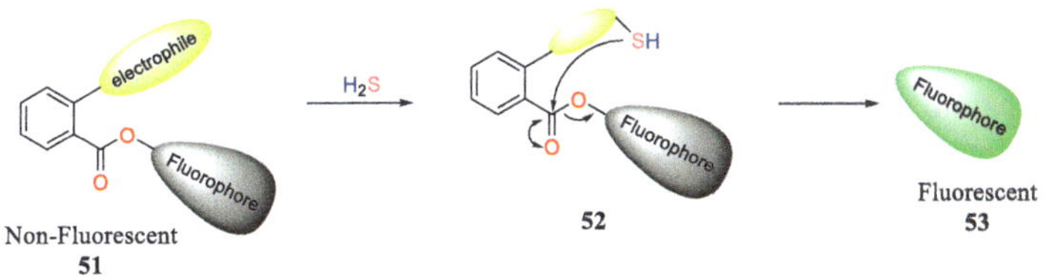

Scheme 10.13 Nucleophilic addition of the H_2S to induce fluorescence variations

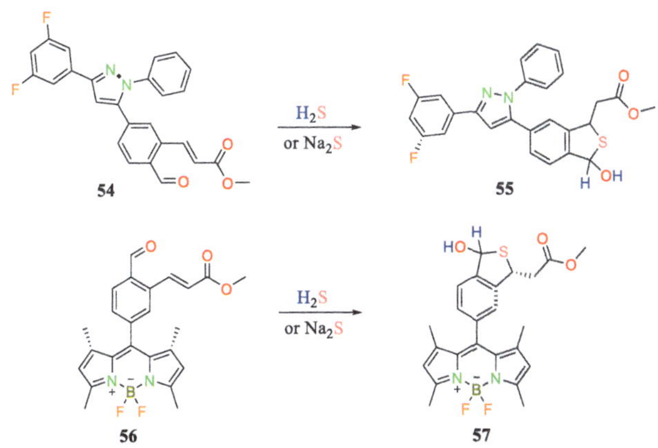

Scheme 10.14 The nucleophilic addition of H_2S gas to induce cyclization products

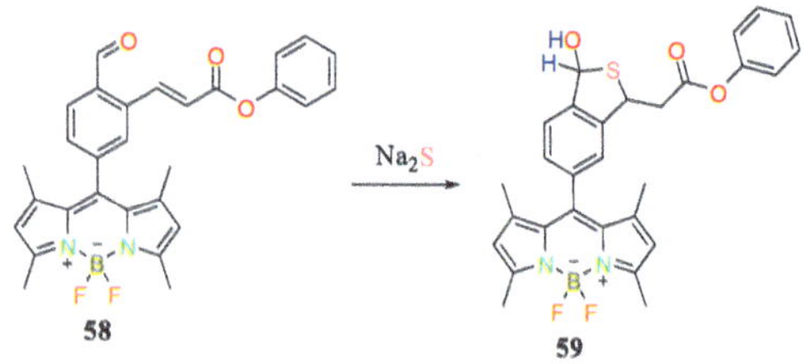

Scheme 10.15 Representation of the cyclization through addition of the H_2S

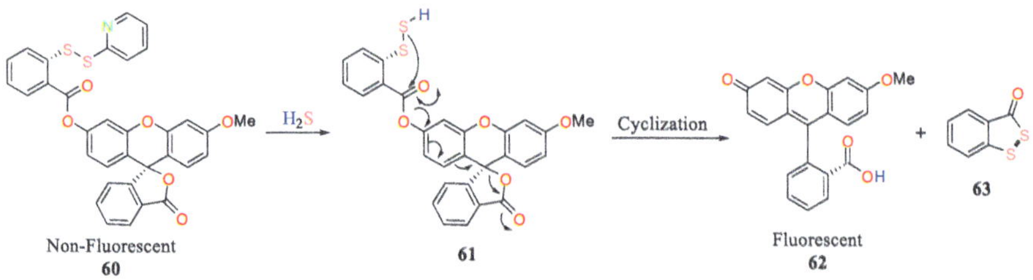

Scheme 10.16 Mechanistic path for nucleophilic addition of H_2S to produce cyclization

(**59**) for the sensitive and explicit identification of the H_2S gas (Scheme 10.15) (Qian et al. 2012). Interestingly, this fluorescent probe displayed prominent fluorescence enhancement toward Na_2S/H_2S in a great thiol contextual as noted in the presence of physiological circumstances and was found worthy for H_2S imaging in living cells.

In a separate work, Liu et al. have reported a fluorescein-based fluorescent probe (**60**) having an ester functionality contiguous to the disulfide (-S-S-) bond (Scheme 10.16) (Liu et al. 2011). As shown in Scheme 10.16, in phosphate-buffered solution, the free-probe is "non-fluorescent" which upon reaction with H_2S leads to the production of S-SH functionality. Next, S-SH group on reaction with eater functionality delivered the fluorescent molecule (**62**) along with a cyclized product **63**.

10.5.2 The H_2S-Induced Azides to Amines Reduction

In this particular section, we will discuss reduction of azides which was carried out in selective fashion utilizing the H_2S gas. The fundamental mechanism of this strategy is exclusive reduction by virtue of H_2S of the azides into the corresponding amines so as to adjust the emission signatures of the fluorescent probe(s) (Scheme 10.17). Interestingly, it has been noticed that the probes following this particular tactic showed high selectivity for the H_2S gas as compared to the other "*biothiols*," for example, glutathione, cysteine, homocysteine, etc. To this regard, two novel fluorescent rhodamine-based probes (**66**) and (**68**), having azide functionalities in their structures were designed and synthesized and employed for the detection of H_2S via fluorescence turn-ON mechanism by the research group of Chang (Scheme 10.18) (Lippert et al. 2011).

Scheme 10.17 Reduction of the azides to amines by H_2S

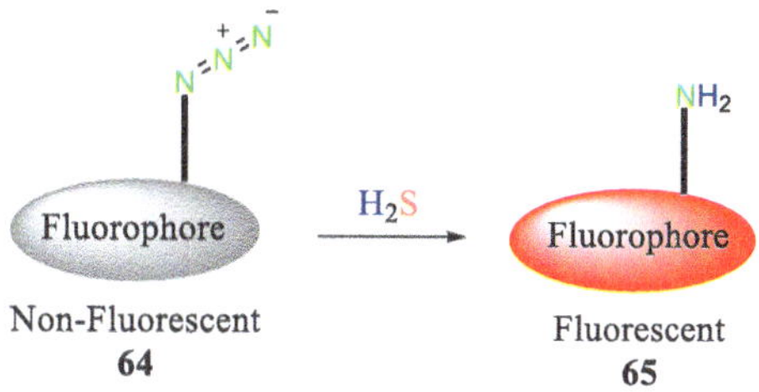

Scheme 10.18 H_2S-induced reduction of rhodamine azides via fluorescence turn-ON

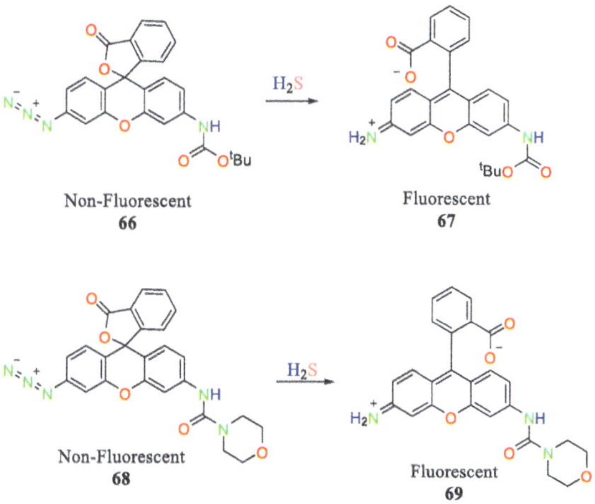

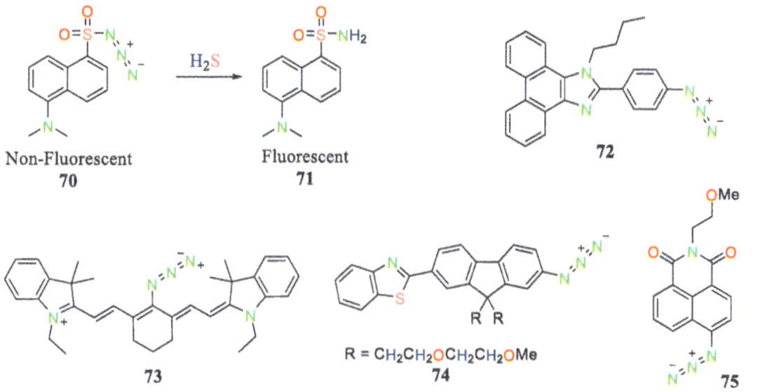

Scheme 10.19 H_2S-induced reduction of dansyl azide **70** and the structures of **72–75**

It has been noticed by Wang and co-workers that sulfonyl azide reduction through H_2S gas into the sulfonamide alters the fluorescent signatures of the dansyl moiety (Peng et al. 2011). From their studies, these authors observed that the dansyl azide-based probe (**70**) is non-emissive which exhibits a strong fluorescence after the reduction with H_2S, as shown in Scheme 10.19. Conversely, a cyanine-based fluorescent probe (**73**) for the cellular H_2S detection was revealed by the research group of Han (Yu et al. 2012b), whereas Das et al. reported a two-photon fluorescent probe

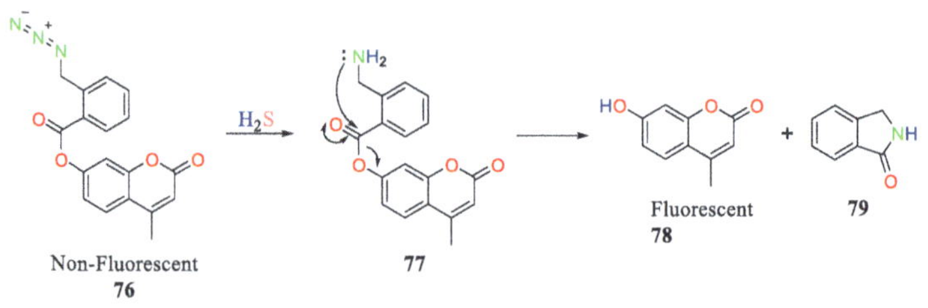

Scheme 10.20 Representation of H_2S-induced reduction to fluorescence turn-on response

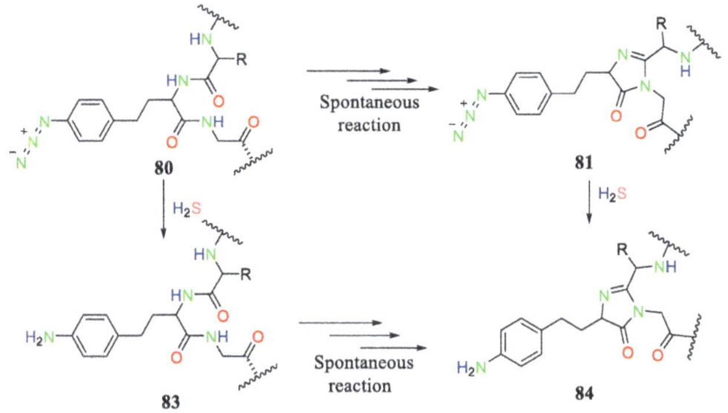

Scheme 10.21 H_2S detection by genetically ciphered fluorescent protein-based probe

(**74**), which showed a 21-fold fluorescence intensification with the H_2S gas (Das et al. 2012). Importantly, this worthy probe was found to be very selective for H_2S gas over other biologically pertinent RNS/ROS or reactive sulfur species (RSS). Moreover, this probe was also used by these workers to detect the H_2S in living cells in addition to the tissues at 90–190 μm depth by means of the two-photon microscopy (TPM). Conversely, two different fluorescent probes based on the naphthalimide and phenanthroimidazole scaffolds (**72** and **74**) for the detection of H_2S were reported by Pluth and Lin research groups (Montoya and Pluth 2012; Zheng et al. 2012).

In another study, Han and his teammates have described a selective H_2S detection with the usage of a fluorescent probe based on the 7-*o*-2′-(azidomethyl) benzoyl-4-methylcoumarin (**76**), whereby an analyte-induced reductive exclusion of the 2′-(azidomethyl)benzoyl crafts a fluorescent species, namely 7-hydroxy-4-methylcoumarin (**78**), displaying fluorescence emission ca. at 450 nm (Scheme 10.20) (Wu et al. 2012). Furthermore, selective detection of the H_2S gas in living cells over a wide range of chemical and/or biological entities was also executed by these authors. Conversely, Chen et al. revealed a "*genetically ciphered fluorescent protein-based probe*" (**80**) through the installation of *p*-azidophenylalanine onto the peptides of fluorescent protein that can explicitly monitor H_2S in the living mammalian cells (Scheme 10.21) (Chen et al. 2012). It was noticed by these authors that the azide functionality after reduction into the amine group by H_2S showed fluorescence turn-ON response in a selective manner. Noticeably, this vital genetically ciphered probe confers several advantages encompassing the add-on of the cell localization tags to administer the probe to unambiguous cell subdomains.

10.5.3 Reduction of Nitro Group to the Amino Functionality

Reduction of nitro functionality into the amino group by means of H_2S is based upon the generation of fluorescence turn-ON response. To this context, the research team of Chen has reported a NIR-based fluorescent probe (**85**) for the selective recognition of H_2S by means of nitro-to-amine reduction approach (Scheme 10.22) (Wang et al. 2012). From their studies in aqueous media, these authors noticed that the photoinduced electron transfer (donor-PET or d-PET) is reliable for the quenching in the fluorescence emission spectra. In a separate report, Cao et al. detailed a NIR-BODIPY based turn-on fluorescent probe (**87**) for the identification of the H_2S gas through H_2S-induced dinitrophenyl ether moiety thiolysis, as depicted in Scheme 10.23 (Cao et al. 2012). Conversely, Palm's research group has reported a cobalt-metal based peptide deformylase (**89**) as a fluorescence resonance-energy-transfer (FRET) sensing ensemble for the detection of H_2S in which active metal site is coordinated with the imidazoles and thiolate atom (Scheme 10.24) (Strianese et al. 2012).

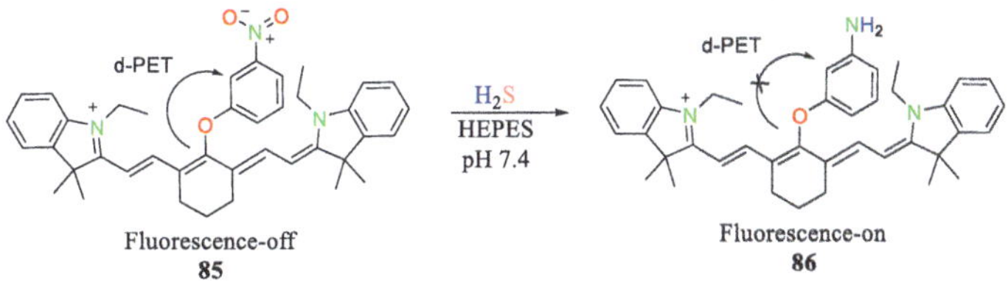

Scheme 10.22 Schematic representation of the H_2S-induced reduction of nitro-to-amine group

Scheme 10.23 The H_2S-induced thiolysis of BODIPY-based system **87**

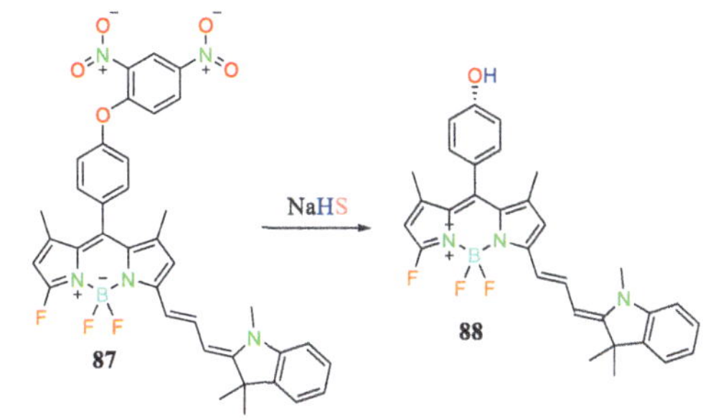

Scheme 10.24 Fluorescence-based H_2O replacement by H_2S molecule

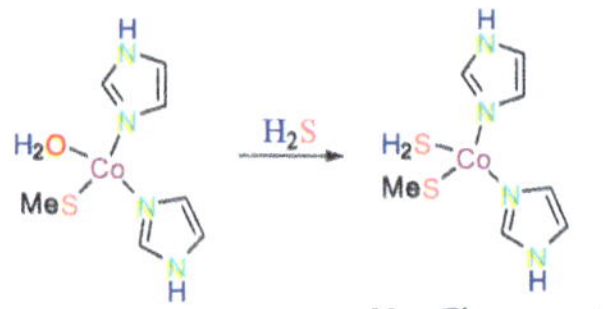

Scheme 10.25 The turn-ON in fluorescence through the H_2S gas

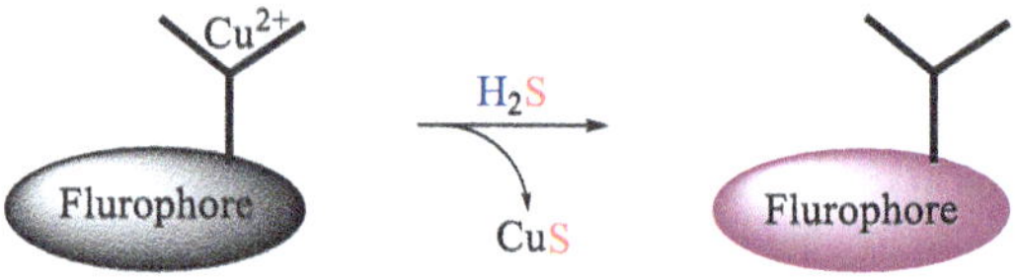

Scheme 10.26 H_2S-induced CuS precipitation via turn-on fluorescence

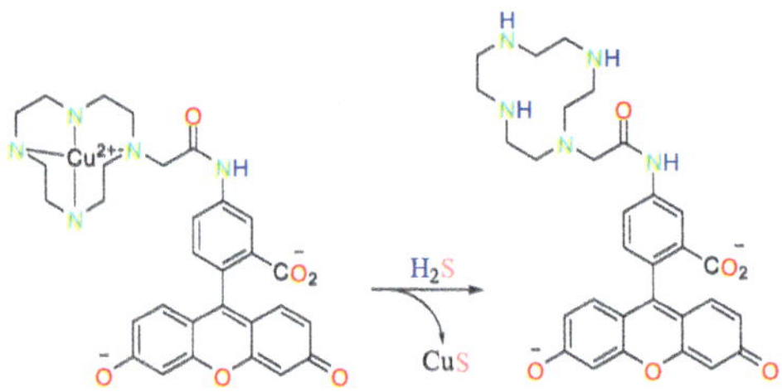

10.5.4 Copper Displacement Technique

In this particular technique, Cu^{2+} (paramagnetic ion) is being displaced from the surrounding of the fluorophore, thereby causing the fluorescence "turn-ON" by means of precipitation of the CuS from H_2S and Cu^{2+}. General schematic representation for this tactic is shown in Scheme 10.25. In this regard, Sasakara et al. have revealed a H_2S-displacement assay, utilizing the fluorescein-based system (Scheme 10.26) (Sasakura et al. 2011). From the study, they observed a great selectivity for the H_2S than other biothiols. As can be inspected from Scheme 10.26, while still binding of Cu^{2+} to the receptor, fluorescence is being quenched, which upon reaction with H_2S releases the Cu^{2+} ion and precipitates the CuS, thereby displaying the intensified fluorescence emission in the solution as well as intracellular systems. Likewise, diverse fluorescent systems have also been reported by various research groups as displayed in Fig. 10.4 (Alday et al. 2020).

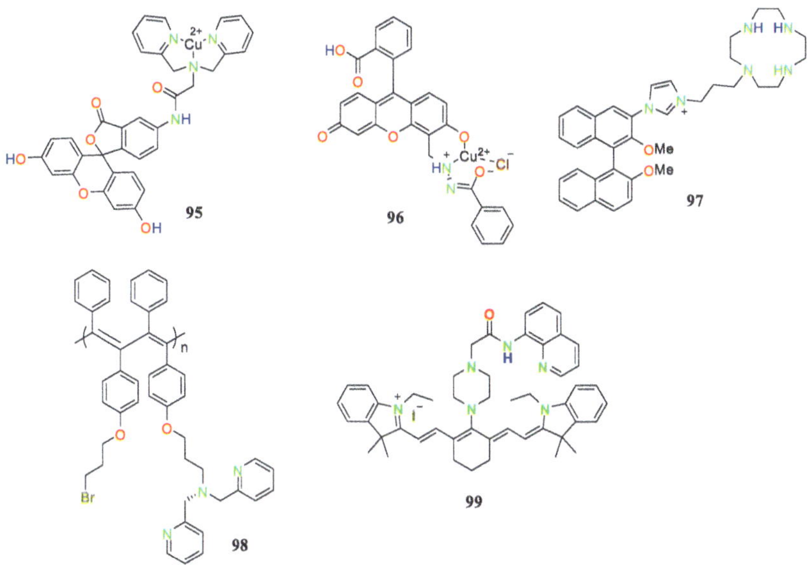

Fig. 10.4 Chemical structures of some important probes reported by different groups

10.6 Conclusions

In this particular chapter, we have exposed a variety of fluorescence probe-based probes employed by the various researchers to detect the NO, CO, and H_2S. As detailed above, a diversity of fluorophores, for instance, rhodamine, fluorescein, cyanine, BODIPY, dansyl, 1,8-naphthalimide, etc., have been introduced with different types of reaction sites for the detection process based on the various photophysical mechanisms encompassing ICT, FRET, PET, etc. We hope that this chapter will add new values to the fluorescent probe in identification of vital biological entities.

Acknowledgements Dr. Rashid Ali is highly thankful to the DST-SERB, New Delhi (Project File no. ECR/2017/000821). S. A. thanks UGC for providing SRF fellowship. And we are also obliged to Jamia Millia Islamia for providing the necessary infrastructure for conducting research.

References

Alday J, Mazzeo A, Suarez S (2020) Selective detection of gasotransmitters using fluorescent probes based on transition metal complexes. Inorganica Chim Acta 510:119696

Brown GC, Bal-Price A (2003) Inflammatory neurodegeneration mediated by nitric oxide, glutamate, and mitochondria. Mol Neurobiol 27(3):325–355. https://doi.org/10.1385/MN:27:3:325

Cao X, Lin W, Zheng K, He L (2012) A near-infrared fluorescent turn-on probe for fluorescence imaging of hydrogen sulfide in living cells based on Thiolysis of dinitrophenyl ether. Chem Commun 48(85):10529. https://doi.org/10.1039/c2cc34031c

Carrola A, Romão CC, Vieira HLA (2023) Carboxyhemoglobin (COHb): unavoidable bystander or protective player? Antioxidants 12(6):1198. https://doi.org/10.3390/antiox12061198

Chen Y, Guo W, Ye Z, Wang G, Yuan J (2011) A europium(Iii) chelate as an efficient time-gated luminescent probe for nitric oxide. Chem Commun 47(22):6266. https://doi.org/10.1039/c0cc05658h

Chen S, Chen Z, Ren W, Ai H (2012) Reaction-based genetically encoded fluorescent hydrogen sulfide sensors. J Am Chem Soc 134(23):9589–9592. https://doi.org/10.1021/ja303261d

Coburn RF (2022) Carbon monoxide (CO), nitric oxide, and hydrogen sulfide signaling during acute CO poisoning. Front Pharmacol:12. https://doi.org/10.3389/fphar.2021.830241

Das SK, Lim CS, Yang SY, Han JH, Cho BR (2012) A small molecule two-photon probe for hydrogen sulfide in live tissues. Chem Commun 48(67):8395. https://doi.org/10.1039/c2cc33909a

Doeller JE, Isbell TS, Benavides G, Koenitzer J, Patel H, Patel RP, Lancaster JR, Darley-Usmar VM, Kraus DW (2005) Polarographic measurement of hydrogen sulfide production and consumption by mammalian tissues. Anal Biochem 341(1):40–51. https://doi.org/10.1016/j.ab.2005.03.024

Fukuto J, Collins M (2007) Interactive endogenous small molecule (gaseous) signaling: implications for teratogenesis. Curr Pharm Des 13(29):2952–2978. https://doi.org/10.2174/138161207782110525

Gallio AE, Fung SS-P, Cammack-Najera A, Hudson AJ, Raven EL (2021) Understanding the logistics for the distribution of heme in cells. JACS Au 1(10):1541–1555. https://doi.org/10.1021/jacsau.1c00288

Hilderbrand SA, Lim MH, Lippard SJ (2004) Dirhodium tetracarboxylate scaffolds as reversible fluorescence-based nitric oxide sensors. J Am Chem Soc 126(15):4972–4978. https://doi.org/10.1021/ja038471j

Hosogi S, Marunaka Y, Ashihara E, Yamada T, Sumino A, Tanaka H, Puppulin L (2021) Plasma membrane anchored nanosensor for quantifying endogenous production of H2O2 in living cells. Biosens Bioelectron 179:113077. https://doi.org/10.1016/j.bios.2021.113077

Hu X, Wang J, Zhu X, Dong D, Zhang X, Wu S, Duan C (2011) A copper(ii) rhodamine complex with a tripodal ligand as a highly selective fluorescence imaging agent for nitric oxide. Chem Commun 47(41):11507. https://doi.org/10.1039/c1cc14032a

Ignarro LJ, Edwards JC, Gruetter DY, Barry BK, Gruetter CA (1980) Possible involvement of S -nitrosothiols in the activation of guanylate cyclase by nitroso compounds. FEBS Lett 110(2):275–278. https://doi.org/10.1016/0014-5793(80)80091-3

Kabil O, Weeks CL, Carballal S, Gherasim C, Alvarez B, Spiro TG, Banerjee R (2011) Reversible heme-dependent regulation of human cystathionine β-synthase by a flavoprotein oxidoreductase. Biochemistry 50(39):8261–8263. https://doi.org/10.1021/bi201270q

Kojima H, Nakatsubo N, Kikuchi K, Kawahara S, Kirino Y, Nagoshi H, Hirata Y, Nagano T (1998) Detection and imaging of nitric oxide with novel fluorescent indicators: diaminofluoresceins. Anal Chem 70(13):2446–2453. https://doi.org/10.1021/ac9801723

Kojima H, Urano Y, Kikuchi K, Higuchi T, Hirata Y, Nagano T (1999) Fluorescent indicators for imaging nitric oxide production. Angew Chem Int Ed 38(21):3209–3212. https://doi.org/10.1002/(SICI)1521-3773(19991102)38:21<3209::AID-ANIE3209>3.0.CO;2-6

Lim MH, Lippard SJ (2005) Copper complexes for fluorescence-based NO detection in aqueous solution. J Am Chem Soc 127(35):12170–12171. https://doi.org/10.1021/ja053150o

Lin L-Y, Lin X-Y, Lin F, Wong K-T (2011) A new spirobifluorene-bridged bipolar system for a nitric oxide turn-on fluorescent probe. Org Lett 13(9):2216–2219. https://doi.org/10.1021/ol200463m

Lippert AR, New EJ, Chang CJ (2011) Reaction-based fluorescent probes for selective imaging of hydrogen sulfide in living cells. J Am Chem Soc 133(26):10078–10080. https://doi.org/10.1021/ja203661j

Liu C, Pan J, Li S, Zhao Y, Wu LY, Berkman CE, Whorton AR, Xian M (2011) Capture and visualization of hydrogen sulfide by a fluorescent probe. Angew Chem 123(44):10511–10513. https://doi.org/10.1002/ange.201104305
Machha A, Schechter AN (2012) Inorganic nitrate: a major player in the cardiovascular health benefits of vegetables? Nutr Rev 70(6):367–372. https://doi.org/10.1111/j.1753-4887.2012.00477.x
Maines MD (1997) THE HEME OXYGENASE SYSTEM: a regulator of second messenger gases. Annu Rev Pharmacol Toxicol 37(1):517–554. https://doi.org/10.1146/annurev.pharmtox.37.1.517
Michel BW, Lippert AR, Chang CJ (2012) A reaction-based fluorescent probe for selective imaging of carbon monoxide in living cells using a palladium-mediated carbonylation. J Am Chem Soc 134(38):15668–15671. https://doi.org/10.1021/ja307017b
Montoya LA, Pluth MD (2012) Selective turn-on fluorescent probes for imaging hydrogen sulfide in living cells. Chem Commun 48(39):4767. https://doi.org/10.1039/c2cc30730h
Motterlini R, Foresti R (2017) Biological signaling by carbon monoxide and carbon monoxide-releasing molecules. Am J Physiol Physiol 312(3):C302–C313. https://doi.org/10.1152/ajpcell.00360.2016
Peng H, Cheng Y, Dai C, King AL, Predmore BL, Lefer DJ, Wang B (2011) A fluorescent probe for fast and quantitative detection of hydrogen sulfide in blood. Angew Chem 123(41):9846–9849. https://doi.org/10.1002/ange.201104236
Pénicaud C, Peyron S, Gontard N, Guillard V (2012) Oxygen quantification methods and application to the determination of oxygen diffusion and solubility coefficients in food. Food Rev Int 28(2):113–145. https://doi.org/10.1080/87559129.2011.595021
Platt JL, Nath KA (1998) Heme oxygenase: protective gene or Trojan horse. Nat Med 4(12):1364–1365. https://doi.org/10.1038/3947
Qian Y, Karpus J, Kabil O, Zhang S-Y, Zhu H-L, Banerjee R, Zhao J, He C (2011) Selective fluorescent probes for live-cell monitoring of sulphide. Nat Commun 2(1):495. https://doi.org/10.1038/ncomms1506
Qian Y, Zhang L, Ding S, Deng X, He C, Zheng XE, Zhu H-L, Zhao J (2012) A fluorescent probe for rapid detection of hydrogen sulfide in blood plasma and brain tissues in mice. Chem Sci 3(10):2920. https://doi.org/10.1039/c2sc20537h
Rather IA, Ali R (2021) Indicator displacement assays: from concept to recent developments. Org Biomol Chem 19(27):5926–5981. https://doi.org/10.1039/D1OB00518A
Rather IA, Ali R (2023) Indicator displacement assays (idas): an innovative molecular sensing approach, ISBN: 978–981-5165-91-3
Rose P, Moore PK, Zhu YZ (2017) H2S biosynthesis and catabolism: New insights from molecular studies. Cell Mol Life Sci 74(8):1391–1412. https://doi.org/10.1007/s00018-016-2406-8
Roughton FJW, Darling RC (1944) The effect of carbon monoxide on the oxyhemoglobin dissociation curve. Am J Physiol Content 141(1):17–31. https://doi.org/10.1152/ajplegacy.1944.141.1.17
Ryter SW, Choi AMK (2013) Carbon monoxide: present and future indications for a medical gas. Korean J Intern Med 28(2):123–140. https://doi.org/10.3904/kjim.2013.28.2.123
Sasakura K, Hanaoka K, Shibuya N, Mikami Y, Kimura Y, Komatsu T, Ueno T, Terai T, Kimura H, Nagano T (2011) Development of a highly selective fluorescence probe for hydrogen sulfide. J Am Chem Soc 133(45):18003–18005. https://doi.org/10.1021/ja207851s
Shibuya N, Tanaka M, Yoshida M, Ogasawara Y, Togawa T, Ishii K, Kimura H (2009) 3-mercaptopyruvate sulfurtransferase produces hydrogen sulfide and bound sulfane sulfur in the brain. Antioxid Redox Signal 11(4):703–714. https://doi.org/10.1089/ars.2008.2253
Strianese M, Palm GJ, Milione S, Kühl O, Hinrichs W, Pellecchia C (2012) A FRET enzyme-based probe for monitoring hydrogen sulfide. Inorg Chem 51(21):11220–11222. https://doi.org/10.1021/ic301363d
Tsuge K, DeRosa F, Lim MD, Ford PC (2004) Intramolecular reductive nitrosylation: reaction of nitric oxide and a copper(II) complex of a cyclam derivative with pendant luminescent chromophores. J Am Chem Soc 126(21):6564–6565. https://doi.org/10.1021/ja049444b

Wang R (2002) Two's company, Three's a crowd: can H 2 S be the third endogenous gaseous transmitter? FASEB J 16(13):1792–1798. https://doi.org/10.1096/fj.02-0211hyp

Wang R, Yu F, Chen L, Chen H, Wang L, Zhang W (2012) A highly selective turn-on near-infrared fluorescent probe for hydrogen sulfide detection and imaging in living cells. Chem Commun 48(96):11757. https://doi.org/10.1039/c2cc36088h

Wu Z, Li Z, Yang L, Han J, Han S (2012) Fluorogenic detection of hydrogen sulfide via reductive unmasking of O-Azidomethylbenzoyl-Coumarin conjugate. Chem Commun 48(81):10120. https://doi.org/10.1039/c2cc34682f

Yang Y, Seidlits SK, Adams MM, Lynch VM, Schmidt CE, Anslyn EV, Shear JB (2010) A highly selective low-background fluorescent imaging agent for nitric oxide. J Am Chem Soc 132(38):13114–13116. https://doi.org/10.1021/ja1040013

Yu H, Xiao Y, Jin L (2012a) A lysosome-targetable and two-photon fluorescent probe for monitoring endogenous and exogenous nitric oxide in living cells. J Am Chem Soc 134(42):17486–17489. https://doi.org/10.1021/ja308967u

Yu F, Song P, Li P, Wang B, Han K (2012b) Development of reversible fluorescence probes based on redox oxoammonium cation for hypobromous acid detection in living cells. Chem Commun 48(62):7735. https://doi.org/10.1039/c2cc33264g

Zheng K, Lin W, Tan L (2012) A phenanthroimidazole-based fluorescent chemosensor for imaging hydrogen sulfide in living cells. Org Biomol Chem 10(48):9683. https://doi.org/10.1039/c2ob26956b

Plant-Mediated Synthesis of Nanomaterials for Biosensing

11

Romana Naaz, Weqar Ahmed Siddiqi, and Mohd. Mohsin

Abstract

Plant-mediated synthesis has become a pivotal research area, especially in biosensing applications, for producing nanomaterials. This chapter explores the importance of plant-mediated synthesis in creating diverse nanomaterials for biosensing. It emphasizes the environmentally friendly, cost-effective, and sustainable aspects of these processes within nanotechnology. Nanoparticles (NPs) can be generated in different sizes and shapes through physical, chemical, or biological routes. By examining various nanomaterial types like gold, silver, zinc, titanium, and palladium synthesized through plant-mediated methods, this chapter showcases the wide range of benefits and applications associated with these eco-friendly synthesis techniques. The mechanisms in plant-mediated synthesis, such as reduction, stabilization, nucleation, aggregation, and capping, are discussed to highlight the distinctive properties of plant-synthesized nanomaterials that make them well-suited for biosensing. Through an in-depth exploration of nanomaterials produced using plant extracts, this chapter offers valuable insights into how plant-mediated synthesis can drive advancements in biosensing technologies.

Keywords

Nanomaterial · Plant-based · Metal nanoparticles · Nanocomposites

---

R. Naaz · W. A. Siddiqi
Department of Applied Sciences and Humanities, Faculty of Engineering and Technology, Jamia Millia Islamia, New Delhi, India

M. Mohsin (✉)
Department of Biosciences, Jamia Millia Islamia, New Delhi, Delhi, India
e-mail: mmohsin1@jmi.ac.in

M. Mohsin, N. Soleja (eds.), *Nanoscale Sensors and their Applications in Biomedical Imaging*, https://doi.org/10.1007/978-981-97-3144-2_11

Abbreviations

AgNPs	Silver nanoparticles
AuNPs	Gold nanoparticles
MNPs	Metal nanoparticles
QDs	Quantum dots
SPR	Surface plasmon resonance

11.1 Introduction

Green chemistry encompasses a wide array of elements, including catalysis, synthesis methods, solvents, raw materials, end products, and streamlined processes. Within green chemistry, green synthesis is a crucial pillar, and finding effective synthetic routes is essential for achieving sustainable industrial production, especially in industries like healthcare. Traditionally, chemists, especially those in organic chemistry, have been educated to formulate synthetic reactions with a primary focus on optimizing both yield and purity. However, numerous chemical fabrication procedures exhibit inefficiency in their utilization of raw material and lead to the generation of significant amounts of side products. The field of nanotechnology has witnessed remarkable advancements, leading to the development of nanomaterials with diverse applications. Among these, the synthesis of nanomaterials through plant-mediated routes has emerged as a promising and eco-friendly approach. The unique properties of plants, combined with the principles of nanotechnology, offer a sustainable and cost-effective pathway to create functional nanomaterials with potential implications in disease diagnosis, environmental monitoring, and food safety, among others. Nano-biotechnology has indeed made significant advancements in the field of cancer diagnosis and treatment in recent years. The use of nanomaterials, including quantum dots, silver, gold, iron oxide, and silica nanoparticles, has been at the forefront of research in developing new strategies for cancer therapy and diagnosis. These enhanced nanomaterials have versatile applications, encompassing diagnosis, drug delivery, and monitoring therapeutic response. Moreover, earlier research has demonstrated the promise of utilizing metal nanoparticles in various fields, such as drug or gene delivery, photoimaging, and sensors, among others (Srivastava and Pandey 2021). Researchers conduct thorough investigations into fine-tuning the surface plasmon resonance (SPR) of metal nanoparticles (MNPs) to unlock their potential applications across a wide range of fields, including sensor technology, biodevices, data storage, spectroscopic methods, and catalytic processes, among numerous others. The exceptional properties of plasmonic metal nanoparticles have garnered significant interest, making them a fantastic nanoscale platform for applications in biological and chemical sensing. Due to their ability to exhibit SPR, plasmonic metal nanoparticles can interact with light in unique ways, resulting in strong and tunable optical responses. This property is particularly advantageous for sensing applications. In biological sensing, plasmonic MNPs are employed to detect and analyze biomolecules, cells, and other biological entities. Their interaction with light can be influenced by changes in the local environment, such as the binding of target molecules, enabling sensitive and specific detection. These nanoparticles find applications in various fields, including medical diagnostics, environmental monitoring, and bioanalytical assays. Similarly, in chemical sensing, plasmonic MNPs are utilized to detect and quantify different chemical species. Their response to the presence of specific chemicals can be monitored optically, offering a rapid and sensitive approach to detect various analytes. This makes them valuable tools for monitoring environmental pollutants, detecting hazardous substances, and characterizing chemical reactions.

Overall, plasmonic MNPs serve as versatile and powerful tools for biological and chemical sensing due to their unique optical properties and tunability. This makes them an exciting and promising area of research and application in nanotechnology and sensing technology. The synthesis of metal nanoparticles (MNPs) utilizing plant extracts is a complex procedure, primarily because it involves a multitude of phytochemicals with the capability to both reduce metal salts and inhibit the aggregation of the resulting nanoparticles. These phytochemicals, including flavonoids, sugars, amino acids, and other compounds, possess functional groups that play essential roles in the bioreduction and stabilization processes of the nanoparticles (Abegunde et al. 2020; del Carmen Travieso Novelles et al. 2021; Shahzadi et al. 2022). Several reports have proposed biochemical reduction mechanisms to explain this phenomenon. It has been noted that the composition of the plant extract plays a pivotal role in shaping and sizing the resultant metal nanoparticles (Naaz et al. 2023). Different plant extracts will have varying concentrations and types of phytochemicals, leading to variations in the final properties of the synthesized nanoparticles. This variability in composition contributes to the complexity of the synthesis mechanism and necessitates careful optimization to obtain nanoparticles with desired characteristics. Biosensors, which integrate elements of molecular biology, microfluidics, and nanomaterials, find a wide range of applications in agriculture, food processing, and environmental monitoring (Bharali et al. 2019). They detect pesticides, pathogens, toxins, proteins, and more in soil, air, water, food, plants, and animals. In 2004, the world witnessed the creation of the inaugural soil microorganism biosensor, which predicts soil diseases by analyzing the oxygen consumption in microbial respiration. It offers an innovative soil diagnostic technique based on numerical data (Moraskie et al. 2021).

11.2 Plant-Mediated Nanomaterial Synthesis

The plant-mediated fabrication of nanomaterials involves using various parts of plants, such as leaves, stems, roots, or seeds, as bioreactors for nanoparticle formation (Pramila et al. 2022). The reduction of metal ions to nanoparticles is mediated by the plant's inherent bioactive compounds, such as phytochemicals, flavonoids, saponins, tannins, and enzymes. Nanoparticles of various sizes, shapes, morphologies, and surface areas can be derived from different parts of plants, such as seeds, leaves, barks, fruits, and roots. This diversity is attributed to the fact that each plant part possesses a unique phytochemical profile with varying concentrations and compositions, which in turn influence the characteristics of the resulting nanoparticles (Pramila et al. 2022). This bioreduction process not only aids in the formation of nanomaterials but also provides a stable and biocompatible environment for their growth.

11.2.1 Advantages of Plant-Based Synthesis

Eco-Friendly One of the most significant advantages of plant-mediated nanomaterial synthesis is its eco-friendliness. Unlike conventional chemical methods that frequently utilize toxic chemicals and produce harmful byproducts, plant-based synthesis employs natural, renewable resources and involves minimal environmental impact. This green approach aligns with sustainable development goals and promotes the synthesis of nanomaterials without contributing to pollution.

Biocompatibility The biomolecules present in plant-mediated nanomaterials impart biocompatibility, making them highly suitable for biomedical applications. In contrast to some synthetic nanomaterials, plant-fabricated nanoparticles exhibit low cytotoxicity, which is crucial for their use in biological systems. This biocompatibility is especially valuable in biosensing applications where the interaction between the nanomaterials and biological molecules must be precise and non-destructive.

Cost-Effectiveness and Scalability Plant-mediated synthesis also offers economic benefits. Plant materials are abundant and can be sourced easily, often economical. The bioreduction activity takes place under ambient conditions, at room temperature and atmospheric pressure, obviating the requirement for costly equipment or specialized facilities. Consequently, this method is scalable, making it feasible for large-scale production of nanomaterials for commercial applications.

Size and Shape Control Plant extracts contain a diverse array of biomolecules, each with its unique properties, enabling the fine-tuning of nanoparticle characteristics. By controlling the reaction parameters and varying the plant extracts, researchers can precisely regulate the size, shape, and surface properties of the synthesized nanomaterials. This level of control is advantageous in biosensing, as it allows tailoring the nanomaterials to specific analytes and sensor requirements, leading to improved sensitivity and selectivity.

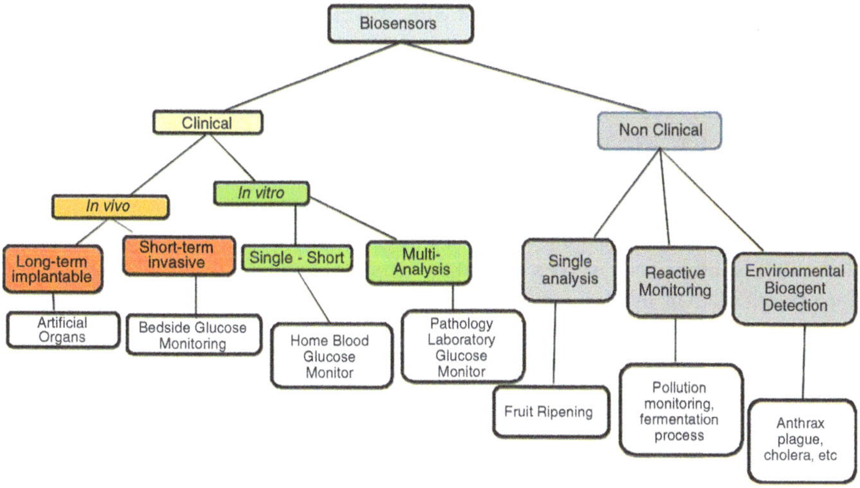

Fig. 11.1 Potential application of biosensors

11.3 Emerging Developments in Environmentally Friendly Nanomaterial Synthesis

In recent decades, there has been substantial attention in the development of nanomaterials with specific characteristics. Researchers have dedicated considerable effort to modifying raw devices in accordance with green chemistry principles to enhance their biosensing capabilities. Besides focusing on the properties of the final materials, there is also a strong emphasis on achieving additional characteristics, including cost-effectiveness, easy setup, low toxicity, and scalability. These desired features are crucial in creating competitive materials that find widespread applications in various fields, as shown in Fig. 11.1.

11.3.1 Silver-Based Nanomaterials

Silver nanoparticles (AgNPs) possess excellent antibacterial and catalytic properties, making them valuable candidates for biosensing applications (Sharanappa et al. 2020). Plant-mediated AgNPs have shown promise in detecting heavy metals, pesticides, and various biomolecules, showcasing their potential in environmental monitoring and healthcare (Naaz et al. 2023). Plant extracts containing various reducing agents, such as terpenoids, tannins, and proteins, effectively convert silver ions into AgNPs. Extensive research is conducted to fine-tune the surface plasmon resonance (SPR) of metal nanoparticles (MNPs), with the aim of unlocking their practical utility in sensors, biodevices, data storage, spectroscopic methods, catalysis, and various other fields. Silver nanoparticles (AgNPs) are particularly appealing for biosensing applications due to their susceptibility to surface-adsorbed molecules, enabling the modulation of SPR wavelength through the concentration of these adsorbed molecules. Hence, researchers have explored the sensing capabilities of AgNPs with various analytes, including glucose, triacylglycerides, and other substances (Varghese Alex et al. 2020). These biofabricated AgNPs have been used in biosensors for detecting heavy metals, pesticides, and environmental toxins. Furthermore, they show great potential for swiftly and sensitively detecting pathogens, including bacteria and viruses, in clinical and environmental samples. *Alex* et al. examined the biosensing and photocatalytic properties of the silver nanoparticles they synthesized. AgNPs served as sensing element, and mancozeb (MCZ) agro-fungicide was used as a model (Varghese Alex et al. 2020).

11.3.2 Cutting-Edge Advancement in Carbon-Based Nanomaterials

Carbon nanomaterials represent distinctive carbon structures created by the rolling of carbon atom sheets. They possess remarkable thermal, optical, mechanical, and electrical characteristics, rendering them exceedingly promising for a wide array of applications. For instance, they are excellent candidates in gas sensing applications, electrochemical biosensing (Zaytseva and Neumann 2016). These exceptional properties contribute to their versatility and potential for an extensive range of practical implementation in different area. Kumar et al. successfully synthesized a plant-derived carbon sample from monkey ladder (Entada gigas) plant seeds using a combustion method. The resulting carbon sample demonstrates excellent electrochemical performance, making it suitable for real-time sensor applications (Sudheer Kumar et al. 2022).

11.3.3 Gold-Based Nanomaterials

Gold nanoparticles (AuNPs), owing to their unique optical properties, have been extensively studied for biosensing applications. The surface plasmon resonance (SPR) effect exhibited by AuNPs results in a characteristic color change when they interact with target analytes. By functionalizing the surface of AuNPs with specific ligands, such as antibodies, DNA, or aptamers, they can be employed for the detection of a broad spectrum of biomolecules, including proteins, nucleic acids, and small molecules. Plant-mediated synthesis of AuNPs offers several advantages over conventional methods. The phytochemicals present in plant extracts, such as flavonoids and phenolic compounds, act as both reducing and stabilizing agents, ensuring the formation of stable and biocompatible AuNPs. Moreover, the tunable SPR properties of these biofabricated AuNPs allow for the development of extremely sensitive and discerning biosensors. Srivastava et al. fabricated the gold through the utilization of *Syngonium podophyllum* leaf extract and subsequently assessed their biosensing capabilities on an indium tin oxide (ITO) electrode (Srivastava and Pandey 2021).

11.3.4 Quantum Dots (QDs)

Quantum dots are semiconductor nanocrystals characterized by their distinctive fluorescent properties, such as size-dependent emission and high photostability. Traditional methods of QD synthesis often involve toxic materials and high temperatures, making them less suitable for biological applications. Plant-mediated synthesis of QDs overcomes these limitations and provides a greener and more biocompatible alternative. By using plant extracts as bioreduction agents, researchers can synthesize QDs with controlled sizes and surface properties, making them suitable for cellular imaging and intracellular sensing. The bright and stable fluorescence of plant-fabricated QDs has paved the way for their application in detecting biomarkers and cellular processes in biological systems. Jacob et al. synthesized ZnS quantum dots (QDs) under ambient conditions using the fungus *Aspergillus* sp. These quantum dots were used as biosensing agent for heavy metals, to check the mechanism behind the antibacterial action, and for the photodegradation of dyes (Jacob et al. 2019).

11.3.5 Iron Oxide Nanomaterials

Iron oxide nanoparticles have gained popularity in biosensing due to their magnetic properties and biocompatibility. Plant-mediated synthesis of IONPs offers a simple and efficient method to obtain magnetic nanomaterials without the need for high-energy inputs or toxic chemicals. The IONPs fabricated through plant-based synthesis have demonstrated promise across a range of biosensing applications, encompassing their use as magnetic resonance imaging (MRI) contrast agents, drug delivery systems, and tools for environmental monitoring. The capability to functionalize the surface of IONPs with specific ligands or biomolecules enhances their versatility and enables targeted detection of analytes in complex biological samples.

11.3.6 Nanocomposite-Based Materials

Nanocomposites for biosensing refer to hybrid materials that incorporate nanoparticles or nanomaterials into a matrix to create a sensing platform with enhanced properties and capabilities for detecting specific biological analytes or events. These nanocomposites have garnered significant attention in the field of biosensing due to their unique characteristics. Bharath et al. synthesized ternary nanocomposite $Au@\alpha\text{-}Fe_2O_3@RGO$ using chlorophyll as both the reducing agent and stabilizer. The nanohybrid exhibits remarkable catalytic efficiency and chemical electrode stability, demonstrating exceptional selectivity toward caffeic acid when utilized as a biosensor in complex coffee samples (Bharath et al. 2018). Pramila et al. achieved the successful synthesis of a multifunctional heterogeneous photocatalyst comprising nanoparticles of bismuth tungstate (Bi2WO6). During this procedure, jackfruit extract served as the fuel, playing a pivotal role in enabling the formation and

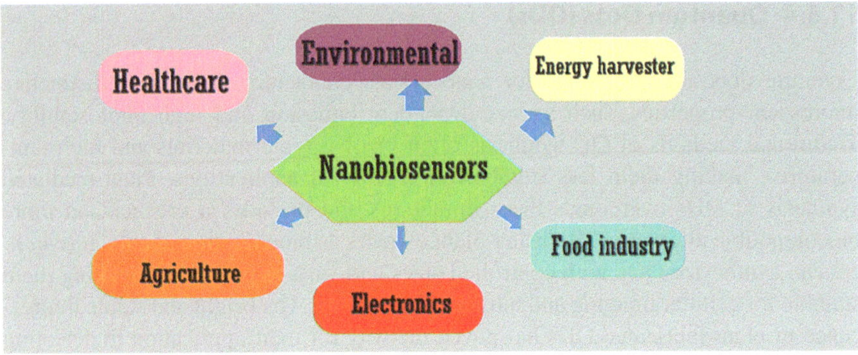

Fig. 11.2 Application of nanobiosensors

growth of the nanoparticles. The resulting Bi_2WO_6 nanoparticles hold potential for various applications due to their unique photocatalytic properties, supercapacitor capabilities, biosensor functionality, and antioxidant activity (Pramila et al. 2022).

11.4 Applications of Plant-Mediated Nanomaterials in Biosensing

Biosensors offer affordable, user-friendly, and precise detection tools suitable for a wide range of research and commercial purposes. Several illustrative applications are outlined in Fig. 11.2.

11.4.1 Environmental Sensing

Scientists are actively working on developing sensors to detect even trace amounts of pathogens by harnessing the unique characteristics of nanostructured materials, aiming to eliminate the presence of harmful chemicals. Environmental monitoring focuses on sensing and identifying pollutants in the air and wastewater, making use of advanced sensor technologies (Moraskie et al. 2021). Plant-fabricated nanomaterials have demonstrated promise in environmental sensing and monitoring. They can be employed to detect pollutants, heavy metals, and toxins in soil, water, and air samples. The use of green biosensing technologies in environmental monitoring ensures more sustainable practices and contributes to safeguarding ecosystems and public health. Biosensors find extensive use in assessing the quality of air and water. These devices are capable of detecting traces of organophosphates from pesticides in the environment and gauging the toxicity levels of wastewater. They play a vital role in environmental monitoring, helping to identify and mitigate potential pollutants, thereby contributing to the preservation and protection of ecosystems and public health.

11.4.2 Biomedical Sensing

Plant-mediated nanomaterials find extensive biomedical applications, especially in the domains of diagnostics and therapeutics. In biosensing, these nanomaterials can be harnessed for the detection of disease-specific biomarkers, thereby facilitating early diagnosis and personalized treatment. For example, functionalized gold nanoparticles have been used in rapid and sensitive tests for infectious diseases, cancer biomarkers, and genetic disorders. Furthermore, biofabricated nanomaterials can serve as drug delivery carriers, enhancing drug solubility, stability, and targeting capabilities. Their biocompatibility and amenability to functionalization with specific ligands render them ideal contenders for targeted drug delivery to precise cells or tissues, thereby mitigating systemic toxicity and enhancing treatment effectiveness. A widely recognized instance of a clinically utilized biosensor is the glucose monitor, employed regularly by individuals with diabetes to monitor their blood sugar levels. These devices accurately detect the concentration of glucose in undiluted blood samples, enabling convenient self-testing and monitoring that has significantly transformed diabetes management. Lin et al. developed an innovative immobilization matrix composed of silver nanoparticles, carbon nanotubes, and a chitosan film (Ag/CNT/Ch) for the creation of an exceptionally sensitive glucose biosensor. This innovative approach offers a promising platform for efficiently immobilizing enzymes, enhancing their performance, and improving the overall efficiency of the biosensor in detecting glucose levels with heightened accuracy (Lin et al. 2009).

11.4.3 Food Safety Sensing

Plant-mediated nanomaterials have been explored for food safety applications, enabling the rapid detection of foodborne pathogens and contaminants, ensuring safer food consumption. In the food industry, biosensors play a crucial role in measuring carbohydrates, alcohols, and acids as part of quality control processes (Thakur et al. 2018). Additionally, they are instrumental in monitoring fermentation during the production of beverages like beer, yogurt, and soft drinks. Another significant application involves the detection of pathogens in fresh meat, poultry, or fish, ensuring food safety and preventing potential health hazards.

11.4.4 Agriculture Sensing

Biosensors play a crucial role in modern agriculture, offering innovative ways to monitor and improve various agricultural processes. These devices utilize biological components, such as enzymes, antibodies, or living cells, to detect and quantify specific target compounds. They have the potential to revolutionize the agricultural industry by providing real-time, accurate, and cost-effective information to farmers and agronomists. Here are some key applications of biosensors in agriculture:

Soil health and nutrient monitoring: Biosensors can assess soil quality by measuring key parameters like pH, moisture, and nutrient levels (nitrogen, phosphorus, potassium). This data helps farmers make informed decisions about fertilization and irrigation, optimizing crop yield and reducing environmental impacts.

Crop disease and pest detection: Biosensors have the capability to identify the presence of pathogens such as bacteria, viruses, and fungi, as well as pests in crops, even before visible symptoms manifest. Early detection facilitates prompt interventions and precise treatments, ultimately diminishing the reliance on broad-spectrum chemical pesticides (Thakur et al. 2018).

Water quality monitoring: Biosensors can be used to assess the quality of irrigation water and detect contaminants that may affect crop health. This ensures that crops are not exposed to harmful substances, preventing potential yield losses and food safety issues.

Detection of toxins and pollutants: Biosensors can identify harmful substances, such as mycotoxins in grains or heavy metals in soil, ensuring that agricultural products meet safety standards and regulations (Thakur et al. 2018).

Monitoring plant growth and development: Biosensors can track plant growth parameters like hormone levels, enabling precise timing for pruning, harvesting, and other management practices.

Precision agriculture and smart farming: Integrating biosensors with other technologies like GPS and remote sensing allows for the implementation of precision agriculture techniques. This means applying inputs only where and when they are needed, reducing waste, and optimizing resource use.

GMO detection and compliance: Biosensors can aid in the identification and quantification of genetically modified organisms (GMOs) in crops, helping farmers adhere to labeling regulations and consumer preferences.

On-site food quality and safety assessment: Biosensors enable rapid on-site testing of food products for contaminants, allergens, and spoilage indicators, enhancing food safety and quality control.

Livestock health monitoring: Biosensors can be used to monitor the health and well-being of livestock by detecting specific biomarkers related to diseases or stress.

Environmental monitoring: Biosensors are valuable tools for assessing the impact of agricultural practices on the environment, helping farmers adopt sustainable practices and comply with environmental regulations.

The adoption of biosensors in agriculture is continuously evolving, and ongoing research and technological advancements are expected to further enhance their capabilities and make them more accessible to farmers worldwide.

11.5 Conclusion and Future Prospects

The utilization of plant for fabrication of nanomaterials has emerged as an innovative and environmentally sustainable approach within the field of biosensing. The distinctive characteristics of these biologically fabricated nanomaterials have paved the way for the development of sensitive, selective, and economical biosensors, spanning applications from environmental monitoring to personalized healthcare. As research in this field continues to progress, it holds the promise of revolutionizing the biosensing landscape, addressing global challenges, and contributing to a sustainable future. Continued research in plant-mediated synthesis is expected to lead to improved methods for nanoparticle fabrication. As the field advances, we can anticipate the development of more efficient and sensitive biosensors for diverse applications in healthcare, environmental monitoring, food safety, and more.

References

Abegunde SM, Idowu KS, Sulaimon AO (2020) Plant-mediated iron nanoparticles and their applications as adsorbents for water treatment–a review. J Chem Rev 2:103–113

Bharali B, Chetia H, Kalita JJ, Mosahari PV, Chhillar AK, Bora U (2019) Engineered nanomaterials in plants: sensors, carriers, and bio-imaging, 1st edn. Elsevier B.V

Bharath G, Alhseinat E, Madhu R, Mugo SM, Alwasel S, Harrath AH (2018) Facile synthesis of Au@α-Fe2O3@RGO ternary nanocomposites for enhanced electrochemical sensing of caffeic acid toward biomedical applications. J Alloys Compd 750:819–827. Elsevier B.V

del Carmen Travieso Novelles M, Ortega AR, Pita BA, López MC, Pérez LD, Medina EA, Pérez OP (2021) Biosynthesis of fluorescent silver nanoparticles from Leea coccinea leaves and their antibacterial potentialities against Xanthomonas phaseoli pv phaseoli. Bioresour Bioprocess 8. Springer Singapore

Jacob JM, Rajan R, Tom TC, Kumar VS, Kurup GG, Shanmuganathan R, Pugazhendhi A (2019) Biogenic design of ZnS quantum dots—insights into their in-vitro cytotoxicity, photocatalysis and biosensing properties. Ceram Int 45:24193–24201. Elsevier Ltd and Techna Group S.r.l.

Lin J, He C, Zhao Y, Zhang S (2009) One-step synthesis of silver nanoparticles/carbon nanotubes/chitosan film and its application in glucose biosensor. Sensors Actuators B Chem 137:768–773

Moraskie M et al (2021) Microbial whole-cell biosensors: current applications, challenges, and future perspectives. Biosens Bioelectron 191:113359

Naaz R, Siddiqui VU, Ahmad A, Qadir SU, Siddiqi WA (2023) Study of antibacterial and antioxidant activities of silver nanoparticles synthesized from Tradescantia pallida (purpurea) leaves extract. J Dispers Sci Technol:1–11. Taylor & Francis

Pramila S, Ranganatha VL, Soundarya TL, Ramu R, Nagaraju G, Mallikarjunaswamy C (2022) Eco-mediated synthesis of visible active Bi2WO6 nanoparticles and its performance towards photocatalyst, supercapacitor, biosensor, and antioxidant activity. J Clust Sci 33:2233–2248. Springer US

Shahzadi I, Aziz Shah SM, Shah MM, Ismail T, Fatima N, Siddique M, Waheed U, Baig A, Ayaz A (2022) Antioxidant, cytotoxic, and antimicrobial potential of silver nanoparticles synthesized using Tradescantia pallida extract. Front Bioeng Biotechnol 10:1–13

Sharanappa A, Shet AR, Patil LR, Hombalimath VS, Kadapure S (2020) Biosynthesis of silver nanoparticles using citrus sinensis peel extract and their application as antibacterial agent. Int J Res Pharm Sci 11:4726–4732

Srivastava S, Pandey A (2021) Syngonium podophyllum leaf extract mediated synthesis and characterization of gold nanoparticles for biosensing potential: a sustainable approach. Curr Nanosci 17:81–89

Sudheer Kumar KH, Yashaswini YHJ, Pratibha S, Hareesh K, Manohara SR (2022) Entada Gigas seeds mediated synthesis of carbon for dielectric and sensing applications. Sens Int 3:100162, The Authors

Thakur S, Thakur S, Kumar R (2018) Bio-nanotechnology and its role in agriculture and food industry. J Mol Genet Med 12:1–5

Varghese Alex K, Tamil Pavai P, Rugmini R, Shiva Prasad M, Kamakshi K, Sekhar KC (2020) Green synthesized ag nanoparticles for bio-sensing and photocatalytic applications. ACS Omega 5:13123–13129

Zaytseva O, Neumann G (2016) Carbon nanomaterials: production, impact on plant development, agricultural and environmental applications. Chem Biol Technol Agric 3:1–26. Springer International Publishing

Biotechnology for Gas Sensing

12

M. Gorshkov, A. Gorshkova, Prachi Sharma, Prabhash Mishra, V. Pavelyev, and Nishant Tripathi

Abstract

In this chapter, the main operation mechanisms of the gas-sensitive elements made on the basis of biotechnologies are briefly considered. Enzymatic gas biosensors, antibody gas biosensors, aptamer gas biosensors, molecularly imprinted polymer sensors, and receptor gas biosensors are covered. The main recognition methods of the input signal are also considered, examples of transducers are given, the advantages and disadvantages of using various biotechnologies as a part of biosensors are discussed. The possibility of new materials (such as carbon nanotubes, graphene, etc.) used in the role of transducers is highlighted separately as well. The prospective development directions of new generation sensors are considered.

Keywords

Nanomaterials · Gas sensor · Biosensor

M. Gorshkov · A. Gorshkova · N. Tripathi (✉)
Samara National Research University, Samara, Russia

P. Sharma
Samara National Research University, Samara, Russia

School of Electronics Engineering (SENSE), Vellore Institute of Technology (VIT), Vellore, India

P. Mishra
Samara National Research University, Samara, Russia

Department of Applied Science & Humanities, Faculty of Engineering and Technology, Jamia Millia Islamia (Central University), New Delhi, India

V. Pavelyev
Samara National Research University, Samara, Russia

IPSI RAS – Branch of the FSRC "Crystallography and Photonics" RAS, Samara, Russia

M. Mohsin, N. Soleja (eds.), *Nanoscale Sensors and their Applications in Biomedical Imaging*, https://doi.org/10.1007/978-981-97-3144-2_12

12.1 Introduction: Stepping into the History of Biogas Sensors

There are many applications of biological science that have been invented in the ancient time, such as agriculture, fermentation, medicine, breeding, genetics, cell biology, DNA recombination, immunity research, development of organic polymers, and much more (Yerlanova et al. 2021). By that time, biological science began to be introduced in many countries, received an even wider scope of applications, and became an interdisciplinary subject. There are so many ideas of the industrial bio-future such as biocomputers etc. have been reported by Prasad et al. (Prasad et al. 2018).

For example, research in the field of organic chemistry has discovered the phenomenon of polymerization (Anon 2014). Polymerization is the formation of a high molecular weight substance or polymer. A polymer consists of monomeric units or monomers connected to each other by the covalent or non-covalent chemical bonds. The prerequisites for the fabrication of polymers were the need to replace natural materials with their cheaper counterparts with similar physical and chemical properties. For example, a material that can conduct electric current and has high strength as well as durability and will not corrode and decompose. Such characteristics are primary requirement for the manufacture of any measuring instruments. Often, the main material of any sensor is metal or semiconductor, which imposes some restrictions on the design. It is known that metals are subject to corrosion, heating. Moreover, they have low elasticity. Therefore, metals have short life time, while electrically conductive polymers are essentially plastics and can combine the mechanical properties of plastics, such as flexibility, strength, malleability, plasticity with high electrical conductivity. The impetus for the study of this direction was that, depending on the synthesis conditions, organic polymers can exhibit superconductivity at significantly higher temperatures than similar metal conductors. In 2000, Alan J. Heeger, Alan McDiarmid, and Hideki Shirakawa were awarded the Nobel Prize in chemistry for the discovery and development of conductive polymers.

Although the study of superconductivity of polymers has not found practical application, the progress in the direction of conductive polymers has contributed to the establishment of other industries. The developments used in the field of conductive polymers allowed H. Koezuka et al. (in 1986) to present the world's first organic field-effect transistor made on the basis of polythiophene (Tsumura et al. 1986). Polythiophene in this scheme is a p-type conductor and is applied in a thin layer between the drain and the source. A year later in 1987, W. Ching et al. presented an organic diode made by vacuum spraying of a thin layer of diamine and tris(8-hydroxyquinolinato)aluminum (Tang 1987), which formed a whole layer of OLED display development. It is a well-known fact that organic chemistry is mainly represented by carbon compounds; and carbon has the ability to form the branched chains of various configurations with different properties. As guided by the rules formed by Butlerov (Butlerow 1861; Бутлеров 1862), it is possible to model such a material that will in the future have such characteristics that will be required to solve a specific problem. The mentioned property has found extremely wide application in the microelectronics section studying chemical sensors and sensor systems. The fact is that the main task of any sensor is to extract the necessary information from the incoming signal to the input not only with high accuracy but also with high selectivity. The signal in this case can be charge, current, potential, voltage, electric field (amplitude, phase, polarization, spectrum), conductivity, permittivity. Classical semiconductor sensors identify a substance by measuring its electrical potential or charge, which significantly limits the range of measured substances (Fraden 2016). Chemical sensors are able to detect a substance based on its chemical properties. For example, whether a substance is an oxide or a base, whether it exhibits metallic or non-metallic properties, etc. To do this, the measured substance must send a signal to the input of the sensor transducer, which will uniquely identify the substance in the volume of the measured mixture. To do this, the substance, for example, is subjected to a redox reaction near the surface of the sensitive layer, which in this case acts as either a donor or an electron acceptor. By measuring the change in the resistance of the sensitive layer, it is possible to identify the substance if you know in advance how many electrons will be required for the redox reaction. Sometimes a catalyst is added to this scheme, which significantly accelerates the reaction and reduces the detection time (Fraden 2016). In the updated version of redox reaction based sensors, researchers used to detect heat produced by redox reaction by using a heat detector. Such sensors are also called thermistors or calorimetric sensors. Their essence lies in the fact that the sensitive layer is covered with a catalyst that accelerates the reaction process. The heat generated changes the resistance of the sensitive layer, which in turn can be measured. In sensors based on organic materials, the polymer usually performs the function of a sensitive layer and the function of a catalyst. In addition, polymers are additionally doped to increase selectivity, so the addition of acid groups increases selectivity to bases, amino groups to acids, etc.

Sometimes we need to determine a substance that is largely identical to another one but is different by, for example, one radical or is isomer. In biochemistry and molecular biology, there is such a thing as complementarity. Complementarity is the property of biopolymers to form intermolecular bonds between spatially complementary fragments of molecules through supramolecular interactions (formation of hydrogen bonds, ionic interaction, Van der Waals interaction, etc.). This property underlies cellular interaction and, in particular, immunity. The basis of any cellular interaction is the reaction of the cellular receptor to changes occurring in the external environment. The receptor molecule binds complementarily to a molecule of a certain chemical that transmits an external regulatory signal and, in turn, transmits this signal inside the cell by changing the conformation of the receptor and stimulating secondary intermediaries or by depolarization of the cell membrane (Alberts et al. 2019). A substance that specifically interacts with a receptor is called a ligand. Depending on the type of receptor, there are chemoreceptors, osmoreceptors, mechanoreceptors, photoreceptors, thermoreceptors, and electroreceptors. The type of receptor depends on the type of stimulus stimulating the receptor. So chemoreceptors react to the addition of a molecule, osmoreceptor to ions that determine osmotic concentration, photoreceptors react to electromagnetic radiation in a certain wavelength range, etc. (Alberts et al. 2019). In complex multicellular animal organisms in embryonic development, cells differentiate into tissues that perform a certain highly specific function. Tissues, in turn, form the body's sensory organs (Hyttel et al. 2009). For example, olfaction is the process of stimulating the receptors of the olfactory epithelium, which excites neurons to transmit a signal to the olfactory bulb. Vision is the process of excitation of the retina, which is represented by five layers of neurons and contains specific receptors, rods and cones, which are excited by visible radiation (Carlson 2019). All cells of the body have receptors, but the cells of the sensory organs perform the most effective function. Of course, such a unique tool had to be used, which eventually led to the appearance of such devices as biosensors.

A biosensor is a device that uses a specific biochemical interaction of an analyte and a bio-recognizing sensitive layer as a recognition component. The sensing process can include the ligand attachment reaction to the receptor as well as catalytic redox reactions taking place with enzymes, immunochemical reactions or reactions taking place in organelles, cells, or in tissues. The biosensor, like any other sensor, consists of a sensitive layer and a transducer that converts the incoming signal into an electrical one. Biorecognition elements are represented proteins, enzymes, antibodies, membrane, nucleic acids, and the whole cells. Polymer films, graphene, carbon nanotubes, piezocrystals, semiconductors doped with organic molecules, etc. are used as transducers. This is due to the fact that various functional groups can be added to the surface of organic polymer materials due to functionalization, which increase the adhesion of the sensitive layer or contribute to the amplification of the output signal. It has been reported that functionalized graphene allows covalent immobilization of immunoglobulins, and graphene demonstrates high sensitivity to changes in electronegativity of adsorbed molecules, which makes this scheme quite simple and effective (Tian et al. 2018). It is well-known fact that in a living cell, signal transmission is carried out through a cascade of complex physiological intracellular processes. Therefore, after examining this process in more detail, researchers concluded that there is absolutely no need to completely repeat the entire reaction chain to form the output signal. Moreover, there is no need to use the whole receptor since a variable site responsible for complementarity is enough. For example, in biosensors in which antibodies are used as a sensitive layer, it turned out that for detection it is sufficient to use only a variable portion of the antibody, which is responsible for interaction with the detected molecule. Such molecules are called aptamers and are oligonucleotide or peptide chains specifically binding to certain target molecules (Nozaki et al. 2020). Aptamers are made by genetic engineering methods. A natural step was the invention of molecularly imprinted polymers, which are a biopolymer obtained by molecular imprinting (Chen et al. 2015). The ligand molecule is imprinted in the polymer matrix, then extracted, after which the polymer molecule self-assembles in the resulting cavity. The molecule obtained in this way completely repeats the receptor in terms of its functions and has complementarity to the target molecules. An additional advantage is that the mentioned molecule can be easily immobilized on a polymer substrate (Lowdon et al. 2020). Another important benefit of artificial receptors over receptors of natural origin is the freedom of choosing the material from which the polymer can be made. For example, in addition to proteins, the frame material can be carbon chains, fused aromatic rings, lipids, etc. Thus, stability, flexibility, and other properties can be freely modulated as per requirements. The manufacture of such devices became possible owing to the development of biochemistry, genetic engineering, and polymer chemistry.

12.2 Modern Nanobiotechnologies Used for Gas Sensing Application

In this section, we are going to cover main working principles of modern nano biogas sensors. As described above, the fundamental difference between biosensors is the presence of sensitive bio-elements in this category of sensors (Naresh and Lee 2021). Basically bio-sensitive elements are represented by enzymes, proteins, nucleic acid, cell receptor, antibodies, membranes, organelles, cells, tissue, microorganisms, and bacteriophages (Karunakaran et al. 2015). And depending on this, the classification of biosensors based on the method of recognition of the stimulus was built. Depending on the measured stimulus, sensors could be classified into acoustic, chemical, biological, electric, magnetic, optical, radiation, thermal, mechanical, etc. (Knopf and Bassi 2018). The type of transducer depends on the input stimulus. Based on this, sensors can be classified as electrochemical, optical, thermal, gravimetric, electromagnetic (see Fig. 12.1).

12.2.1 Enzymatic Gas Biosensors

The working mechanism of enzymatic gas sensors is based on the redox reaction of the analyte and the enzyme. Such reactions are accompanied by either the release of heat (calorimetric) or radiation (optical) or the presence of current carriers or an

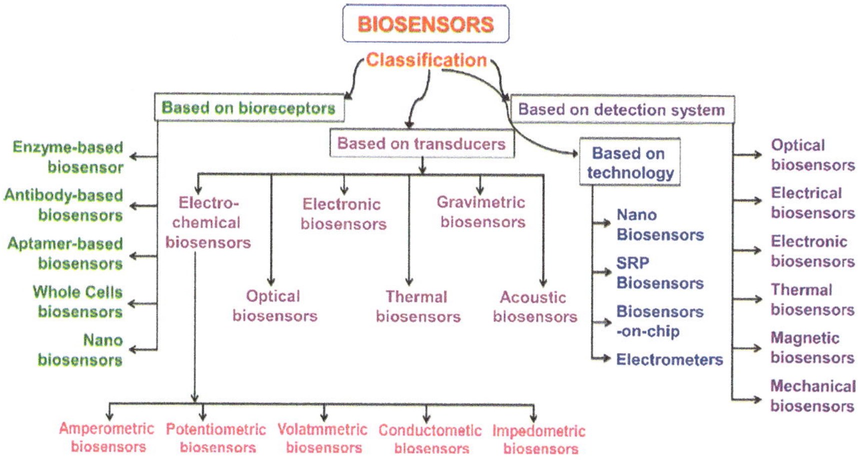

Fig. 12.1 The classification of biosensors. Reprinted from Naresh and Lee (2021)

active radical (electrochemical). The transducer is thus selected depending on the type of stimulus. Most enzyme reactions are accompanied by the release of heat in the range of 5–100 kJ/mol with respect to change in temperature in the range of 2–10 °C. The amount of heat generated correlates with the amount of the reacted substance. Devices that react to changes in heat are called thermistor or thermophile, and the measurement method is calorimetric. An example of such a device is the enzyme thermistor, which is an insulated cylinder containing a heat exchanger unit in which two electrodes are located (see Fig. 12.2a).

Out of two electrodes, one used to coat with an enzyme alcohol oxidase and the other electrode acts as a reference electrode. The sensitivity of the electrode is required as 100 mV with a temperature change of 10^{-3} °C. The concentration of the analyte being measured is 0.01–100 mM. 0.1–1 ml of the mixture is sufficient for measurement. Generally, the substance utilized for study (ethyl alcohol) is cleaved by the action of the enzyme alcohol oxidase and catalase to hydrogen peroxide with the release of heat. It has been reported the conductivity of Al made electrodes reduced with respect to increment in the temperature, which can be measured as an electrical quantity (Danielsson et al. 1988).

In addition to heat, enzymatic reactions proceed with the formation of a reaction byproduct, which can also participate in the redox reaction on the surface of the transducer. For example, the product of enzymatic oxidation of phenol is a catechol reaction product (Kaisheva et al. 1997). In highly sensitive biosensors, the working electrode consists of Ag/AgCl. Catechol, which is formed after the oxidation of phenol by the enzyme polyphenol oxidase, is oxidized to quinone using molecular oxygen (Fig. 12.2b). Quinones can be electrochemically reduced at approximately

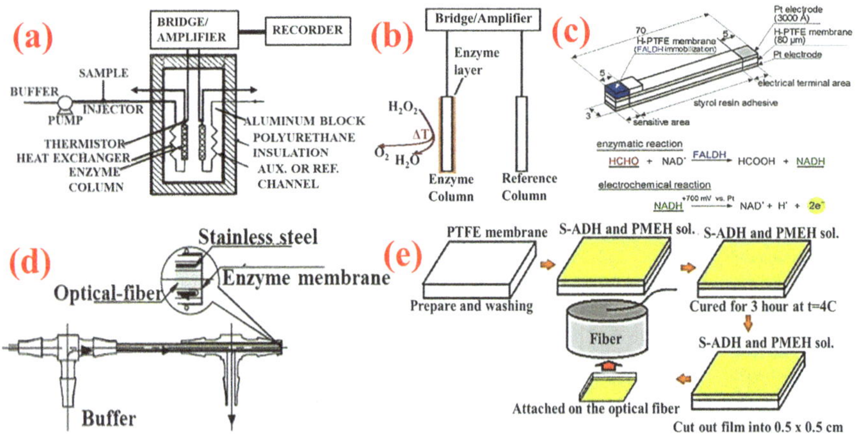

Fig. 12.2 (**a**) A calorimetric ethanol gas-phase biosensor, (**b**) a calorimetric phenol gas-phase biosensor, (**c**) a calorimetric ethanol, phenol, aldehyde coenzyme used gas-phase biosensor, (**d**) an optical ethanol used ruthenium organic complex gas-phase biosensor, (**e**) an optical ethanol used NADH gas-phase biosensor. Reprinted from Danielsson et al. (1988) and Mitsubayashi et al. (2008, 2003)

−150 mV on the electrode surface (Kaisheva et al. 1997). The change in charge carriers is recorded.

A classification of this sensor can be called as a "biosensor based on the reaction of the analyte with coenzyme nicotinamide adenine dinucleotide." This coenzyme in the body is a substrate for the implementation of redox reactions in the cell and is capable of oxidizing substances such as ethanol, phenol, aldehyde in the presence of the corresponding enzyme alcohol dehydrogenase, formaldehyde dehydrogenase, and aldehyde dehydrogenase (Fig. 12.2c). During the oxidation of the substance under study (ethanol, phenol, aldehyde), nicotinamide adenine dinucleotide is reduced to its reduced form, and then oxidized back at the electrode, while losing two electrons and a proton, changing the conductivity of the electrode (Mitsubayashi et al. 2008). The sensitivity of this sensor has been found as 40–3000 ppb.

Enzyme sensors also belong to biofluorometric sensors. Biofluorometry is a method for determining the concentration of a substance by the fluorescence intensity of an organic compound. For example, an optical biosensor for registering gaseous ethanol is reported. The essence of the method is to apply oxygen-sensitive ruthenium organic complex to the substrate (substrate or optical fiber). Also, the enzyme alcohol oxidasa is applied to the sensitive part of the sensor, which oxidizes ethanol to form aldehyde and hydrogen peroxide. Peroxide reacting with ruthenium organic complex causes it to fluoresce in the presence of oxygen with a wavelength of 600 nm (Fig. 12.2d). The sensitivity of this sensor has been reported as 0.71–51.49 ppm (Mitsubayashi et al. 2003). An improved version of this scheme is the use of nicotinamide adenine dinucleotide (NADH) instead of ruthenium organic complex, which, by directly reacting with ethanol, is able to fluoresce with a wavelength of 490 nm (Fig. 12.2e) (Ye et al. 2017).

12.2.2 Antibody Gas Biosensors

An antibody or immunoglobulin is a protein molecule consisting of two light fragments and one heavy fragment that performs an immune function in a multicellular organism by recognizing a specific antibody and binding to it (Schroeder and Cavacini 2010). The recognition mechanism consists in a special spatial-structure of the light chain of the antibody. Antibody considers as a lock which can attach with the antigen. Antigen acts a key which repeats the shape of the light chain of the antibody (Janeway Jr et al. 2001). There are specific antibodies for detecting the gas molecules. In such sensors, the signal detection mechanism can be optical and gravimetric (Naresh and Lee 2021).

Optical sensors occupy a fairly wide range of various devices operating on the optical principle of signal detection. Such sensors include the chemiluminescence biosensors, surface plasmon resonance biosensors (Naresh and Lee 2021).

Chemiluminescent sensors use fluorescent dye-labeled antibodies. The fluorescence resonance energy transfer to the antigen acts as an acceptor. An example of such a sensor is a gas-solid biosensor. This biosensor is used to recognize airborne protein antigen. This category includes specific proteins, allergens, spores, etc. To manufacture the affinity purified monoclonal anti-GST sensor, fluorophore-labeled europium(lll) trisbipyridine cryptate was deposited on a polypropylene electret non-woven filter, which was then placed in a gas installation in which a gas mixture was supplied for 2 s at a temperature of 80 °C. The air flow was 3 l/min. In the process of feeding the air mixture, the antibody specifically interacted with the antigen. After that, the filter is placed in the EnVision microplate reader. Fluorescence resonance energy from the fluorophore donor is transferred to the fluorescence resonance energy acceptor (Fig. 12.3a). The intensity of fluorescence indicates the quantitative composition of the analyte (Iwanaga et al. 2009).

Another type of optical sensor is a sensor based on surface plasmon resonance. An example of a surface plasmon resonance biosensor is a sensor detecting gaseous trinitrotoluene (TNT). The sensor is a multilayer structure made of a glass substrate, a silane-thiol layer, and a metal layer of Ag or Au to excite the surface plasmon. The sensitive layer is a monoclonal antibody for TNT which is covalently bound to 11,119-dithio-bis(succinimidyl undecanoate) (DSU) (Fig. 12.3b). The measurement takes place using surface-enhanced resonance Raman (SERS). The sensitivity of this sensor is about 730 ppt (Bowen et al. 2003).

A fairly common type of antibody sensors are gravimetric or mass-sensitive sensors. The sensitive element of a gravimetric sensor is a piezoelectric sensor. The

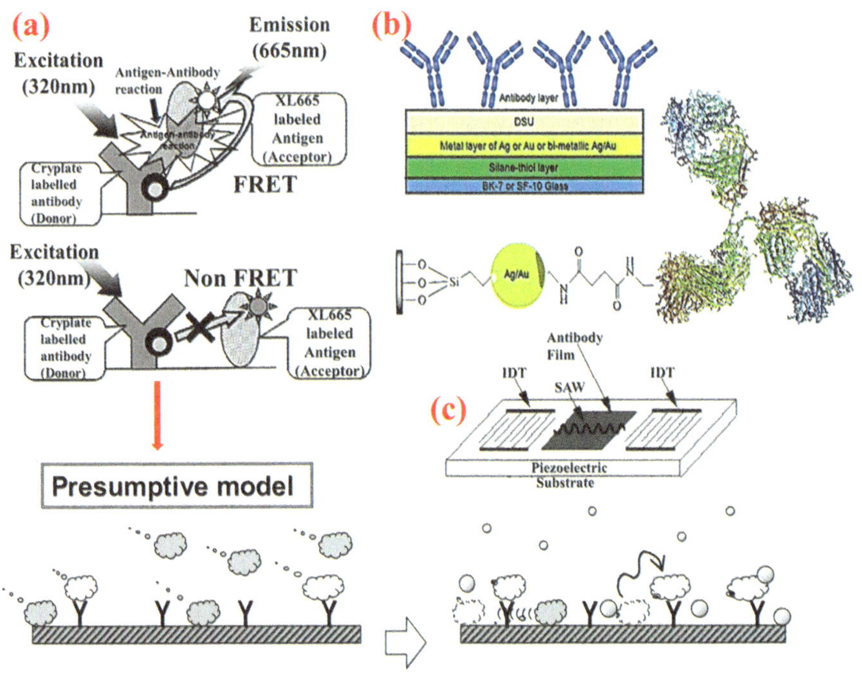

Fig. 12.3 (**a**) A chemiluminescent sensor based monoclonal anti-GST, (**b**) an optical TNT sensor based antibody, (**c**) a gravimetric sensor based antibody. Reprinted from Iwanaga et al. (2009) and Stubbs et al. (2002)

action of such sensors is based on the static bending of the piezocrystal surface, which is proportional to the mass of the analyte (Fig. 12.3c). The piezocrystal is covered with a sensitive layer that selectively reacts with the test substance (Stubbs et al. 2002).

12.2.3 Aptamer Gas Biosensors

Aptamers are oligonucleotide or peptide molecules that are used for attaching target molecules. Aptamers are manufactured synthetically from large arbitrary libraries by the SELEX method (Stoltenburg et al. 2012). Aptamers mainly copy the functions of receptors, antibodies, DNA components, complementary proteins sections, etc. The stimulus formation options in sensors based on aptamers are in many ways similar to those ones based on antibodies. The examples of such sensors are: optical, gravimetric, and electrochemical sensors. Unlike antibody sensors, electrochemical gas sensors are in many ways more common due to easier signal recognition mechanisms and higher cost-efficiency.

For example, there is a chemoresistive biosensor that detects the concentration of the explosive 2,4-dinitrotoluene (DNT) in the air (Stoltenburg et al. 2012). For selective recognition, a peptide is being made from the following amino acid sequences: His-Pro-Asn-Phe-Ser-Lys-Tyr-IleLeu-HisGln-Arg-Cys. The transducer in this scheme is reduced graphene oxide (rGO). This peptide is covalently attached via 1-ethyl-3-(3-dimethyla-minopropyl)carbodiimide and N-succinimide (EDC/NHS) to the mentioned transducer. Reduced graphene oxide, due to its electronic properties, reacts very fast to the various groups of substances by the electron transfer, which affects the conductivity of the transducer. When DNT is attached to the aptamer, its NO_2 group strongly withdraws electrons, which increases the resistance of the rGO surface. Such sensor had response of 2.426 ppb for DNT (Stoltenburg et al. 2012).

A DNA aptamer with the sequence 5′-CGACCAGCTCATTCCTCAGGAGA-AACATGGAGTCTCGATGATAGTAGGAGCGGCGGAACGTAGGAAGAGAG-GATGACGGAGGATCCGAGCTCACCAGTC-3′ (Van_74) can be synthesized from a fragment of nucleic acid unlike peptide aptamers. This aptamer is used to bind vanillin on the surface of the ISFET sensor. Vanillin binds complementarily to the oligopeptide chain near the −OCH3, −CHO, −OH groups and the benzene ring and pulls the electrons toward itself, resulting in an increase in the surface potential of the FET surface (Fig. 12.4). The change in the surface potential can be measured. The sensitivity of such a sensor is 2.7 ppt for vanillin vapor (Kuznetsov et al. 2017).

12.2.4 Molecularly Imprinted Polymer Sensors

Molecularly imprinted polymer is a biopolymer manufactured by molecular imprinting (Lowdon et al. 2020). Molecular imprinting is a method of obtaining "molecular imprints" based on the polymerization of functional monomers in the presence of specially introduced target template molecules. The manufacturing method includes three stages. In the first step, the monomer and the

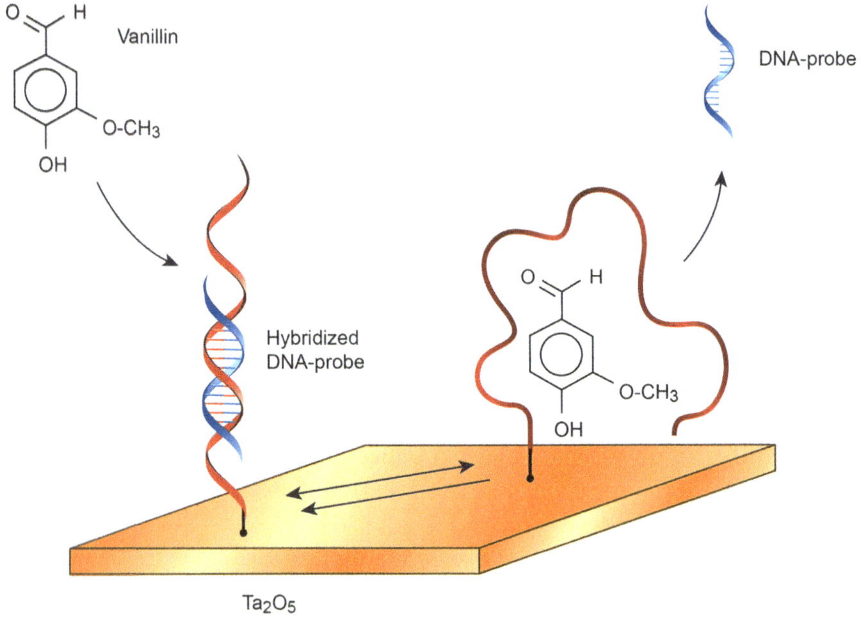

Fig. 12.4 The aptamer gas biosensors and the mechanism of complimentary interaction of DNA aptamer with the vanillin. Reprinted from Kuznetsov et al. (2017)

molecular template are mixed in a suitable solvent before polymerization begins. In this case, the so-called pre-organization occurs, i.e., the formation of a stable pre-polymerization complex of monomer molecules and a template substance, owing to which the functional monomer molecules are being arranged and fixed in a certain way around the template molecule. At the second stage, as a result of polymerization of pre-polymerization complexes with a large excess of cross-linking agent, a strongly crosslinked polymer with a rigid structure is being formed. At the third stage, the resulting polymer is being crushed, sieved to the desired particle size and the molecular template is being removed from the polymer mesh. As a result, cavities are being formed in the material. These cavities interact with target template molecules or compounds close to them in structure. Thus, the sensitivity of molecular imprinted polymers is based on the shape correspondence of the prints and specific functional groups within them to the template molecules. Depending on the bond type (covalent or non-covalent) between a monomer and template molecules, covalent and non-covalent imprinting are distinguished. In accordance with the type of signal recognition, MIP sensors can be gas biosensors of antibodies and gas biosensors of aptamers. MIPs are used in the GUI. MIP sensors consist of a piezoelectric quartz crystal coated with a polymer with an imprint of 2-meth-ilizoborneol (MIB). This MIP-based QCM sensor is capable of measuring geosmin at a concentration of 200 micrograms/l (Ji et al. 1999). Such a sensor can detect formaldehyde as well (Feng et al. 2005).

Optical sensors have a huge variety of constructions. The optical fiber is being coated with a MIP layer using electropolymerization technique. Such a sensor is capable of detecting formaldehyde at a concentration of 1 ppm (Fig. 12.5) (Gonzalez-Vila et al. 2017).

12.2.5 Receptors of Gas Biosensors

A receptor is a sensory ending of a neuron or a specialized cell capable of perceiving stimuli from the external and internal environment. Depending on the type of receptor, we can distinguish the following: chemoreceptors, osmoreceptors, mechanoreceptors, photoreceptors, thermoreceptors, electroreceptors. As a result of evolution, each type of receptor has been adapted to a certain type of stimulus. That is why chemoreceptors react to changes in the external and internal environment, photoreceptors react to electromagnetic radiation in a certain wavelength range, thermoreceptors react to temperature changes, etc. Each of the receptors reacts to a certain group of stimuli. The most differentiated receptors have maximum selectivity to external stimuli. This is called "olfactory system." The olfactory system is represented by the olfactory epithelium. The

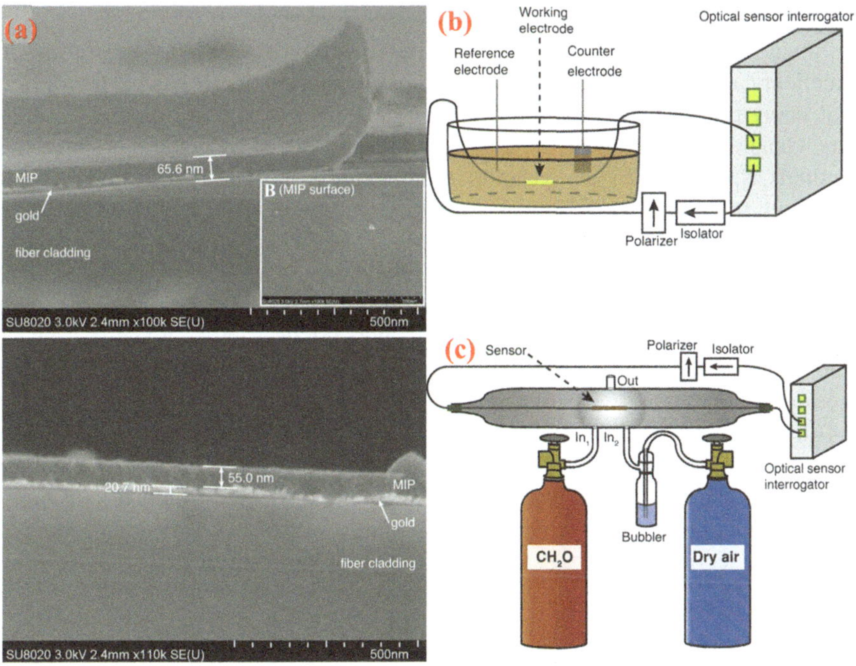

Fig. 12.5 A molecularly imprinted polymer sensor. (**a**) The MIP-coated optical fiber: A—the structure of fiber, B—detail of the coating surface, (**b**) the preparation for polymerization stage, (**c**) the formaldehyde gas detection process. Reprinted from Gonzalez-Vila et al. (2017)

olfactory epithelium cells are special olfactory receptors, which belong to the group of heptaspiral receptors. They have seven transmembrane alpha-helical domains in the structure. The signaling function of this structure is provided by conjugation with heterotrimeric G-proteins (G-protein coupled receptors—GPCR). Olfactory receptors were first identified in a rat olfactory epithelium in 1991 with the help of molecular cloning. In this experiment, researchers used a reverse transcriptase PCR reaction with degenerate primers constructed on the basis of conservative sites identified by GPCR receptor sequences. Olfactory receptor genes have the largest family in mammalian genomes, the number of which varies in different species (Buck and Axel 1991; Glusman et al. 2001; Zhang and Firestein 2007). The amino acid sequence of the synthesized receptor includes 320 ± 25 residues. The difference in residues length is mainly attributed to the consequence of the receptors N- and C-ends.

The mechanism of smell is quite complex. The odorant molecule interacts complementarily with the chemoreceptor located in the olfactory epithelium in accordance with the key-lock principle. After initiation, the receptor conformation starts to change. The receptor opens a site for the attachment of a mediator protein called G-protein and adenylate cyclase and triggers the production of cyclic AMP (cAMP). Sodium channels are being activated in this process. The receptor membrane is being depolarized due to the movement of sodium ions into the cell. The potential difference of the membrane increases and an action potential which excites the neuron occurs.

Since the main target cells are volatile compounds, olfactory receptors are excellent candidates for the manufacture of gas biosensors. It has been reported that many types of biosensors have been developed based on olfactory receptors. Generally, Olfactory receptors used to be produced by genetic engineering methods. Specific OR gene is included into expression vectors (plasmid or virus designed for gene expression in cells). Then the vectors are transfected to expression cells, which results in the expression of ORs in the cells (Fig. 12.6a, b). The receptor thus obtained is immobilized on the surface of the transducer (Fig. 12.6c, d). For example, the olfactory receptor of a meat fly is able to detect 1,4-diaminobutane, butanoic acid, and 1-hexanol in a chemoresistive sensor. The odorant deposited on the receptor causes a potential change that can be measured. The sensitivity of such a sensor is of few ppb to 100 ppm for 1,4-diaminobutane odor, 20–200 ppm for butanoic acid odor, and 8–500 ppm for 1-hexanol (Pietrantonio et al. 2013). The combination of different olfactory biosensors leads to the emergence of a new type of device—an artificial nose. This instrument detects a wide range of volatile substances with high accuracy and high selectivity.

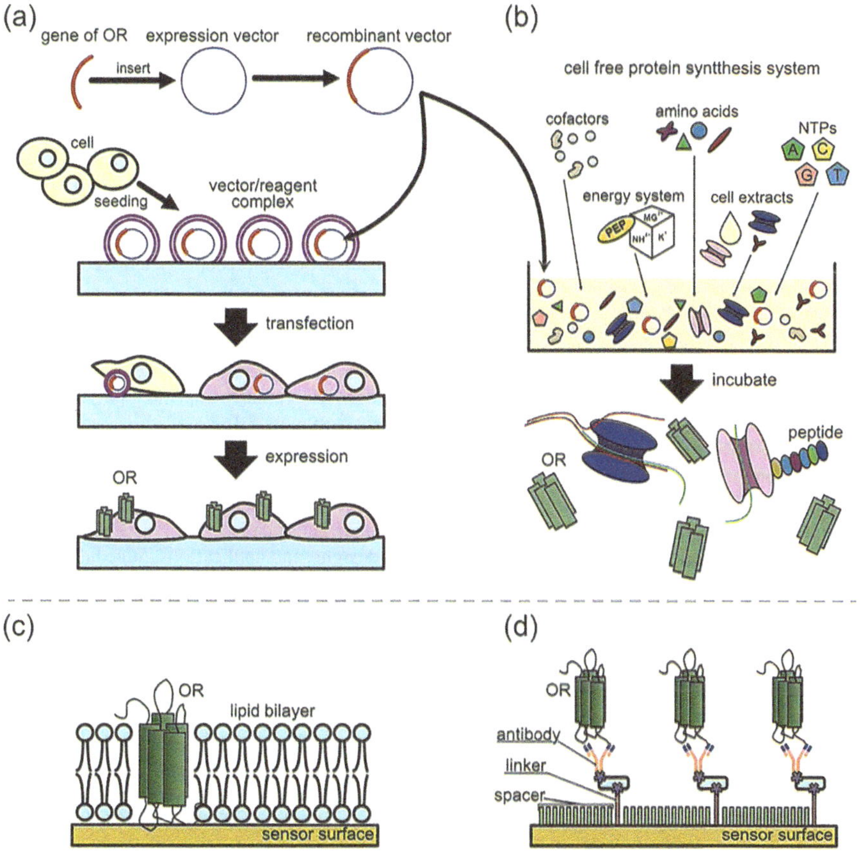

Fig. 12.6 Steps to prepare olfactory based biosensors: (**a**) the synthesis of ORs receptors in bacterial cells, (**b**) the free-cell ORs receptor synthesis, (**c**) a non-covalent adsorption ORs on transducer, (**d**) a covalent adsorption ORs on transducer. Reprinted from Mitsubayashi et al. (2022)

12.3 Conclusion

Sensorics, and especially biosensorics, is a rapidly developing industry that has every prospect of moving to the next stage of development in the future. The study of wildlife opens a new technological branch—synthetic biochemistry. Owing to discoveries in the field of genetic engineering and organic chemistry, we already can synthesize artificial antibodies and artificial receptors. The combination of various sensitive elements makes it possible to manufacture more complex gas analyzers with high accuracy, selectivity, and small dimensions. The needs of the rapidly developing industry for compact and cheap methods of group and personal security can be met in the near future.

References

Alberts K, Hopkin A, Johnson D, Morgan M, Raff K, Roberts PW (2019) Essential cell biolog, 5th edn. W.W. Norton & Company, New York

Anon (2014) "Polymerization". The IUPAC compendium of chemical terminology. doi:https://doi.org/10.1351/goldbook.P04740

Bowen J, Noe L, Sullivan B, Morris K, Martin V, Donelly G (2003) Gas-phase detection of trinitrotoluene utilizing a solid- phase antibody immobilized on a gold film by means of surface plasmon resonance spectroscopy. Appl Spectrosc 57:8

Buck L, Axel R (1991) A novel multigene family may encode odorant receptors: a molecular basis for odor recognition. Cell 65:175–187

Butlerow (1861) Einiges über die chemische Structur der Körper (A. Butlerov, some facts about the chemical structure of substances). Zeitschrift für Chemie und Pharmacie : magazin 4:549–560

Carlson MB (2019) Human embryology and developmental biology, 6th edn. Elsevier

Chen W, Ma Y, Pan J, Meng Z, Pan G, Sellergren B (2015) Molecularly imprinted polymers with stimuli-responsive affinity: progress and perspectives. Polymers 7:1689–1715

Danielsson B, Mosbach K, Winquest F, Lundstrom I (1988) Biosensors based on thermistors and semiconductors and their bioanalytical applications. Sensors Actuators 13:139–146

Feng L, Liu Y, Zhou X, Hu J (2005) The fabrication and characterization of a formaldehyde odor sensor using molecularly imprinted polymers. J Colloid Interface Sci 284:378–382

Fraden J (2016) Handbook of modern sensors physics, designs, and applications, 5th edn. Springer International Publishing Switzerland

Glusman G, Yanai I, Rubin I, Lancet D (2001) The complete human olfactory subgenome. Genome Res 11:685–702

Gonzalez-Vila, Debliquy M, Lahem D, Zhang C, Megret P, Caucheteur C (2017) Molecularly imprinted electropolymerization on a metal-coated optical fiber for gas sensing applications. Sens Actuators B Chem 244:1145–1151

Hyttel P, Sinowatz F, Vejlsted M, Betteridge K (2009) Essentials of domestic animal embryology, 1st edn. Saunders Ltd.

Iwanaga H, Tsuzuki H, Kamiyama Y, Ueda H (2009) Verification of a specific reaction between an airborne antigen and an immobilized antibody at a gas-solid interface. Anal Sci 25:1101–1106

Janeway Jr, Travers P, Walport M et al (2001) Immunobiology: the immune system in health and disease, 5th edn. Garland Science, New York

Ji H-S, McNiven S, Ikebukuro K, Karube I (1999) Selective piezoelectric odor sensors using molecularly imprinted polymers. Anal Chim Acta 390:93–100

Kaisheva, Iliev I, Christov S, Kazareva R (1997) Electrochemical gas biosensor for phenol. Sensors Actuators B 44:571–577

Karunakaran, Bhargava K, Benjamin R (2015) Biosensors and bioelectronics. Elsevier Science

Knopf G, Bassi AS (2018) Smart biosensor technology, 2nd edn. CRC Press Taylor & Francis Group, Boca Raton

Kuznetsov N, Komarova M, Andrianova V, Grudtsov E (2017) Kuznetsov, Aptamer based vanillin sensor using an ion-sensitive field-effect transistor. Mikrochim Acta 185(1):3

Lowdon J, Dilien H, Singla P, Peeters M, Cleij T, van Grinsven B, Eersels K (2020) MIPs for commercial application in low-cost sensors and assays—an overview of the current status quo. Sens Actuators B Chem 325:128973

Mitsubayashi K, Kon T, Hashimoto Y (2003) Optical bio-sniffer for ethanol vapor using an oxygen-sensitive optical fiber. Biosens Bioelectron 19:193–198

Mitsubayashi K, Nishio G, Sawai M, Saito T, Kudo H, Saito H, Otsuka K, Noguer T, Marty J (2008) A bio-sniffer stick with FALDH (formaldehyde dehydrogenase) for convenient analysis of gaseous formaldehyde. Sens Actuators B Chem 130:32–37

Mitsubayashi K, Toma K, Iitani K, Arakawa T (2022) Gas-phase biosensors: a review. Sens Actuators B Chem 367:132053

Naresh V, Lee N (2021) A review on biosensors and recent development of nanostructured materials-enabled biosensors. Sensors 21:1109

Nozaki R, Ikuta T, Ueno K, Tsukakoshi K, Ikebukuro K, Maehashi K (2020) Ethanol detection at the parts per billion level with single-stranded-DNA-modified graphene field-effect transistors. Phys Status Solidi Basic Res 257:1–6

Pietrantonio, Cannata D, Benetti M, Verona E, Varriale A, Staiano M, D'Auria S (2013) Detection of odorant molecules via surface acoustic wave biosensor array based on odorant-binding proteins. Biosens Bioelectron 41:328–334

Prasad BD, Sahni S, Kumar P, Siddiqui MW (2018) PLANT BIOTECHNOLOGY, VOLUME 1, principles, techniques, and applications. Apple Academic Press, Inc

Schroeder H, Cavacini L (2010) Structure and function of immunoglobulins. Allergy Clin Immunol 125(202):41–52

Stoltenburg R, Nikolaus N, Strehlitz B (2012) Capture-SELEX: selection of DNA aptamers for aminoglycoside antibiotics. Anal Methods Chem 2012:415697

Stubbs W, Hunt S, Lee D, Doyle (2002) Gas phase activity of anti-FITC antibodies immobilized on a surface acoustic wave resonator device. Biosens Bioelectron 17:471–477

Tang S (1987) VanSlyke, organic electroluminescent diodes. Appl Phys Lett 51:913

Tian W, Liu X, Yu W (2018) Research progress of gas sensor based on graphene and its derivatives: a review. Appl Sci 8(7):1118

Tsumura, Koezuka H, Ando T (1986) Macromolecular electronic device: field-effect transistor with a polythiophene thin film. Appl Phys Lett 49(18):1210–1212

Ye M, Arakawa T, Chien P-J, Suzuki T, Toma K, Mitsubayashi K (2017) Acetone biosensor based on fluorometry of reduced nicotinamide adenine dinucleotide consumption in reversible reaction by secondary alcohol dehydrogenase. IEEE Sensors J 17:5419–5425

Yerlanova G, Serik M, Kopyltsov A (2021) High performance computers: from parallel computing to quantum computers and biocomputers. J Phys Conf Ser 1889:032032

Zhang X, Firestein S (2007) Comparative genomics of odorant- and pheromone receptor genes in rodents. Genomics 89:441–450

Бутлеров А (1862) О химическом строении веществ (A. Butlerov, on the chemical structure of substances). Учёные записки Казанского университета (отд физ-мат и мед наук) 1(1):1–11

Nanobiosensors as Biomedical Diagnostic Tools

13

Zainab Ahmed and Mohd. Mohsin

Abstract

The development of new methods capable of highly sensitive detection and quantification of particular parameters has contributed significantly to the advancement of medical diagnosis. Nanobiosensors are becoming increasingly important in the field of biomedical diagnostics as a result of their high levels of sensitivity and specificity as well as their ability to perform monitoring in real time. Various types of nanobiosensors have been designed and they are capable of diagnosing a range of diseases like diabetes, neurodegenerative diseases, autoimmune disorders. The purpose of this chapter is to offer a broad overview of nanobiosensors, encompassing various types and their applications in the field of medical diagnostics.

Keywords

Nanobiosensors · Biomarker · Quantum dot · Nanoparticles · Graphene · Biodetection

Abbreviations

AFP	Alpha-fetoprotein
AGA	Antigliadin antibodies
AuNP	Gold nanoparticle
CA-125	Cancer antigen 125
CNT	Carbon nanotubes
CRP	C-reactive protein

Z. Ahmed · M. Mohsin (✉)
Department of Biosciences, Jamia Millia Islamia, New Delhi, Delhi, India
e-mail: mmohsin1@jmi.ac.in

M. Mohsin, N. Soleja (eds.), *Nanoscale Sensors and their Applications in Biomedical Imaging*, https://doi.org/10.1007/978-981-97-3144-2_13

EDC/NHS	1-Ethyl-3-(3-dimethylaminopropyl)carbodiimide/N-hydroxysuccinimide
EGFR	Epidermal growth factor receptor
GO	Graphene oxide
GO/pPG	Graphene oxide/polypropylene glycol
GO/PTCNH2	Graphene oxide paired with amino-terminated perylene derivative
HER2/neu	Human epidermal growth factor receptor 2
LOD	Limit of detection
MBP	Myelin basic protein
MWCNT	Multiwalled carbon nanotube
MWCNT-PS	Multiwalled carbon nanotube polystyrene
NEEs	Nanoelectrode ensembles
PSA	Prostate-specific antigen
PSA NCAM	Polysialylated-neural cell adhesion molecule
p53	Tumour protein P53
QD	Quantum dot
SPR	Surface plasmon resonance
SQUIDs	Superconducting quantum interference devices

13.1 Introduction

In the recent past, there have been significant advancements in medical diagnosis. These advancements can be attributed to developing new techniques that are capable of detecting and quantifying specific molecules and components. Valuable insights into the physiological state of a living cell can be acquired by ascertaining the presence or absence of these molecules and components. The rapid, accurate, and specific nature of the medical diagnostic process necessitates minimising the occurrence of "false positives" or erroneous outcomes (Nagraik et al. 2021). Prior to the advancement of these techniques, diagnostic procedures relied solely on observable parameters. The ability to conduct more accurate evaluations and comprehensive examinations of the patient's condition now relies on data pertaining to the existence or non-existence of substances or other parameters that contribute to the diagnosis. Accurate and reliable patient information can be obtained by quantifying specific molecular constituents. Nonetheless, it is imperative to couple this approach with a comprehensive comprehension of the fundamental principles governing the patient's physiological condition as well as the underlying mechanisms of the disorder or disease. Biomarkers are objective parameters that can be measured or evaluated to gather information regarding a physiologic/pathologic condition or the response to a therapeutic intervention. The most prevalent biomarkers are molecular biomarkers, which are biochemically measured. Molecular biomarkers are significant molecules that play a crucial role in the functioning of cells, either intercellular or intracellular. Any deviation from normal cellular conditions or activities can lead to alterations in the expression or presence of these biomarkers (Golubnitschaja and Flammer 2007). The interest in biomarkers in medicine stems from the ability to link a varied parameter to a specific clinical implication, some of the biomarkers used in diagnostics are listed in Table 13.1.

Due to the availability of sophisticated equipment in laboratories, currently, there exist a variety of techniques to identify an extensive range of molecules with high sensitivity. Researchers can detect even trace amounts of biomolecules using methods like isotope-coded affinity tags, mass spectroscopy (Jiang et al. 2015), and 2D gel electrophoresis (Hanash 2000). Also, the techniques mentioned can do more than just find biomolecules. They can also give useful information about other

Table 13.1 List of different biomarkers used in the medical diagnosis

Biomarker	Type	Application	References
C-reactive protein (CRP)	Protein	Inflammation, cardiovascular risk, infection	Zimmerman et al. (2003)
Prostate-specific antigen (PSA)	Protein	Prostate cancer screening and monitoring	Khan et al. (2022)
Troponin	Protein	Myocardial infarction (heart attack)	Chaulin (2022)
Haemoglobin A1c (HbA1c)	Protein	Diabetes management and monitoring	Kowsar and Mansouri (2022)
CA-125	Protein	Ovarian cancer screening and monitoring	Yumashev et al. (2022)
HER2/neu	Protein	Breast cancer diagnosis and treatment	Gupta et al. (2023)
Alpha-fetoprotein (AFP)	Protein	Liver cancer and foetal defects	Kim et al. (2018), Mizejewski (2004)
Procalcitonin	Protein	Bacterial infection and sepsis	Foushee et al. (2012)
Thyroid-stimulating hormone (TSH)	Hormone	Thyroid function and disorders	Bensenor et al. (2012)
Oestrogen	Hormone	Menstrual cycle, menopause, and fertility	Rettberg et al. (2014)
Human chorionic gonadotropin (hCG)	Hormone	Pregnancy testing and monitoring	Szczerba et al. (2016)
Glucose	Metabolite	Diabetes diagnosis and management	Roden (2016)
Creatinine	Metabolite	Kidney function and renal health	Zhao (2013)
PSA-NCAM	Cell surface marker	Neural development and brain tumours	Amoureux et al. (2010)
p53	Genetic marker	Tumour suppressor gene in cancer	Hayman et al. (2019)
Epidermal growth factor receptor (EGFR)	Genetic marker	Lung cancer diagnosis and treatment	Sigismund et al. (2018)
MicroRNAs	Genetic marker	Various diseases, including cancer and neurodegenerative diseases	Takousis (2018), Christodoulatos and Dalamaga (2014)

properties of the biomolecules being studied, such as their conformation, modification states, or interactions with other molecules. Due to the fragile nature of the aforementioned techniques, the presence of the analyte in the appropriate form for detection is a prerequisite. Prior to detection, it becomes necessary to treat samples; these treatments consist primarily of purification as well as concentration processes, which are typically performed by chromatography. These additional steps limit the application of these methods in real-world situations outside of the laboratory, as they severely restrict the number of samples that can be processed, and that is how they add to the overall cost as well. Furthermore, the execution of the experimental protocol demands trained personnel with extensive knowledge of the field. Currently, there is a need to detect fluctuations in the levels of biomolecules in real time, enabling early diagnosis of potential diseases.

The field of nanotechnology has drawn considerable global interest in the context of medical applications like diagnosis followed by the management of diseases. Nanoscience refers to the scientific investigation of structures and molecules at the nanoscale that typically ranges from 1 to 100 nm. The corresponding field of technology, which applies this knowledge to practical applications such as the development of devices, is commonly known as nanotechnology (Mansoori and Soelaiman 2005). Nanotechnology promotes the integration of diagnostic and therapeutic modalities, constituting a crucial element of a personalised approach to addressing various disorders. The application of nanotechnology has found extensive utilisation across various scientific and medical domains owing to the distinctive quantum mechanical characteristics exhibited by specific nanomaterials (Zhong 2009).

13.2 Nanobiosensors

For the detection of biologically active target molecules in a sample, scientists use analytical devices called biosensors, which consist of a transducer with an active biological molecule to generate a signal. Biosensors have two incorporated parts (Rackus et al. 2015; Fig. 13.1):

1. The receptor, commonly constituted by biological entities such as microorganisms, DNA, enzymes, tissues, antibodies or even synthetic molecules.
2. The transducer, an integral element of the sensor, possesses the ability to convert the recognition event into a measurable signal that can be utilised for the quantification of the analyte.

Upon the occurrence of the recognition event within the biosensor, a notable alteration happens in the characteristics of the setup, thereby enabling the transducer to perceive and convert the disturbance into a signal (electrical signal). In certain instances, the inclusion of signal labels within the system is imperative in order to attain a signal that is easily comprehensible. These sensors are commonly known as label-based biosensors. Typically, labels employed in various applications consist of nanoparticles (NPs), organic molecules, or proteins (such as enzymes or

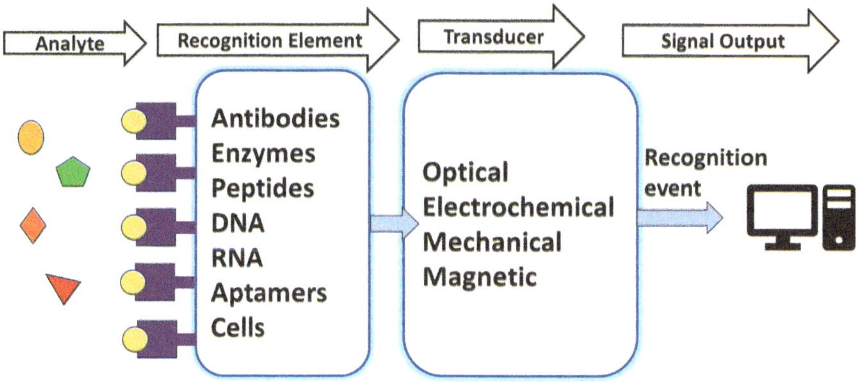

Fig. 13.1 The schematic representation of the components comprising a biosensor. This phenomenon illustrates the manner in which a receptor captures an analyte, resulting in a modification of the transducing element. The transducer is responsible for converting the captured event into an electrical signal

receptors) that have the ability to produce an electrochemical or fluorescent signal. Another option available is known as label-free biosensors. In these biosensors, the interaction between the analyte and the recognition element results in a change, either physical or chemical. The transducer is capable of directly detecting this alteration, thereby enabling the biosensor to monitor interactions without the requirement of supplementary labels or markers (Luo and Davis 2013).

A biosensor built with nanoscale components is called a nanobiosensor, they are made up of nanomaterials. The biological recognition element is the analyte of interest that is targeted, the transduction mechanism converts the modifications occurred by recognition element into a measurable quantity, and the bioreceptor allows for selective binding of the target analyte to the sensor.

Nanobiosensors are widely regarded as highly impactful advancements in the field of biosensing technology. There have been numerous nanostructure-based biosensing device studies conducted globally. Nanomaterials are employed in detection capabilities through the utilisation of electron microscopes. Metallic nanoparticles have demonstrated exceptional properties that make them highly suitable for a wide range of optical and electronic applications. Moreover, they have been effectively utilised in the nucleic acid sequence analysis. Carbon nanotubes possess several significant advantages, including enhanced enzyme loading capabilities, higher aspect ratio, functionalisation potential, and improved electrical communication (Davis et al. 2003). Nanoparticles facilitate immobilisation, improve bioanalyte loading, and exhibit excellent catalytic properties (Katz et al. 2004). Quantum dots are very good at fluorescence, have band energy that can be changed based on their size, and keep charge carriers in good quantum confinement (Mansur 2010).

13.2.1 Various Types of Nanobiosensors

13.2.1.1 Optical Nanobiosensor

Optical sensors refer to biosensing platforms that employ optical principles for the purpose of detecting and quantifying biomolecular interactions. The utilisation of nanoscale materials or structures is commonly observed in these sensors, wherein they interact with specific biomolecules, leading to detectable alterations in their optical characteristics, such as absorbance, light scattering, fluorescence, or surface plasmon resonance (SPR). Optical biosensors possess a distinct advantage over electrochemical biosensors as they enable the detection of multiple analytes through the utilisation of diverse monitoring wavelengths. The multiplexing of optical methods is facilitated by the capacity to choose distinct samples using different wavelengths of light. Among the different optical detection methods, the devices that are most frequently favoured include those that rely on the interferometry, surface plasmon resonance, fluorescence spectroscopy, and guided modes spectroscopy in optical waveguides. The methods employed above evaluate a range of parameters, including polarisation, decay time, energy, amplitude, phase, and other relevant factors. Among these, amplitude measurement is commonly used due to its convenient correlation with analyte density (Damborský et al. 2016).

13.2.1.2 Electrochemical Nanobiosensor

Electrochemical sensors simply employ methods that are electrical in nature to facilitate or monitor biological reactions. These nanosensors consist predominantly of metal-based nanoparticles. Nanoparticles facilitate the biochemical reaction between molecules, which aids by inhibiting one of the analytes. Therefore, the biochemical process becomes highly targeted, and the potential for producing unwanted byproducts is greatly reduced. For instance, colloidal gold nanoparticles are used for the purpose of immobilising DNA on gold electrodes, which significantly decreases the detection limit and enhances the biosensor's overall performance (Cai et al. 2001). Biosensors have been developed to detect glucose, hydrogen peroxide, and xanthine by utilising gold nanoparticles that are linked with enzymes (Xu et al. 2003; Zhao et al. 1996). Xu et al. (2003) have found that horseradish peroxidase (HRP) can be immobilised on electroplated gold electrodes that are loaded with nanoparticles of carbon. This immobilisation technique enables a more rapid amperometric response and improved electrocatalytic reduction capability. Consequently, biosensors containing nanoparticles exhibited lower detection limit and are more sensitive than those that did not contain nanoparticles.

Photochemical reactions can be made to work better with the help of nanoscale semiconductor crystals. When making a new photoelectrochemical system, biological molecules like enzymes and precursors can be labelled in these systems for detecting biomolecule. For example, Curri et al. (2002) used immobilised cadmium sulphide (CdS) to make a system for enzymatic detection of formaldehyde by immobilising formaldehyde dehydrogenase on gold electrodes that catalyse oxidation of formaldehyde (Curri et al. 2002). Numerous other studies have demonstrated that metal-based nanoparticles can be linked to biological probes to find a specific biomolecule in a mixture (González-Garcıa et al. 2000).

13.2.1.3 Magnetic Nanobiosensor

Magnetic nanobiosensors are fabricated through the utilisation of specifically engineered magnetic nanoparticles. The nanoparticles under consideration predominantly consist of iron-based materials. Magnetic sensors are widely suggested for use in biomedical applications due to their versatility and diverse range of analytical capabilities. Magnetic materials used in detection are composed of various transition metal compounds combined with iron, giving the blend unique characteristics. The integration of magnetic nanoparticles has significantly enhanced the efficacy and sensitivity of conventional biodetection devices. The magnetic behaviour of iron blends containing transition metals and d-block metals with unpaired electrons is extremely versatile. The utilisation of a magnetic bioassay technique, which involves the selective isolation of magnetically labelled biological targets through the employment of a magnetometer, has been increasingly gaining recognition and acceptance in various scientific fields (Richardson et al. 2001). Superconducting quantum interference devices (SQUIDs) apply the superparamagnetic characteristic of magnetised nanoparticles to achieve rapid identification of biological analytes. These tools are employed in biomedical research for the purpose of detecting specific analyte through the utilisation of antibodies that are bound to magnetic nanoparticles (Chemla et al. 2000). The phenomenon of screening through the utilisation of superparamagnetic properties is notably observed in magnetic nanoparticles.

13.2.1.4 Piezoelectric Nanobiosensor

Piezoelectric nanobiosensors refer to a specific category of nanobiosensors that exploit the piezoelectric phenomenon to transform mechanical stress or pressure arising from biomolecular interactions into electrical signals. The piezoelectric phenomenon refers to the production of an electric current or voltage within specific materials upon exposure to mechanical strain. In piezoelectric nanobiosensors, specific biomolecules (like antibodies, enzymes, or DNA probes) that only interact with target analytes are added to a piezoelectric material, which is often a thin film or nanowire made of zinc oxide (ZnO) or lead zirconate titanate (PZT). When these biomolecules bind to the target analytes, they cause the piezoelectric material to deform or be stressed, which changes its electrical properties.

For the purpose of detecting *M. tuberculosis*, a piezoelectric sensor utilising gold nanoparticle (AuNP)-mediated enzyme-assisted enhancement of signals has been created by Zhang and He (2022). The sensor uses a 3′-protruding hairpin-shaped DNA duplex. The hairpin probe unfolds in the presence of the *M. tuberculosis* 16S rDNA fragment, facilitating the precise cleavage of the hairpin by exonuclease III. Target and probe DNA are both released in this action. The freed target DNA then creates a hybrid with another hairpin-shaped DNA duplex, starting another digestion cycle and producing a large number of uncut DNA probes. Hybridisation with an electrode-modified capture probe pulls uncut DNA to the surface. Subsequently, an AuNP-labelled detection probe then crosses the electrodes and hybridises with untrimmed DNA. The piezoelectric sensor detects *M. tuberculosis* through conductivity changes. This sensor could diagnose tuberculosis quickly and accurately (Zhang and He 2022).

13.2.1.5 Acoustic Wave Nanobiosensor

Acoustic wave nanobiosensors represent a specific category of nanobiosensors that employ acoustic waves for the purpose of detecting and quantifying biomolecular interactions. The sensors are constructed using materials or structures at the nanoscale level, which have the capability to detect alterations caused by biochemical events in the form of acoustic properties such as changes in frequency, velocity, or impedance. They can enhance the sensing reaction during detection, thereby enhancing the accuracy of the detection boundaries. This category of nanobiosensors exhibits a range of stimulation-based effects. The acoustic wave nanobiosensors in the mass-based variant employ sol particles that have been modified with antibodies. These antibodies are capable of binding to the coating of the electrode. The surface of electrode is modified with analyte-conjugated elements, resulting in the immobilisation of antibody molecules on the electrode covering. The presence of a large number of confined sol fragments of the antibody changes the vibration frequency of the piezoelectric-based detecting system. The alteration in vibrational frequency serves as the fundamental basis for detection. It is preferable for the diameter of the sol-based antibody particles to fall within the range of 5–100 nm. In general, gold, titanium dioxide, cadmium sulphide, and platinum nanoparticles are commonly favoured (Liu et al. 2004; Su et al. 2000).

13.2.1.6 Quantum Dot-Based Nanobiosensor

Quantum dot (QD)-based nanobiosensors are nanoscale devices that utilise quantum dots, which are semiconductor nanocrystals, as the sensing element. Quantum dots have exceptional optical properties, such as tunable fluorescence emission, a high quantum yield, and resistance to photobleaching (Reiss et al. 2009). The fluorescence of these nanocrystals can be adjusted in size, ranging from larger diameter QDs that exhibit red fluorescence within the visible light spectrum to smaller diameter QDs that emit blue fluorescence. Due to this rationale, QDs possess a wide excitation range combined with a narrow and symmetrical emission spectrum. This characteristic allows QDs to be excited by a single wavelength and operate in a multicolour mode, facilitating their application in disease imaging and mapping. This makes them ideal candidates for the development of highly sensitive and multiplexed biosensors. QDs can be classified into two distinct categories based on their chemical composition. The first category includes elements belonging to groups III (boron, aluminium, gallium) to V (nitrogen, phosphorus, arsenic) of the periodic table, while the second category comprises elements from subgroup II (zinc, cadmium) and main group VI (oxygen, sulphur, selenium) of the periodic table. The second category of quantum dots exhibits a higher optimised quantum yield, which renders them superior and consequently preferable compared to the first category of quantum dots (Ghaderi et al. 2011). Extensive study has been conducted on nanobiosensors based on quantum dots for various applications, including bioimaging, disease diagnosis, and environmental monitoring.

13.2.1.7 Graphene-Based Nanobiosensor

Graphene is a honeycomb-structured monolayer of carbon atoms. Each of these carbons forms three intralayer sp^2 or sigma (σ) bonds with its three carbon atom neighbours (Georgakilas et al. 2012). There has been a major interest in the utilisation of sensors that incorporate graphene-based nanocomposites. Graphene oxide (GO), a nanomaterial derived from the chemical exfoliation of graphite using potent oxidising agents, can be subjected by incorporating additional materials such as macromolecules, metal oxides, and metals. These modifications aim to enhance the electrochemical performance of biosensors (Pourmadadi et al. 2020; Wang and Dai 2015). The optical, electrical, and electrochemical behaviour of GO differs from that of graphene, primarily attributed to the presence of oxygen in its structure (Chimene et al. 2015). In recent years, there have been proposals for novel sensing platforms utilising both original and modified graphene nanoparticles. Numerous platforms have been employed for the immobilisation of biomolecules such as antibodies, DNA, and enzymes, with the aim of developing biosensors that exhibit exceptional sensitivity and selectivity.

Transducers for biosensors are manufactured using graphene-based nanomaterials. Transducers are known to transform the bioreceptor-target biomolecule interaction into detectable signals. In order for the interaction to occur, target molecules must bind to the transducer surface. Target molecules such as antibodies and ssDNA can be attached to graphene using EDC/NHS chemistry, while enzymes are typically immobilised using physisorption. Because of its high electrical conductivity, large surface area, high electron transfer rate, and ability to immobilise a variety of molecules, graphene has been used to create a variety of biosensors (Rao et al. 2009).

13.2.1.8 Carbon Nanotube-Based Nanobiosensor

Carbon nanotube-based nanosensors are a specific category of nanosensors that employ carbon nanotubes (CNTs) as the primary sensing component. Carbon nanotubes are nanostructures with a cylindrical shape that are formed by rolling up sheets of graphene. These nanotubes exhibit distinctive properties in terms of their electrical, mechanical, and chemical characteristics, which render them highly suitable for a wide range of sensing applications. CNT-based nanosensors possess notable attributes such as enhanced sensitivity, rapid response rates, and the capacity to detect a diverse array of analytes (Sotiropoulou et al. 2003; Zhao et al. 2002). Consequently, they have emerged as invaluable instruments in various domains, encompassing biomedical diagnostics, environmental monitoring, and gas sensing.

CNT-based biosensors have the capability to identify and measure biomolecular interactions, including but not limited to DNA hybridisation, protein–protein interactions, and enzymatic reactions. The utilisation of functionalised CNTs facilitates the binding of targeted biomolecules, leading to alterations in their electrical characteristics. This, in turn, enables the detection of biomolecular interactions without the need for labelling and in a real-time manner. The presence of glucose in various bodily fluids is quantified using an enzyme called glucose oxidase. The enzyme is used in the traditional sensing method to identify glucose in basic body tissues like blood. The determination of glucose from small amounts of bodily fluids like tears, saliva, etc. is made possible by the incorporation of nanotubes as immobilisation assemblies. Single-walled nanotubes have been successfully modified for enzyme-based glucose estimation, and the activity of the enzyme has increased noticeably (Azamian et al. 2002).

13.2.1.9 Nanowire-Based Nanobiosensor

Nanowire sensors have demonstrated significant promise as a viable detection platform for use in biological and medical contexts. Nanowires are nanostructures that exhibit a cylindrical morphology, bearing resemblance to carbon nanotubes. The lengths of these objects span from a few micrometres to centimetres, whereas their diameters fall within the nanometre range. Nanowires and nanotubes exhibit dissimilarities in their aspect ratios, which refer to the ratio between their lengths and diameters. The aspect ratio of nanotubes exhibits a significantly high value, whereas the aspect ratio of nanowires falls within the range of 1000. Nanowires exhibit exceptional electron transport characteristics, with significantly enhanced charge carrier mobility when compared to their bulk counterparts.

Nanowires are considered highly suitable components for the detection of pathogens, as well as biological and chemical species, due to their compact dimensions and impressive functionality. Nanowires are designed to operate within specific dimensions, making them suitable for in vivo applications and enabling their utilisation in restricted environments such as within living cells. In a study conducted by Wang et al., a fabrication technique was utilised to create biosensor optical fibres integrated with nanowires coated with antibodies. The purpose of this biosensor was to detect the presence of toxicants within living cells (Wang 2002).

13.2.1.10 Cantilever-Based Nanobiosensor

Biosensors based on cantilevers are the best example of nanotransducers. This biosensor employs a cantilever manufactured micromechanically. Due to interaction with biosensing components, a nearly 100 nm-thick cantilever is deformed. Cantilever-based nanobiosensors have been widely recognised as a robust platform for label-free and time-resolved detection of molecular interactions, owing to their exceptional sensitivity (Backmann et al. 2010).

Cantilever-based sensors can detect vapours, ions, DNA, proteins, antibiotics, and pathogenic microorganisms. In static mode, these sensors can find a process by figuring out how much the cantilever bends because molecules on one side of the cantilever cause stress. When there is a mass on the surface of the cantilever, it changes the frequency of resonance in a cantilever that is oscillating. This shows that a process is happening in dynamic mode. In this mode of operation, the mechanical sensitivity is caused by changes in surface stress caused by molecular interactions with the surface and by side-to-side interactions with the molecular layer (Fritz 2008).

13.3 Applications of Nanobiosensors in Medical Diagnostics

Nanobiosensors exhibit a remarkable degree of versatility and multifunctionality, hence offering a wide range of potential applications. In the field of medical diagnostics, nanobiosensors are employed for the purpose of detecting serum antigens, carcinogens, and other metabolic diseases. Serum analysis is a crucial method for the detection of several ailments such as diabetes, cancer, and allergic reactions, among others. This section aims to cover commonly acknowledged clinical applications of nanobiosensors.

13.3.1 Diagnosis of Cancer

The objective of targeting cancer cells in vivo has been accomplished using multifunctional nanoparticle probes based on quantum dots. The enhanced permeability of quantum dot probes and the presence of cancer cell-specific biomarkers contribute to their accumulation in tumour sites. This phenomenon has resulted in the visualisation of cancer cells that exhibit sensitivity and display many fluorescent colours under in vivo circumstances. The aforementioned technique has been effectively utilised in the study of human prostate cancer in murine models (Gao et al. 2004).

Another class of nanobiosensors, specifically the electrochemical nanobiosensors which encompass amperometric, potentiometric, conductometric, impedimetric, and field-effect devices, has garnered significant attention in the field of early-stage cancer diagnosis (Freitas et al. 2018). These are employed in the identification of biomarkers associated with breast cancer, including particular genes, microRNA, proteins, circulating tumour cells, breast cancer cell lines, and biomarkers derived from exosomes or exosome-related sources (Gajdosova et al. 2020).

The utilisation of magnetic nanoparticles for in vivo detection of cancer cells is a significant advancement in cancer detection. A significant elevation in telomerase activity is commonly reported in several types of malignancies, indicating its potential as a viable target for detection purposes. The approach developed by Grimm et al. employs magnetic nanoparticles that undergo a change in their magnetic state when subjected to annealing with telomerase-synthesised TTAGGG repeats. The utilisation of MRI for the rapid processing of numerous images demonstrates the potential of this approach in the field of in vivo sensing (Grimm et al. 2004).

13.3.2 Detection of Virus

The timely and accurate identification of viral pathogens is crucial in facilitating prompt and efficient patient treatment, hence mitigating the risk of widespread outbreaks or pandemics. Due to this rationale, the utilisation of biosensors has been prevalent in facilitating the process of diagnosis, hence circumventing the need for complex protein or DNA identification methods in the context of certain viral infections (Castillo-Henríquez et al. 2020).

In a study, Hassanpour et al. devised an innovative optical biosensor consisting of citrate-capped silver nanoparticles (AgNPs) linked with plasmid DNA (pDNA). This biosensor was specifically designed to detect *Haemophilus influenza* in human biofluids with high selectivity and sensitivity (Hassanpour et al. 2018). The researcher group Shafiee et al. conducted a study on a photonic crystal biosensor with the aim of detecting various subvariants of HIV-1 (namely, subvariants A, B, and D) through the binding of the analyte to the biosensor (Shafiee et al. 2014). Furthermore, Gong et al. conducted reverse-phase polymerisation to synthesise a nanocomposite consisting of polyaniline/graphene (PAN/GN). This nanocomposite was applied in the fabrication of an electrical DNA-biosensor, which exhibited remarkable sensitivity and selectivity in detecting the HIV-1 gene fragment (Gong et al. 2019).

13.3.3 Pathogenic Bacteria Detection

Bacterial detection is accomplished by the utilisation of culture techniques and biochemical assays that are laborious. Therefore, it is imperative to provide an expeditious, dependable, and discerning approach for the identification of bacteria in clinical specimens. The biosensor developed by Suaifan et al. has the capability to rapidly detect *Staphylococcus aureus* within a short span of time. The sensing mechanism relies on the proteolytic activity shown by pathogen proteases towards a particular peptide substrate that is positioned between two magnetic nanobeads. The colour shift seen in this scenario is attributed to the dissociation of magnetic nanobeads–peptide moieties (Suaifan et al. 2017).

Ahari et al. employed an alternative methodology whereby they developed a potentiometric nanobiosensor capable of discerning the presence of bacteria by detecting a specific exotoxin released by the bacterium. The approach is mainly employed for the analysis of food samples that have been contaminated; however, it may also be utilised for clinical detection purposes (Ahari et al. 2017).

The study conducted by Elahi et al. revealed a novel approach for the early identification of pathogenic *Shigella*. The objective of this investigation was to develop an optical genosensing device by hybridising gold nanoparticles (AuNPs)-DNA probes with the *Spa* gene sequence. In the absence of the corresponding target, the biosensor induces a colour change in the sample solution, resulting in a purple hue. Conversely, in the presence of the specified gene sequence, the solution retains its original red coloration (Elahi et al. 2019).

13.3.4 Detection of Glucose

The detection of glucose holds significant importance within the food and pharmaceutical sectors. The monitoring of blood glucose levels represents a significant breakthrough in the field of nanobiosensors. This section provides an extensive overview of different nanosensors designed for glucose detection. These nanosensors have been effectively demonstrated to function in various biological samples such as blood, plasma, or serum.

Xu et al. designed a colorimetric nanosensor in which the sol-gel transition of acrylic acid-modified gold nanoparticles (AuNPs) induced by hydrogen peroxide (H_2O_2) can be utilised to detect glucose in diluted human serum. The limit of detection (LOD) for this method is 1 μM (Xu et al. 2012). The process of AuNPs aggregation is facilitated by cascade reactions. Specifically, the enzyme glucose oxidase (GOx) catalyses the production of hydrogen peroxide (H_2O_2). In the presence of Fe^{2+}, this hydrogen peroxide triggers the aggregation of AuNPs, resulting in an apparent change in colour (Xu et al. 2012).

Furthermore, fluorescent nanobiosensors have been designed for glucose detection, with their principle grounded in creating a correlated fluorescence signal by converting the glucose signal. There is a need for three essential elements for the process to occur: (a) biomolecules or molecules that can specifically recognise glucose, (b) a signal transducer, and (c) a signal detector.

Fluorescent nanomaterials act as effective transducers for converting glucose signals into fluorescent signals, which can be measured in terms of peak shift, peak intensity, or fluorescent lifetime. Fluorescence resonance energy transfer (FRET) is recognised as a highly sensitive and dependable analytical technique that finds extensive application in fluorescence nanobiosensing systems. A glucose nanobiosensor was developed using quantum dots and incorporating glucose catalytic reactions. Quantum dots containing glucose dehydrogenase were modified with CdSe/ZnS to detect glucose. FRET was employed as a method of sensing. Prior to the addition of glucose, the fluorescence emitted by the quantum dots was reduced by the presence of methylene blue. Following the introduction of glucose, biocatalytic reactions were initiated, resulting in the restoration of fluorescence in the quantum dots. The biocatalytic reaction involved the conversion of glucose into gluconic acid. This reaction was facilitated by the glucose dehydrogenase biocatalyst, which utilised the NADH–NAD^+ pair. Specifically, NADH reduced methylene blue to its colourless form, MBH. Subsequently, the fluorescence resonance energy transfer (FRET) between quantum dots and methylene blue ceased. The fluorescent glucose nanobiosensor exhibited a detection limit of 10^{-5} M (Bahshi et al. 2009).

13.3.5 Rapid Diagnosis of Autoimmune Disease

Autoimmune diseases encompass a range of chronic conditions that can be categorised as either organ-specific, such as Graves' disease, thyroid diseases, primary biliary cirrhosis, and type 1 diabetes, or systemic, including rheumatoid arthritis, antiphospholipid syndrome, and systemic lupus erythematosus (Tozzoli 2008). These disorders typically arise from a malfunction of immunological self-tolerance, resulting in the production of autoantibodies with high affinity. These autoantibodies erroneously target the cells, tissues, and organs of the individual's own body.

Electrochemical biosensors, particularly those utilising amperometry and voltammetry, are commonly employed techniques for the detection of autoimmune disorders. Several advantages of these techniques include their reusable and rapid detection capability, the requirement of low sample volume, ease of miniaturisation, and feasibility of quantitative analysis. Numerous scholarly articles have documented the utilisation of diverse biosensors and nanobiosensors for the detection of various biomarkers associated with autoimmune diseases. These sensors are listed in Table 13.2.

13.3.6 Neurodegenerative Disease Detection

Parkinson's and Alzheimer's diseases are two widely recognised examples of neurodegenerative disorders, characterised by the degeneration of nerve cells and the onset of dementia. The probability of developing a neurodegenerative disease significantly increases with advancing age. Recent studies show that neurodegeneration will affect a large number of people in the coming decades. This shows how important it is to diagnose, what causes it, and find new ways to treat and cure it.

The majority of scholarly investigations pertaining to the utilisation of nanoparticles (NPs) as diagnostic agents for Parkinson's disease (PD) have focused on dopamine (DA) nanobiosensors (Sim et al. 2020). Shin and colleagues conducted a study in which they developed a new type of nanoparticles consisting of silver and molybdenum disulfide (Ag/MoS_2). These nanoparticles were utilised for the detection of dopamine (DA) and the researchers observed an amplified electrochemical signal from the Ag/MoS_2 electrochemical biosensor they synthesised. This finding

Table 13.2 List of nanobiosensors having application in autoimmune disease

Detection method/sensor	Biomarker	Biofluid	Nanomaterial	References
Electrochemical (amperometric)	Anti-tTG (IgA)	Serum	AuNPs	Giannetto et al. (2014)
Electrochemical (amperometric)	ACPAs (rheumatoid arthritis)	Serum	MWCNT–PS composite	Villa Mde et al. (2011)
Electrical impedance spectroscopy (EIS)	Anti-MBP	Serum and CSF	TiO2 nanoparticles	Derkus et al. (2013)
Cyclic voltammetric	AGA (IgA) AGA (IgG)	Serum	MWCNTs/AuNPs	Neves et al. (2012)
Differential pulse voltammetry	Anti-tTG	Serum	QDs	Martín-Yerga et al. (2017)
Differential pulse voltammetry	MBP	Serum	GO/pPG nanocomposite	Derkus et al. (2017)
Photoelectrochemical	TNFα	Serum	GO/PTCNH2	Hu et al. (2018)
Optical (fluorescent)	microRNA-145 (MS)	Serum	Silver nanoclusters (NC)	Mansourian et al. (2017)
Optical (electrochemiluminescence)	Anti-tTG	Serum	Membrane-template gold NEEs	Habtamu et al. (2015)

indicates that the biosensor has promising applications in the field of Parkinson's disease (Shin et al. 2019). A variety of nanobiosensors, encompassing both electrochemical and optical approaches, have been developed with the aim of facilitating the early detection of neurodegenerative diseases. The diagnostic tools are designed to exhibit specificity and sensitivity by taking into account the distinct biomarkers like alpha-synuclein, dopamine, dopamine receptor, tau tangles, and amyloid plaques associated with neurodegenerative disorders such as Parkinson's disease and Alzheimer's disease (Fazlali et al. 2022).

13.4 Conclusion and Future Perspectives

Nanobiosensors are an evolution in biomedical diagnostics, providing unprecedented sensitivity and specificity for detecting biomolecular interactions. Nanobiosensors make it possible to monitor biomolecular interactions in real time, giving immediate results and facilitating quick decision-making in medical conditions. Many nanobiosensors provide label-free detection, eliminating the need for time-consuming and complex labelling steps. This simplifies the diagnostic process and reduces the possibility of false-negative or false-positive outcomes. The miniaturisation of nanobiosensors enables the development of point-of-care devices, allowing medical diagnosis to take place at the patient's bedside or in remote areas. This allows for more rapid and convenient testing, which is especially useful in resource-constrained environments. Nanobiosensors can be used in non-invasive diagnostic procedures such as breath analysis or saliva testing. These non-invasive methods are less invasive and less painful for patients. Nanobiosensors detect specific genetic markers or biomarkers that help with real-time analysis and personalised medicine. This enables healthcare providers to tailor treatments based on the characteristics of individual patients, resulting in more effective therapies.

Despite the considerable potential they hold, the integration of nanobiosensors into medical environments encounters various obstacles. In order to fully harness the potential of nanobiosensors in the field of diagnostics, it is imperative to address various concerns pertaining to biocompatibility, stability, reproducibility, and cost.

References

Ahari H, Hedayati M, Akbari-adergani B, Kakoolaki S, Hosseini H, Anvar A (2017) Staphylococcus aureus exotoxin detection using potentiometric nanobiosensor for microbial electrode approach with the effects of pH and temperature. Int J Food Prop 20(sup2):1578–1587

Amoureux MC, Coulibaly B, Chinot O, Loundou A, Metellus P, Rougon G, Figarella-Branger D (2010) Polysialic acid neural cell adhesion molecule (PSA-NCAM) is an adverse prognosis factor in glioblastoma, and regulates olig2 expression in glioma cell lines. BMC Cancer 10:91. https://doi.org/10.1186/1471-2407-10-91

Azamian BR, Davis JJ, Coleman KS, Bagshaw CB, Green ML (2002) Bioelectrochemical single-walled carbon nanotubes. J Am Chem Soc 124(43):12664–12665. https://doi.org/10.1021/ja0272989

Backmann N, Kappeler N, Braun T, Huber F, Lang HP, Gerber C, Lim RY (2010) Sensing surface PEGylation with microcantilevers. Beilstein J Nanotechnol 1(1):3–13

Bahshi L, Freeman R, Gill R, Willner I (2009) Optical detection of glucose by means of metal nanoparticles or semiconductor quantum dots. Small 5(6):676–680

Bensenor IM, Olmos RD, Lotufo PA (2012) Hypothyroidism in the elderly: diagnosis and management. Clin Interv Aging 7:97–111

Cai H, Xu C, He P, Fang Y (2001) Colloid Au-enhanced DNA immobilization for the electrochemical detection of sequence-specific DNA. J Electroanal Chem 510(1–2):78–85

Castillo-Henríquez L, Brenes-Acuña M, Castro-Rojas A, Cordero-Salmerón R, Lopretti-Correa M, Vega-Baudrit JR (2020) Biosensors for the detection of bacterial and viral clinical pathogens. Sensors 20(23):6926

Chaulin AM (2022) The importance of cardiac troponin metabolism in the laboratory diagnosis of myocardial infarction (comprehensive review). Biomed Res Int 2022:6454467

Chemla YR, Grossman HL, Poon Y, McDermott R, Stevens R, Alper MD, Clarke J (2000) Ultrasensitive magnetic biosensor for homogeneous immunoassay. Proc Natl Acad Sci USA 97(26):14268–14272

Chimene D, Alge DL, Gaharwar AK (2015) Two-dimensional nanomaterials for biomedical applications: emerging trends and future prospects. Adv Mater 27(45):7261–7284. https://doi.org/10.1002/adma.201502422

Christodoulatos GS, Dalamaga M (2014) Micro-RNAs as clinical biomarkers and therapeutic targets in breast cancer: quo vadis? World J Clin Oncol 5(2):71–81

Curri ML, Agostiano A, Leo G, Mallardi A, Cosma P, Della Monica M (2002) Development of a novel enzyme/semiconductor nanoparticles system for biosensor application. *Mater Sci Eng C* 22(2):449–452

Damborský P, Švitel J, Katrlík J (2016) Optical biosensors. Essays Biochem 60(1):91–100

Davis JJ, Coleman KS, Azamian BR, Bagshaw CB, Green ML (2003) Chemical and biochemical sensing with modified single walled carbon nanotubes. Chem Eur J 9(16):3732–3739

Derkus B, Emregul E, Yucesan C, Cebesoy Emregul K (2013) Myelin basic protein immunosensor for multiple sclerosis detection based upon label-free electrochemical impedance spectroscopy. Biosens Bioelectron 46:53–60

Derkus B, Bozkurt PA, Tulu M, Emregul KC, Yucesan C, Emregul E (2017) Simultaneous quantification of myelin basic protein and tau proteins in cerebrospinal fluid and serum of multiple sclerosis patients using nanoimmunosensor. Biosens Bioelectron 89:781–788

Elahi N, Baghersad MH, Kamali M (2019) Precise, direct, and rapid detection of Shigella Spa gene by a novel unmodified AuNPs-based optical genosensing system. J Microbiol Methods 162:42–49

Fazlali P, Mahdian A, Soheilifar MS, Amininasab SM, Shafiee P, Wani IA, AL-Mokaram AM (2022) Nanobiosensors for early detection of neurodegenerative disease. J Compos Compd 4(10):24–36

Foushee JA, Hope NH, Grace EE (2012) Applying biomarkers to clinical practice: a guide for utilizing procalcitonin assays. J Antimicrob Chemother 67(11):2560–2569

Freitas M, Nouws HP, Delerue-Matos C (2018) Electrochemical biosensing in cancer diagnostics and follow-up. Electroanalysis 30(8):1584–1603

Fritz J (2008) Cantilever biosensors. Analyst 133(7):855–863

Gajdosova V, Lorencova L, Kasak P, Tkac J (2020) Electrochemical nanobiosensors for detection of breast cancer biomarkers. Sensors 20(14):4022

Gao X, Cui Y, Levenson RM, Chung LW, Nie S (2004) In vivo cancer targeting and imaging with semiconductor quantum dots. Nat Biotechnol 22(8):969–976

Georgakilas V, Otyepka M, Bourlinos AB, Chandra V, Kim N, Kemp KC, Hobza P, Zboril R, Kim KS (2012) Functionalization of graphene: covalent and non-covalent approaches, derivatives and applications. Chem Rev 112(11):6156–6214

Ghaderi S, Ramesh B, Seifalian AM (2011) Fluorescence nanoparticles "quantum dots" as drug delivery system and their toxicity: a review. J Drug Target 19(7):475–486

Giannetto M, Mattarozzi M, Umiltà E, Manfredi A, Quaglia S, Careri M (2014) An amperometric immunosensor for diagnosis of celiac disease based on covalent immobilization of open conformation tissue transglutaminase for determination of anti-tTG antibodies in human serum. Biosens Bioelectron 62:325–330

Golubnitschaja O, Flammer J (2007) What are the biomarkers for glaucoma? Surv Ophthalmol 52(6):S155–SS61

Gong Q, Han H, Yang H, Zhang M, Sun X, Liang Y, Liu Z, Zhang W, Qiao J (2019) Sensitive electrochemical DNA sensor for the detection of HIV based on a polyaniline/graphene nanocomposite. J Mater 5(2):313–319

González-Garcıa MB, Fernandez-Sanchez C, Costa-Garcıa A (2000) Colloidal gold as an electrochemical label of streptavidin–biotin interaction. Biosens Bioelectron 15(5–6):315–321

Grimm J, Perez JM, Josephson L, Weissleder R (2004) Novel nanosensors for rapid analysis of telomerase activity. Cancer Res 64(2):639–643

Gupta A, Patil BU, Gangane NM (2023) Role of p53 and Her2/Neu as a prognostic biomarker in breast carcinoma. Med J Dr DY Patil Univ 16(2):191–196

Habtamu HB, Sentic M, Silvestrini M, De Leo L, Not T, Arbault S, Manojlovic D, Sojic N, Ugo P (2015) A sensitive electrochemiluminescence immunosensor for celiac disease diagnosis based on nanoelectrode ensembles. Anal Chem 87(24):12080–12087

Hanash SM (2000) Biomedical applications of two-dimensional electrophoresis using immobilized pH gradients: current status. Electrophoresis 21(6):1202–1209

Hassanpour S, Baradaran B, Hejazi M, Hasanzadeh M, Mokhtarzadeh A, de la Guardia M (2018) Recent trends in rapid detection of influenza infections by bio and nanobiosensor. TrAC Trends Anal Chem 98:201–215

Hayman L, Chaudhry WR, Revin VV, Zhelev N, Bourdon JC (2019) What is the potential of p53 isoforms as a predictive biomarker in the treatment of cancer? Expert Rev Mol Diagn 19(2):149–159

Hu H, Pan D, Xue H, Zhang M, Zhang Y, Shen Y (2018) A photoelectrochemical immunoassay for tumor necrosis factor-α using a GO-PTCNH2 nanohybrid as a probe. J Electroanal Chem 824:195–200

Jiang L, Chughtai K, Purvine SO, Bhujwalla ZM, Raman V, Pasa-Tolic L, Heeren RM, Glunde K (2015) MALDI-mass spectrometric imaging revealing hypoxia-driven lipids and proteins in a breast tumor model. Anal Chem 87(12):5947–5956

Katz E, Willner I, Wang J (2004) Electroanalytical and bioelectroanalytical systems based on metal and semiconductor nanoparticles. Electroanalysis: An International Journal Devoted to Fundamental and Practical Aspects of Electroanalysis 16(1–2):19–44

Khan R, Arshad F, Hassan IU, Naikoo GA, Pedram MZ, Zedegan MS, Pourfarzad H, Aljabali AAA, Serrano-Aroca Á, Haggag Y, Mishra V, Mishra Y, Birkett M, Tambuwala MM (2022) Advances in nanomaterial-based immunosensors for prostate cancer screening. Biomed Pharmacother 155:113649

Kim DH, Oh HG, Park WH, Jeon DC, Lim KM, Kim HJ, Jang BK, Song KS (2018) Detection of alpha-fetoprotein in hepatocellular carcinoma patient plasma with graphene field-effect transistor. Sensors 18(11):4032

Kowsar R, Mansouri A (2022) Multi-level analysis reveals the association between diabetes, body mass index, and HbA1c in an Iraqi population. Sci Rep 12(1):21135

Liu T, Tang JA, Jiang L (2004) The enhancement effect of gold nanoparticles as a surface modifier on DNA sensor sensitivity. Biochem Biophys Res Commun 313(1):3–7

Luo X, Davis JJ (2013) Electrical biosensors and the label free detection of protein disease biomarkers. Chem Soc Rev 42(13):5944–5962

Mansoori GA, Soelaiman TF (2005) Nanotechnology—*an introduction for the standards community*. ASTM International

Mansourian N, Rahaie M, Hosseini M (2017) A Nanobiosensor based on fluorescent DNA-hosted silver nanocluster and HCR amplification for detection of MicroRNA involved in progression of multiple sclerosis. J Fluoresc 27:1679–1685

Mansur HS (2010) Quantum dots and nanocomposites. Wiley Interdiscip Rev Nanomed Nanobiotechnol 2(2):113–129

Martín-Yerga D, Fanjul-Bolado P, Hernández-Santos D, Costa-García A (2017) Enhanced detection of quantum dots by the magnetohydrodynamic effect for electrochemical biosensing. Analyst 142(9):1591–1600

Mizejewski GJ (2004) Biological roles of alpha-fetoprotein during pregnancy and perinatal development. Exp Biol Med 229(6):439–463

Nagraik R, Sharma A, Kumar D, Mukherjee S, Sen F, Kumar AP (2021) Amalgamation of biosensors and nanotechnology in disease diagnosis: mini-review. Sens Int 2:100089

Neves MM, González-García MB, Santos-Silva A, Costa-García A (2012) Voltammetric immunosensor for the diagnosis of celiac disease based on the quantification of anti-gliadin antibodies. Sensors Actuators B Chem 163(1):253–259

Pourmadadi M, Shayeh JS, Arjmand S, Omidi M, Fatemi F (2020) An electrochemical sandwich immunosensor of vascular endothelial growth factor based on reduced graphene oxide/gold nanoparticle composites. Microchem J 159:105476

Rackus DG, Shamsi MH, Wheeler AR (2015) Electrochemistry, biosensors and microfluidics: a convergence of fields. Chem Soc Rev 44(15):5320–5340

Rao CE, Sood AE, Subrahmanyam KE, Govindaraj A (2009) Graphene: the new two-dimensional nanomaterial. Angew Chem Int Ed 48(42):7752–7777

Reiss P, Protiere M, Li L (2009) Core/shell semiconductor nanocrystals. Small 5(2):154–168

Rettberg JR, Yao J, Brinton RD (2014) Estrogen: a master regulator of bioenergetic systems in the brain and body. Front Neuroendocrinol 35(1):8–30

Richardson J, Hawkins P, Luxton R (2001) The use of coated paramagnetic particles as a physical label in a magneto-immunoassay. Biosens Bioelectron 16(9–12):989–993

Roden M (2016) Diabetes mellitus: definition, classification and diagnosis. Wien Klin Wochenschr 128:37–40

Shafiee H, Lidstone EA, Jahangir M, Inci F, Hanhauser E, Henrich TJ, Kuritzkes DR, Cunningham BT, Demirci U (2014) Nanostructured optical photonic crystal biosensor for HIV viral load measurement. Sci Rep 4(1):4116

Shin JW, Yoon J, Shin M, Choi JW (2019) Electrochemical dopamine biosensor composed of silver encapsulated MoS 2 hybrid nanoparticle. Biotechnol Bioprocess Eng 24:135–144

Sigismund S, Avanzato D, Lanzetti L (2018) Emerging functions of the EGFR in cancer. Mol Oncol 12(1):3–20

Sim TM, Tarini D, Dheen ST, Bay BH, Srinivasan DK (2020) Nanoparticle-based technology approaches to the management of neurological disorders. Int J Mol Sci 21(17):6070

Sotiropoulou S, Gavalas V, Vamvakaki V, Chaniotakis NA (2003) Novel carbon materials in biosensor systems. Biosens Bioelectron 18(2–3):211–215

Su X, Chew FT, Li SF (2000) Design and application of piezoelectric quartz crystal-based immunoassay. Anal Sci 16(2):107–114

Suaifan GA, Alhogail S, Zourob M (2017) Rapid and low-cost biosensor for the detection of Staphylococcus aureus. Biosens Bioelectron 90:230–237

Szczerba A, Białas P, Pięta PP, Jankowska A (2016) hCG—related molecules and their measurement. Ginekol Pol 87(1):65–70

Takousis P (2018) Analysis of micro-RNA expression by qPCR on a microfluidics platform for Alzheimer's disease biomarker discovery. Biomarkers for Alzheimer's Disease Drug Development, 283–292

Tozzoli R (2008) The diagnostic role of autoantibodies in the prediction of organ-specific autoimmune diseases. Clin Chem Lab Med 46(5):577–587

Villa Mde G, Jiménez-Jorquera C, Haro I, Gomara MJ, Sanmartí R, Fernández-Sánchez C, Mendoza E (2011) Carbon nanotube composite peptide-based biosensors as putative diagnostic tools for rheumatoid arthritis. Biosens Bioelectron 27(1):113–118

Wang J (2002) Electrochemical nucleic acid biosensors. Anal Chim Acta 469(1):63–71

Wang Z, Dai Z (2015) Carbon nanomaterial-based electrochemical biosensors: an overview. Nanoscale 7(15):6420–6431

Xu X, Liu S, Ju H (2003) A novel hydrogen peroxide sensor via the direct electrochemistry of horseradish peroxidase immobilized on colloidal gold modified screen-printed electrode. Sensors 3(9):350–360

Xu C, Ren J, Feng L, Qu X (2012) H_2O_2 triggered sol–gel transition used for visual detection of glucose. Chem Commun 48(31):3739–3741

Yumashev AV, Rudiansyah M, Chupradit S, Kadhim MM, Jalil AT, Abdelbasset WK, Suksatan W, Parra RM, Mustafa YF, Abdullaev B, Bidares R (2022) Optical-based biosensor for detection of oncomarker CA 125, recent progress and current status. Anal Biochem 655:114750

Zhang J, He F (2022) Mycobacterium tuberculosis piezoelectric sensor based on AuNPs-mediated enzyme assisted signal amplification. Talanta 236:122902

Zhao YY (2013) Metabolomics in chronic kidney disease. Clin Chim Acta 422:59–69

Zhao J, O'daly JP, Henkens RW, Stonehuerner J, Crumbliss AL (1996) A xanthine oxidase/colloidal gold enzyme electrode for amperometric biosensor applications. Biosens Bioelectron 11(5):493–502

Zhao YD, Zhang WD, Chen H, Luo QM, Li SF (2002) Direct electrochemistry of horseradish peroxidase at carbon nanotube powder microelectrode. Sensors Actuators B Chem 87(1):168–172

Zhong W (2009) Nanomaterials in fluorescence-based biosensing. Anal Bioanal Chem 394(1):47–59

Zimmerman MA, Selzman CH, Cothren C, Sorensen AC, Raeburn CD, Harken AH (2003) Diagnostic implications of C-reactive protein. Arch Surg 138(2):220–224

14 Janus Nano-Assembly Based Sensing Platform for Cancer Theranostics: An Unrivaled Mastering Bioimaging Perspective

Syeda Warisul Fatima, Shahenvaz Alam, and Sunil K. Khare

Abstract

This study aims to explore the design, synthesis, and applications of Janus nanoparticles in the field of cancer theranostics. Janus nanoparticles, named after the two-faced Roman god Janus, possess distinct sides or regions that enable multifunctionality and versatility in cancer therapy and diagnostics. This study investigates the potential of Janus nanoparticles for targeted drug delivery, molecular imaging, and personalized medicine approaches in cancer treatment. The research focuses on addressing challenges related to stability, biocompatibility, scalability, and clinical evaluation, aiming to overcome these barriers for the successful translation of Janus nanoparticles into clinical practice. Targeted cancer theranostics, involving concurrent therapeutic treated and selective imaging of malignant tissue, has recently emerged as a potent method for enhancing patient prognosis. Targeted multimodal molecular imaging has garnered attention by overcoming specific limitations of conventional imagining modalities. In the current era, the quest to develop suitable bio-based materials and nano-scaled

S. W. Fatima (✉)
Enzyme and Microbial Biochemistry Laboratory, Department of Chemistry, Indian Institute of Technology Delhi, New Delhi, India

Department of Radiation Oncology, Emory University School of Medicine, Atlanta, Georgia, USA

S. Alam
Enzyme and Microbial Biochemistry Laboratory, Department of Chemistry, Indian Institute of Technology Delhi, New Delhi, India

Department of Biochemistry and Molecular Biology, University of Georgia, Athens, Georgia, USA

S. K. Khare
Enzyme and Microbial Biochemistry Laboratory, Department of Chemistry, Indian Institute of Technology Delhi, New Delhi, India

M. Mohsin, N. Soleja (eds.), *Nanoscale Sensors and their Applications in Biomedical Imaging*, https://doi.org/10.1007/978-981-97-3144-2_14

colloidal particulates for established application domains has resulted in notable efforts in the advancement of controlled size and shape, anisotropy, and features of nanostructures, in which the building blocks are designed via random self-assembly. It is also noteworthy to mention that salient features including multiple imaging methods to screening and diagnosis, as well as the integration of imaging elements into JNPs (such as quantum dots and fluorescent dyes) could be accomplished. State-of-the-art developments in JNPs, including their classification and the performance of fabric nanoparticles in cancer theranostics, are also covered. This study will outline the primary production strategies as well as major advancements of JNPs paving way in the biomedical and cancer therapy.

Keywords

Janus nanoparticle · Multimodal imaging · Cancer theranostics · Drug delivery · Nanotechnology · Cancer imaging

14.1 Introduction

Janus nanoparticles have gained significant attention in the field of cancer theranostics due to their unique structure and multifunctional properties (Wang et al. 2021a; Li et al. 2021a). These nanoparticles, named after the Roman god Janus who had two faces looking in opposite directions, possess a dual-sided functionality that allows for both therapeutic and diagnostic applications. In this article, we will explore the potential of Janus nanoparticles in cancer theranostics and their impact on improving cancer diagnosis, treatment, and monitoring (Zhang et al. 2021a; Wang et al. 2021b; Liu et al. 2021a; Zhang et al. 2021b). Advanced nanomaterials, i.e., Janus nanoplatform play a central role in theranostics development owing to their distinctive features (Wang et al. 2021c). This is because they have strong stimulus-responsive abilities, customizable structures, are biocompatible, have two surfaces with different hydrophobic qualities, and have distinctive functional groups. Featuring two parts of the Janus amphiphilic particles with differing hydrophobicity has been the most critical aspect. Using imaging agents, such as gold (AU) for photoacoustic imaging processing (PAl), silver for surface-enhanced Raman scattering (SERS), and Fe_30_4 and MnO_2 for magnetic resonance imaging (MRI), has rendered it possible to develop JNPs (Zhang et al. 2021a; Wu et al. 2020; Zhang et al. 2019a; Wang et al. 2019a; Zhang and Liang 2019; Huang et al. 2018). The development of a novel Janus nanoplatform that combines hybrid nanoparticle with face such as core@shell has revolutionized the cancer therapy. The hybrid nanomaterial's anisotropy allows for the orthogonal site-selective alteration of each face, facilitating the integration of a fluorescent dye and a targeting peptide for cancer detection (Zhang et al. 2018a). The resulting hybrid nanocarrier successfully promotes in vivo tumor-targeted multimodal imaging by magnetic resonance, computed tomography, and fluorescent tracking (fluorescent dye loading) in sarcoma-bearing mouse model due to the inherent characteristics of this Janus nanoplatform along with those selectively generated on their surfaces. Janus nanostructures have shown great promise as bioengineering platforms for the transportation of bioactive hydrophilic/hydrophobic materials as well as for use as therapeutic carriers, drug delivery systems, and biosensor equipment (Lee et al. 2007; Huang et al. 2006).

It gives an account of improvements made in the creation of Janus particles utilizing diverse materials and methods. It rates Janus particles as a promising particulate system with the potential to enhance traditional delivery methods by offering a superior loading capacity and targeting specificity while also encouraging multi-drug loading and single-dose drug administration (Wang et al. 2021a, d). The creation of nano-based sensors for diagnostic uses has been facilitated by nanoparticles. However, the incapacity of nanoparticles to detect several analytes with a single type of nanoparticle limits their usage in the creation of sensing devices for multiple analyte detection. The name "JNPs" refers to micro- or nanoscale particles that have been segmented into compartments or sections, each of which has a unique set of chemical or physical properties, creating multifunctional particles endowed with distinguishing traits (Li et al. 2021a, b; Liu et al. 2021b; Wang et al. 2021e). Furthermore, JNPs can carry out numerous tasks simultaneously within a single particle without interference from nearby parts. This review focuses on the development of biosensors using Janus particles as well as the examination of numerous qualities these Janus particles possess that make them suitable for use as biosensors. The several kinds of Janus particle-based biosensors that are now on the market are also covered. Finally, the limitations of Janus particles and their potential future applications of cancer have been covered.

14.1.1 Overview of Janus Nanoparticles

Janus nanoparticles (JNPs) contain two faces that are different from one another in terms of their physical or chemical composition, giving them special qualities. Since the 1990s, JNPs have been the subject of significant scientific investigation because their anisotropy can result in phenomena like self-assembly or interfacial activity (Zhang et al. 2021c). Since then, potent synthetic methods have been created that enable JNPs with virtually infinite combinations of components, traits, and morphologies. As a result, JNPs have been used in a variety of applications, including optics, interface engineering, catalysis, and drug administration. In this viewpoint, the synthesis of these significant anisotropic materials is covered. This chapter specifically defines the main classes of JNP synthetic techniques, provides some historical background, discusses current advancements in this field, and assesses the advantages and disadvantages of the various approaches with an eye toward potential future directions and applications (Wang et al. 2021f). With this work, we anticipate that this viewpoint will aid in orienting people new to the area who wish to create and/or utilize JNPs (Li et al. 2021c; Liu et al. 2021c).

The anisotropic JNPs bestow synergistic effects by concentrating multiple properties on a single carrier. A new and intelligent material called Janus particles has the potential to enhance theranostics, medication delivery, and pharmaceutical formulation. In contrast to other conventional particles, Janus particles feature two distinct partitions that have different functions, physicochemical qualities, and morphological traits (Huang et al. 2006). Since they can accommodate two pharmaceutical agents that are in opposition to one another and can be engineered at the molecular level to achieve better target affinity, lower drug dosage to achieve a therapeutic effect, and control drug release with improved pharmacokinetics and pharmacodynamics, Janus particles have recently attracted a lot of attention as effective particulate drug delivery systems (Wang et al. 2021a; Wu et al. 2020; Huang et al. 2006). The development of Janus particles aims to enhance the precision and efficacy of administering pharmacological agents for cancer treatment (Wang and Li 2021a; Zhang and Li 2021a; Li and Li 2021).

14.1.2 Importance of Janus Nanoparticles in Versatile Applications

Janus nanoparticles have gained significant attention due to their ability to combine therapeutic and diagnostic functionalities in a single platform (Wang et al. 2021a; Liu et al. 2021d; Wang and Li 2021b; Zhang and Li 2021b; Wang and Li 2021c). Their design allows for targeted drug delivery, controlled release of therapeutic agents, and precise imaging of cancer cells. This makes them a promising tool in the field of cancer theranostics, offering improved treatment outcomes and personalized medicine (Zhang et al. 2021b; Liu et al. 2021e).

14.1.3 Focus on the Design and Synthesis of Janus Nanoparticles for Specific Applications

This article focuses on the design and synthesis of Janus nanoparticles, with an emphasis on their applications in cancer therapy (Wang et al. 2021b, 2019b, 2018). We will explore the different considerations involved in designing Janus nanoparticles, including the selection of nanoparticle material, determination of size and shape, and surface functionalization strategies. Additionally, we will discuss various synthesis techniques and characterization methods used to fabricate and analyze Janus nanoparticles (Zhang and Liang 2019; Lee et al. 2007; Liu et al. 2019).

14.2 Design Considerations for Janus Nanoparticles

14.2.1 Selection of Nanoparticle Material

The choice of nanoparticle material depends on the desired properties and applications. Common materials used for Janus nanoparticles include gold, silver, silica, polymer-based nanoparticles, and magnetic nanoparticles (Wang et al. 2021c, f; Chen et al. 2020). Factors such as biocompatibility, stability, and surface functionalization capabilities should be considered during material selection.

14.2.2 Determination of Nanoparticle Size and Shape

The size and shape of Janus nanoparticles play a crucial role in their properties and applications. The size can affect their cellular uptake, biodistribution, and circulation time. The shape can influence their stability, drug loading capacity, and targeting efficiency (Zhang et al. 2019a, 2018b). Techniques such as seed-mediated growth, emulsion-based methods, and solvent evaporation can be employed to control the size and shape of Janus nanoparticles (Zhang et al. 2018a).

14.2.3 Surface Functionalization Strategies

The surface of Janus nanoparticles needs to be functionalized to impart specific properties, such as targeting ligands, therapeutic agents, and imaging agents. Ligand attachment methods, including covalent bonding and non-covalent interactions, can be used to attach targeting ligands to one side of the nanoparticle (Huang et al. 2018; Liu et al. 2021b). Coating techniques and self-assembly approaches can be utilized to achieve functionalization on both sides of the Janus nanoparticle.

14.3 Synthesis Techniques for Janus Nanoparticles

14.3.1 Chemical Synthesis Methods

Chemical synthesis methods are commonly employed to fabricate Janus nanoparticles with controlled growth and composition by manipulating reaction conditions. Seed-mediated growth, emulsion-based methods, and solvent evaporation techniques are commonly employed. These methods allow for precise control over the synthesis of JNPs to achieve desired nanoparticle properties (Wang and Li 2021a, b; Li et al. 2019).

14.3.1.1 Seed-Mediated Growth

Seed-mediated growth involves the use of pre-existing nanoparticles as seeds for further growth. Different materials can be used as seeds to initiate the growth of Janus nanoparticles with desired properties. By carefully controlling the reaction parameters, such as temperature, concentration, and reaction time, the growth of nanoparticles can be directed to form Janus structures with distinct sides (Wang et al. 2019a; Li et al. 2018).

14.3.1.2 Emulsion-Based Methods

Emulsion-based methods utilize emulsions, which are systems consisting of two immiscible liquids, to create Janus nanoparticles. One approach involves the formation of an emulsion containing two different precursors for nanoparticle synthesis (Li et al. 2020). By controlling the emulsion conditions and subsequent reactions,

Janus nanoparticles can be formed with one side composed of one material and the other side composed of another material.

14.3.1.3 Solvent Evaporation Techniques

Solvent evaporation techniques involve the controlled evaporation of solvents containing nanoparticle precursors. One common method is the droplet evaporation technique, where droplets of the precursor solution are deposited onto a substrate. As the solvent evaporates, the nanoparticles form and self-assemble into Janus structures due to differences in surface properties or chemical interactions (Zhang et al. 2019a).

Physical Methods

Physical methods offer alternative approaches to fabricate Janus nanoparticles, relying on physical forces and interactions to create the desired structures.

Electrostatic Assembly

Electrostatic assembly involves the use of electrostatic interactions to guide the assembly of nanoparticles into Janus structures. By modifying the surface charges of the nanoparticles, they can be attracted or repelled by each other, leading to the formation of Janus architectures. Electrostatic assembly allows for precise control over the composition and arrangement of the nanoparticles (Wang et al. 2021e; g).

Layer-by-Layer Deposition

Layer-by-layer deposition is a technique that involves the sequential adsorption of different layers onto a substrate or nanoparticle surface. By alternating the deposition of materials with distinct properties, Janus nanoparticles can be created. This method offers flexibility in choosing the materials and controlling the thickness and composition of each layer.

14.4 Surface Modification of Janus Nanoparticles

14.4.1 Introduction to Surface Modification

Surface modification plays a crucial role in tailoring the properties and functionalities of Janus nanoparticles. It involves the attachment of various molecules or ligands onto the nanoparticle surface, enabling specific interactions with target cells or molecules. Surface modification strategies enhance the targeting capabilities, drug delivery efficiency, and imaging performance of Janus nanoparticles.

14.4.2 Ligand Attachment Strategies

Ligand attachment strategies are employed to functionalize the surface of Janus nanoparticles, allowing for specific interactions with target cells or biomolecules.

14.4.2.1 Covalent Bonding

Covalent bonding involves the formation of strong chemical bonds between the ligand and the nanoparticle surface. This strategy ensures stable attachment and long-term functionality of the ligands (Lee et al. 2007; Liu et al. 2020). Covalent bonding methods may include the use of reactive functional groups on the nanoparticle surface, such as amino groups or carboxyl groups, which can react with corresponding functional groups on the ligands.

14.4.2.2 Non-covalent Interactions

Non-covalent interactions are reversible interactions that do not involve the formation of covalent bonds. These interactions include electrostatic interactions, hydrogen bonding, hydrophobic interactions, and van der Waals forces (Wang et al. 2021a; Huang et al. 2018). Non-covalent attachment methods are relatively simpler and more versatile, allowing for easy modification of the nanoparticle surface with a wide range of ligands.

14.4.3 Functionalization of Each Side of the Janus Nanoparticle

To maximize the potential of Janus nanoparticles in cancer theranostics, each side of the nanoparticle can be functionalized with specific molecules or agents.

14.4.3.1 Targeting Ligands for Specific Cell Recognition

One side of the Janus nanoparticle can be functionalized with targeting ligands, such as antibodies, peptides, or aptamers, that specifically recognize and bind to biomarkers expressed on the surface of cancer cells. These ligands enhance the nanoparticle's ability to selectively target cancer cells while minimizing non-specific interactions with healthy cells, thus improving the efficiency of drug delivery and reducing off-target effects (Wang et al. 2019a; Liu et al. 2021d; Wang et al. 2020).

14.4.3.2 Therapeutic Agents for Drug Delivery

The other side of the Janus nanoparticle can be functionalized with therapeutic agents, such as chemotherapy drugs, gene therapies, or immunotherapies. These agents can be loaded onto the nanoparticle surface or encapsulated within the nanoparticle core. The Janus nanoparticle acts as a carrier, protecting the therapeutic agents from degradation and delivering them directly to the tumor site, increasing their concentration at the desired location and minimizing systemic toxicity (Li et al. 2021a; Zhang et al. 2021a; Wang et al. 2021b; Liu et al. 2021a).

14.4.3.3 Imaging Agents for Diagnostics

Both sides of the Janus nanoparticle can be functionalized with imaging agents, such as fluorescent dyes, quantum dots, or magnetic nanoparticles. These imaging agents provide contrast and enable non-invasive imaging of the tumor site. By incorporating specific imaging agents, such as near-infrared dyes or magnetic resonance contrast agents, Janus nanoparticles can be detected using various imaging techniques, including fluorescence imaging, magnetic resonance imaging (MRI), or computed tomography (CT) scans (Zhang and Liang 2019; Huang et al. 2018; Zhang et al. 2018a; Lee et al. 2007). This facilitates accurate tumor detection, monitoring of treatment response, and real-time visualization of the nanoparticles' distribution within the body.

Thus, surface modification of Janus nanoparticles is crucial for enhancing their targeting capabilities, drug delivery efficiency, and imaging performance. Covalent bonding and non-covalent interactions provide strategies for attaching ligands to the nanoparticle surface. Functionalization of each side of the Janus nanoparticle with targeting ligands, therapeutic agents, and imaging agents enables specific cell recognition, efficient drug delivery, and accurate diagnostics (Li et al. 2021b). These surface modification strategies contribute to the potential of Janus nanoparticles in revolutionizing cancer theranostics by improving treatment outcomes and enabling personalized medicine.

Here is an outline summarizing the design methods and applications of Janus nanoparticles in cancer therapy:

Design Method | Application in Cancer Therapy

1. Surface Functionalization
   - Incorporation of targeting ligands or antibodies on one side of the nanoparticle | Targeted drug delivery to cancer cells.
   - Attachment of imaging agents for cancer imaging and diagnosis | Enhanced cancer imaging and early detection.
   - Conjugation of therapeutic agents for controlled release | Improved therapeutic efficacy and reduced side effects.
2. Core-Shell Structure
   - Encapsulation of chemotherapy drugs or gene therapies in the core | Targeted and controlled drug delivery to cancer cells.
   - Surface modification of the shell for tumor-specific targeting | Enhanced treatment efficacy and reduced off-target effects.
3. Janus Nanoparticle Assembly
   - Self-assembly of Janus nanoparticles into larger structures | Formation of multifunctional platforms for combination therapy.
   - Integration of different therapeutic agents or imaging modalities on different sides of the assembled structure | Synergistic effects and personalized treatment approaches.
4. Surface Charge Manipulation
   - Modification of the surface charge of Janus nanoparticles | Enhanced cellular uptake and intracellular drug delivery.
   - Utilization of electrostatic interactions for targeted drug delivery | Improved treatment response and reduced systemic toxicity.
5. Surface Coating and Modification
   - Coating Janus nanoparticles with biocompatible polymers | Improved stability, biocompatibility, and circulation time.

   - Introduction of stimuli-responsive coatings for triggered drug release | Controlled and on-demand drug release at the tumor site.
6. Integration of Plasmonic Materials
   - Incorporation of gold or other plasmonic materials in Janus nanoparticles | Photothermal therapy for tumor ablation or enhanced drug release.
   - Utilization of plasmonic properties for imaging and diagnosis | Multimodal imaging and precise tumor targeting.

14.5 Characterization Techniques for Janus Nanoparticles

Characterization techniques are essential for assessing the morphology, structure, surface properties, and functional performance of Janus nanoparticles. These techniques provide valuable insights into the quality and efficacy of the synthesized nanoparticles, ensuring their suitability for specific applications.

14.5.1 Morphological Characterization

Morphological characterization techniques allow for the visualization and analysis of the shape, size, and distribution of Janus nanoparticles.

14.5.1.1 Transmission Electron Microscopy (TEM)

TEM provides high-resolution images of the nanoparticles, allowing for detailed examination of their morphology and structure. It enables the determination of particle size, shape, and the presence of any defects or surface modifications. Additionally, TEM can provide information about the distribution and orientation of the two distinct sides of Janus nanoparticles (Wu et al. 2020; Zhang et al. 2019a; Wang et al. 2019a).

14.5.1.2 Scanning Electron Microscopy (SEM)

SEM is another imaging technique that provides surface morphology information of Janus nanoparticles. It offers a larger field of view compared to TEM and can be used to assess particle size, shape, and surface topography (Li et al. 2021c; Liu et al. 2021c; Wang and Li 2021a; Zhang and Li 2021a). SEM can be particularly useful for studying the overall arrangement and distribution of Janus nanoparticles within a sample.

14.5.2 Structural Characterization

Structural characterization techniques are employed to analyze the crystal structure, composition, and chemical bonding of Janus nanoparticles (Wang et al. 2021a; Chen et al. 2020).

14.5.2.1 X-Ray Diffraction (XRD)

XRD is a powerful technique used to determine the crystal structure and phase composition of nanoparticles. By analyzing the diffraction pattern of X-rays scattered by the nanoparticles, information about the lattice parameters, crystal structure, and orientation can be obtained (Zhang et al. 2019a; Zhang and Li 2021a). XRD can provide insights into the arrangement of atoms within the Janus nanoparticles and their potential impact on properties such as stability and reactivity.

14.5.2.2 Fourier-Transform Infrared Spectroscopy (FTIR)

FTIR spectroscopy is used to analyze the chemical composition and functional groups present on the surface of Janus nanoparticles (Zhang et al. 2021a; Liu et al. 2021b). It measures the absorption of infrared radiation by the nanoparticles, providing information about the types of chemical bonds and functional groups present. FTIR can be used to confirm the successful attachment of ligands or functional groups onto the nanoparticle surface, ensuring the desired surface modifications.

14.5.3 Surface Analysis Techniques

Surface analysis techniques allow for the characterization of the surface chemistry, charge, and interaction capabilities of Janus nanoparticles.

14.5.3.1 X-Ray Photoelectron Spectroscopy (XPS)

XPS is a technique used to analyze the elemental composition and chemical state of the surface of Janus nanoparticles. It provides information about the binding energies of the elements present, allowing for the identification of specific chemical species and functional groups. XPS can help confirm the successful functionalization of the nanoparticle surface and assess the chemical environment of the ligands or coatings (Liu et al. 2021d).

14.5.3.2 Dynamic Light Scattering (DLS)

DLS measures the hydrodynamic size and size distribution of Janus nanoparticles in a liquid suspension. By analyzing the intensity fluctuations of scattered light, information about the particle size, polydispersity, and stability can be obtained. DLS is particularly useful for assessing the colloidal stability and aggregation behavior of Janus nanoparticles, which can impact their performance in various applications (Wu et al. 2020; Zhang et al. 2019a).

14.5.4 Functional Characterization

Functional characterization techniques evaluate the performance and behavior of Janus nanoparticles in terms of their intended applications.

14.5.4.1 Drug Release Kinetics

To assess the efficiency and controllability of drug delivery, studies can be conducted to measure the release kinetics of therapeutic agents from Janus nanoparticles (Wang et al. 2019b). Techniques such as UV–Vis spectroscopy, HPLC, or fluorescence spectroscopy can be used to quantify the amount of released drug over time, providing insights into the release profile and potential for sustained or triggered release.

14.5.4.2 Targeting Efficiency

The targeting efficiency of Janus nanoparticles can be evaluated by assessing their ability to bind specifically to target cells or biomarkers (Li et al. 2018, 2020; Zhang et al. 2019b; Wang et al. 2021g; Liu et al. 2020). Techniques such as flow cytometry or confocal microscopy can be used to quantify the binding or internalization of Janus nanoparticles by target cells, providing information about their targeting capabilities and potential for specific cell recognition.

14.5.4.3 Imaging Performance

To evaluate the imaging capabilities of Janus nanoparticles, techniques such as fluorescence imaging, MRI, or CT scans can be utilized. These techniques can assess the contrast enhancement, signal intensity, and distribution of the nanoparticles within a biological system. By analyzing the imaging results, the sensitivity, specificity, and suitability of the Janus nanoparticles for diagnostics and imaging applications can be determined (Liu et al. 2020).

Thus, characterization techniques are crucial for analyzing the morphology, structure, surface properties, and functional performance of Janus nanoparticles. Techniques such as TEM, SEM, XRD, FTIR, XPS, DLS, and functional characterization methods enable the assessment of nanoparticle size, shape, crystal structure, surface chemistry, targeting efficiency, drug release kinetics, and imaging performance (Huang et al. 2018; Zhang et al. 2018b). These techniques provide valuable information for optimizing the design, synthesis, and application of Janus nanoparticles in various fields, including cancer therapy, diagnostics, and imaging.

14.6 Applications of JNPS

The dual-sided functionality of Janus nanoparticles is a significant advantage in cancer theranostics. One side of the nanoparticle can be functionalized with targeting ligands, such as antibodies or peptides, that specifically recognize and bind to cancer cells. This targeted binding enables the nanoparticles to selectively accumulate at the tumor site, increasing the concentration of therapeutic agents and imaging agents in the cancer cells while minimizing their distribution in healthy tissues.

On the other side, Janus nanoparticles can be designed to carry therapeutic agents, such as chemotherapy drugs or gene therapies. These therapeutic agents can be loaded onto the nanoparticle surface or encapsulated within the nanoparticle core, depending on the desired release kinetics (Chen et al. 2020; Zhang et al.

2018b; Li et al. 2019). By delivering the therapeutic agents directly to the cancer cells, Janus nanoparticles enhance treatment efficacy and reduce systemic toxicity compared to conventional chemotherapy drugs.

In addition to their therapeutic capabilities, Janus nanoparticles offer diagnostic functionalities. These nanoparticles can be engineered to incorporate imaging agents, such as fluorescent dyes, quantum dots, or magnetic nanoparticles (Liu et al. 2021a, 2020; Wang et al. 2020). By incorporating these imaging agents, the nanoparticles become visible under different imaging modalities, including fluorescence imaging, magnetic resonance imaging (MRI), or computed tomography (CT) scans. This allows for non-invasive imaging of the tumor site, facilitating cancer diagnosis and monitoring.

The imaging capabilities of Janus nanoparticles enable precise localization and visualization of the tumor site. This is particularly important in cancer diagnosis, where accurate tumor detection and staging are crucial for determining the appropriate treatment strategy. By using Janus nanoparticles as contrast agents, clinicians can obtain high-resolution images of the tumor site, providing valuable information about its size, shape, and location (Liu et al. 2021e; Li et al. 2020).

Furthermore, Janus nanoparticles can be functionalized with specific ligands or antibodies that target biomarkers expressed on the surface of cancer cells. These biomarker-targeted nanoparticles enable molecular imaging, allowing for the detection and characterization of specific cancer subtypes. This molecular imaging approach provides valuable information about the tumor's molecular profile, aiding in personalized treatment planning and monitoring of treatment response (Wang et al. 2021c; Wu et al. 2020).

The ability of Janus nanoparticles to combine therapeutic and diagnostic functionalities in a single platform is a major advantage in cancer theranostics (Zhang et al. 2019b; Wang et al. 2021g; Liu et al. 2020). This integrated approach allows for simultaneous imaging and treatment of cancer cells, enabling real-time monitoring of treatment response. By using the same nanoparticles for both imaging and therapy, the nanoparticles can be tracked and quantified, providing valuable information about their distribution and accumulation at the tumor site.

Janus nanoparticles also offer controlled drug release, which is crucial for optimizing therapeutic outcomes. By modifying the nanoparticle's surface properties or incorporating stimuli-responsive materials, drug release can be triggered by specific conditions, such as changes in pH, temperature, or enzyme activity. This controlled release ensures that therapeutic agents are delivered to the cancer cells in a sustained and controlled manner, maximizing their effectiveness and minimizing off-target effects.

In addition to their therapeutic and diagnostic capabilities, Janus nanoparticles have been explored for other applications in cancer theranostics. For instance, these nanoparticles can be engineered to carry nucleic acids, such as siRNA or miRNA, for gene therapy applications. By delivering these nucleic acids to the cancer cells, Janus nanoparticles can modulate gene expression, targeting specific oncogenes or tumor suppressor genes to inhibit tumor growth or enhance the immune response against cancer cells (Wang et al. 2019b; Li et al. 2018; Wang et al. 2020).

Moreover, Janus nanoparticles can be functionalized with stimuli-responsive materials that can be activated by external stimuli, such as light, heat, or magnetic fields. This opens up opportunities for targeted drug delivery and therapy activation. For example, by using near-infrared (NIR) light, which can penetrate tissues deeply, Janus nanoparticles can be activated to release therapeutic agents or generate heat to induce hyperthermia therapy selectively at the tumor site (Liu et al. 2021a; Wang et al. 2021d).

The versatility of Janus nanoparticles extends beyond therapy and diagnostics. These nanoparticles can also be used for monitoring treatment response and guiding subsequent therapy decisions. For example, by incorporating imaging agents that respond to changes in tumor microenvironment, such as pH or oxygen levels, Janus nanoparticles can provide real-time information about treatment efficacy. Changes in the imaging signal can indicate tumor regression, resistance to treatment, or the development of metastasis, guiding the adjustment of treatment strategies accordingly (Liu et al. 2021b).

Furthermore, Janus nanoparticles can be engineered to respond to specific biomarkers or signals associated with treatment response, such as changes in enzymatic activity or drug metabolism. By incorporating these responsive elements, the nanoparticles can act as smart probes, providing real-time feedback on treatment efficacy and enabling personalized treatment optimization.

14.6.1 Cancer Therapeutics

14.6.1.1 Targeted Drug Delivery

- Janus nanoparticles can be functionalized with targeting ligands or antibodies that specifically recognize cancer cells or tumor biomarkers.
- This enables selective delivery of therapeutic agents, such as chemotherapy drugs or gene therapies, directly to the tumor site, increasing treatment efficacy while minimizing off-target effects on healthy tissues (Li et al. 2021c).

14.6.1.2 Overcoming Drug Resistance

- Janus nanoparticles offer a potential solution to overcome drug resistance mechanisms in cancer cells (Wang and Li 2021b; Zhang and Li 2021b; Wang and Li 2021c).
- By incorporating multiple drugs or therapeutic modalities on different sides of the nanoparticle, Janus nanoparticles can target multiple pathways or vulnerabilities in cancer cells, reducing the likelihood of resistance development.

14.6.1.3 Combination Therapy

- Janus nanoparticles enable the delivery of multiple therapeutic agents simultaneously, facilitating combination therapy approaches (Li et al. 2021a; Lee et al. 2007).

- Different drugs, immunotherapies, or gene therapies can be integrated on each side of the nanoparticle, allowing for synergistic effects and enhanced treatment outcomes.
- Combination therapies using Janus nanoparticles have shown promise in preclinical models by improving treatment response rates and overcoming limitations of single-agent therapies.

14.6.2 Diagnostics and Imaging

14.6.2.1 Molecular Imaging

- Janus nanoparticles can be engineered with imaging agents, such as fluorescent dyes, quantum dots, or magnetic nanoparticles (Zhang and Li 2021b).
- These imaging agents enable the visualization and tracking of Janus nanoparticles in vivo, facilitating molecular imaging for cancer diagnosis and monitoring of treatment response.

14.6.2.2 Theranostic Platforms

- Janus nanoparticles can serve as theranostic platforms, combining therapeutic and diagnostic capabilities in a single system.
- By integrating imaging agents and therapeutic agents on different sides of the nanoparticle, Janus nanoparticles enable simultaneous cancer imaging, diagnosis, and targeted therapy.
- This theranostic approach allows for real-time monitoring of treatment response and personalized adjustments to therapy, leading to improved patient outcomes.

14.6.3 Key JNP Applications Are as Follows

14.6.3.1 Gene Therapy

- Janus nanoparticles can be utilized in gene delivery systems for targeted and efficient delivery of therapeutic genes to cancer cells (Wang et al. 2021b, 2019b).
- By encapsulating or conjugating therapeutic genes on Janus nanoparticles, specific gene targets can be addressed, offering potential solutions for genetic-based cancer therapies.

14.6.3.2 Photothermal Therapy

- Janus nanoparticles can incorporate plasmonic materials, such as gold nanoparticles, which possess strong light-absorbing properties (Zhang and Li 2021a).
- When exposed to near-infrared light, these plasmonic Janus nanoparticles generate localized heat, leading to tumor ablation or enhanced drug release in photothermal therapy.

14.6.3.3 Biosensing

- Janus nanoparticles can be functionalized with biosensing elements, such as antibodies or aptamers, to detect specific cancer biomarkers.
- By utilizing the unique properties of Janus nanoparticles, such as their dual-sided nature, biosensing platforms can be developed for early cancer detection and monitoring of disease progression.

These applications of Janus nanoparticles demonstrate their versatility and potential in various aspects of cancer therapy, diagnostics, and other related fields. Continued research and development in this area hold promise for improving cancer treatment outcomes and advancing personalized medicine approaches (Li et al. 2021a; Zhang et al. 2019a, 2018b; Liu et al. 2020).

14.7 Challenges of JNPs

14.7.1 Stability and Biocompatibility Issues

- One of the challenges in the development of Janus nanoparticles is ensuring their stability and biocompatibility.
- Janus nanoparticles should maintain their structural integrity and functionality during storage, circulation in the body, and interaction with biological systems (Wang et al. 2021c; Zhang and Li 2021b).
- Careful selection of materials, surface coatings, and fabrication methods is required to address stability and biocompatibility concerns.

14.7.2 Scalability and Reproducibility

- The scalability and reproducibility of Janus nanoparticles pose challenges in their large-scale production.
- Achieving consistent particle size, shape, and surface functionalities is crucial for reliable and reproducible results (Zhang and Li 2021b).
- Developing scalable synthesis methods and quality control measures is essential for the translation of Janus nanoparticles into clinical applications.

14.7.3 Preclinical and Clinical Evaluation

- Extensive preclinical evaluation is necessary to assess the safety, efficacy, and toxicity profiles of Janus nanoparticles.
- Animal studies should be conducted to evaluate the biodistribution, pharmacokinetics, and long-term effects of Janus nanoparticles.
- Clinical evaluation, including rigorous clinical trials, is needed to assess the feasibility, effectiveness, and potential side effects of Janus nanoparticles in humans.

14.7.4 Emerging Trends and Future Directions

- The field of Janus nanoparticles is rapidly evolving, and several emerging trends and future directions can be identified.
- Integration of artificial intelligence and machine learning algorithms for precise targeting, therapy guidance, and personalized medicine approaches.
- Development of stimuli-responsive Janus nanoparticles that can release therapeutic agents in response to specific cues in the tumor microenvironment.
- Exploration of nanotheranostics, combining imaging, therapy, and monitoring capabilities within a single Janus nanoparticle system.
- Investigation of Janus nanoparticles for immunomodulation, including their potential in enhancing immune responses and overcoming immunosuppression in the tumor microenvironment (Liu et al. 2021e; Wang et al. 2019b, 2018; Liu et al. 2019; Chen et al. 2020).

Here is a comparison of Janus particles utilized for cancer therapy, along with examples:

1. Liposome-based Janus particles:
   - Liposome-based Janus particles consist of lipid bilayers with different properties on each side, enabling the encapsulation and delivery of therapeutic agents to specific targets.
2. Polymer-based Janus particles:
   - Polymer-based Janus particles are composed of polymers with different properties on each side, allowing for controlled release of drugs and targeted therapy (Liu et al. 2021e).
3. Gold-based Janus particles:
   - Gold-based Janus particles possess different surface coatings, with one side functionalized for targeted drug delivery and the other side for imaging or photothermal therapy (Li et al. 2019).
4. Silica-based Janus particles:
   - Silica-based Janus particles have distinct surface chemistries on each side, allowing for controlled drug release and targeted therapy (Zhang et al. 2021b; Wang et al. 2021e).
5. Magnetic Janus particles:
   - Magnetic Janus particles possess magnetic properties on one side, enabling targeted drug delivery and magnetic hyperthermia for cancer therapy (Liu et al. 2020).

Janus nanoparticles have shown promising potential in various types of cancer. Here is a brief report on the application of Janus nanoparticles in different cancer types:

1. *Breast Cancer:*
   - Janus nanoparticles have been utilized for targeted drug delivery to breast cancer cells, improving therapeutic efficacy and reducing side effects (Lee et al. 2007). They have also been used for imaging and early detection of breast cancer, enhancing diagnostic accuracy.
2. *Lung Cancer:*
   - Janus nanoparticles have been employed for targeted drug delivery and controlled release in lung cancer treatment, improving drug efficacy and reducing toxicity (Wang et al. 2018). They have also been employed for lung cancer, aiding in accurate diagnosis.
3. *Prostate Cancer:*
   - Janus nanoparticles have been utilized for targeted drug delivery to prostate cancer cells, enhancing treatment outcomes and minimizing off-target effects. Help in prognosis of prostate cancer, facilitating early detection and disease monitoring (Zhang et al. 2018b).
4. *Colorectal Cancer:*
   - Janus nanoparticles have been employed for targeted drug delivery and controlled release in colorectal cancer treatment, improving therapeutic efficacy and reducing side effects.
   - They have also been used for imaging and early detection of colorectal cancer, aiding in accurate diagnosis and disease monitoring (Zhang et al. 2019b).
5. *Pancreatic Cancer:*
   - Janus nanoparticles have been utilized for targeted drug delivery and combined therapy in pancreatic cancer, enhancing treatment response and overcoming drug resistance (Li et al. 2020). They have also been used for imaging and early detection of pancreatic cancer, facilitating timely intervention and improved patient outcomes (Wang et al. 2021g).

14.7.5 Mechanism of Action of JNPs

It is important to note that the field of Janus nanoparticles in cancer therapy is rapidly evolving, with ongoing research and advancements. These references provide further insights into the specific applications and advancements in Janus nanoparticles for different cancer types.

The mechanism of action of Janus nanoparticles in cancer therapy can vary depending on the specific design and functionalization of the nanoparticles. However, here are some general mechanisms by which Janus nanoparticles exert their therapeutic effects:

1. *Targeted Drug Delivery:*
   - Janus nanoparticles can be designed to selectively deliver therapeutic agents, such as chemotherapeutic drugs or RNA-based therapies, to cancer cells.
   - The distinct surface properties of Janus nanoparticles allow for specific targeting of cancer cells, minimizing off-target effects.

   - The therapeutic agents can be encapsulated within or conjugated onto the nanoparticle surface, enabling controlled release and enhanced drug efficacy.
   - This targeted drug delivery mechanism helps to increase the concentration of therapeutic agents at the tumor site, improving treatment outcomes (Wang et al. 2019a; Li et al. 2021c).
2. *Enhanced Cellular Uptake:*
   - Janus nanoparticles can facilitate enhanced cellular uptake of therapeutic agents by cancer cells.
   - The unique surface properties of Janus nanoparticles, such as charge or ligand functionalization, can promote receptor-mediated endocytosis or active transport mechanisms.
   - This increased cellular uptake allows for higher intracellular concentrations of therapeutic agents, enhancing their effectiveness against cancer cells (Wang et al. 2018; Li et al. 2018).
3. *Combination Therapy:*
   - Janus nanoparticles can be designed to deliver multiple therapeutic agents simultaneously, enabling combination therapy.
   - Different sides of Janus nanoparticles can carry different therapeutic agents, such as chemotherapy drugs, immunotherapeutic agents, or gene therapies.
   - Combination therapy using Janus nanoparticles can target multiple pathways or vulnerabilities in cancer cells, leading to synergistic effects and improved treatment outcomes (Zhang et al. 2019b; Wang et al. 2021g; Liu et al. 2020).
4. *Imaging and Diagnosis:*
   - Janus nanoparticles can be functionalized with imaging agents, such as fluorescent dyes or contrast agents, enabling cancer imaging and diagnosis.
   - The imaging capabilities of Janus nanoparticles allow for accurate visualization and localization of tumors, aiding in early detection and precise tumor targeting.
5. *Photothermal Therapy*:
   - Janus nanoparticles, particularly those with gold or other plasmonic materials, can exhibit photothermal properties.
   - When exposed to near-infrared light, Janus nanoparticles can generate localized heat, leading to thermal ablation of cancer cells or enhanced drug release.
   - Photothermal therapy using Janus nanoparticles can selectively target and destroy cancer cells while sparing healthy tissues (Zhang and Li 2021b; Liu et al. 2021e).

It is important to note that the exact mechanism of action can vary depending on the specific properties and functionalization of Janus nanoparticles, as well as the type of cancer being targeted. Further research and development are still ongoing to explore and optimize the therapeutic mechanisms of Janus nanoparticles in cancer therapy.

The field of Janus nanoparticles in cancer therapy holds promising future prospects (Wang et al. 2021a). It is important to note that the timeline provided is a generalized overview, and the specific advancements and milestones may vary based on individual research efforts and scientific breakthroughs. The field of Janus nanoparticles in cancer therapy is rapidly evolving, and ongoing research is expected to uncover novel applications and therapeutic strategies in the coming years. Here are some potential areas of development and future prospects for Janus nanoparticles in cancer treatments. Breakthroughs in the utilization of Janus nanoparticles as cancer modalities have shown significant advancements in cancer therapy. Some notable results and discussions regarding these breakthroughs include:

1. *Enhanced Therapeutic Efficacy:*
   - Janus nanoparticles have demonstrated improved therapeutic efficacy compared to conventional therapies.
   - By enabling targeted drug delivery, Janus nanoparticles can increase the concentration of therapeutic agents at the tumor site, leading to enhanced treatment outcomes.
   - Studies have shown that Janus nanoparticles can improve the efficacy of chemotherapy drugs, immunotherapies, and gene therapies, leading to better tumor regression rates and prolonged survival in preclinical models.
2. *Reduced Side Effects:*
   - Janus nanoparticles have the potential to minimize off-target effects and reduce systemic toxicity associated with cancer therapies.
   - The selective targeting of Janus nanoparticles to cancer cells allows for lower doses of therapeutic agents, thereby reducing side effects on healthy tissues (Lee et al. 2007; Wang et al. 2019b).
   - By encapsulating drugs within or conjugating them onto Janus nanoparticles, controlled release can be achieved, further reducing toxicity and improving the safety profile of cancer treatments.
3. *Overcoming Drug Resistance:*
   - Janus nanoparticles offer the opportunity to overcome drug resistance mechanisms in cancer cells.
   - Combination therapies using Janus nanoparticles, incorporating different drugs or therapeutic modalities, can target multiple pathways or vulnerabilities in cancer cells, reducing the likelihood of resistance development.
   - Janus nanoparticles can also enhance the delivery of immunotherapies, which can activate the immune system to overcome drug resistance and enhance treatment response.
4. *Imaging and Diagnosis:*
   - Janus nanoparticles can be designed to combine different therapies, such as chemotherapy, photothermal therapy, or magnetic hyperthermia, with various imaging modalities.
   - By integrating imaging capabilities, such as MRI, fluorescence, or photoacoustic imaging, into Janus nanoparticles, simultaneous therapy and real-time monitoring of treatment can be achieved.
   - This integration of therapies and imaging modalities in Janus nanoparticles has the potential to provide a comprehensive approach to cancer treatment,

enabling precise and personalized therapy guidance (Liu et al. 2019; Wang et al. 2021g).

14.7.5.1 Theranostics

- Janus nanoparticles can be designed to combine therapeutic and diagnostic capabilities, leading to theranostic applications.
- By incorporating imaging agents or contrast agents, Janus nanoparticles can enable simultaneous cancer imaging, diagnosis, and targeted therapy.
- Theranostic Janus nanoparticles have the potential to facilitate real-time monitoring of treatment response, allowing for personalized adjustments to therapy and improved patient outcomes (Li et al. 2021a; Wang et al. 2021e).

14.7.5.2 Combination of Therapies and Imaging Modalities

- Janus nanoparticles have demonstrated remarkable imaging and diagnostic capabilities in cancer.
- By incorporating imaging agents, such as fluorescent dyes or contrast agents, Janus nanoparticles enable accurate visualization and localization of tumors.
- This imaging capability facilitates early detection, precise tumor targeting, and real-time monitoring of treatment response, thereby improving diagnostic accuracy and treatment planning (Liu et al. 2019).

14.7.5.3 Combination Therapies

- Janus nanoparticles can be further optimized to deliver multiple therapeutic agents simultaneously, enabling synergistic combination therapies.
- By incorporating different drugs, immunotherapies, or gene therapies on each side of the Janus nanoparticle, treatment strategies can be tailored to target multiple pathways or vulnerabilities in cancer cells (Wang et al. 2021c).
- Combination therapies using Janus nanoparticles have the potential to overcome drug resistance, improve treatment response rates, and enhance patient outcomes.

14.7.5.4 Immunotherapy Enhancements

- Janus nanoparticles can be utilized to enhance the efficacy and delivery of immunotherapies, such as immune checkpoint inhibitors or cancer vaccines.
- By encapsulating or conjugating immunotherapeutic agents onto Janus nanoparticles, targeted delivery to immune cells or tumor microenvironments can be achieved, enhancing immune responses against cancer cells.
- Janus nanoparticles have the potential to improve the effectiveness of immunotherapies and overcome immune evasion mechanisms in cancer.

14.7.5.5 Personalized Medicine

- Janus nanoparticles can be tailored to specific cancer types or even individual patients, enabling personalized medicine approaches. The design flexibility of Janus nanoparticles allows for personalized medicine approaches in cancer therapy.

- By incorporating tumor-specific targeting ligands or biomarkers, Janus nanoparticles can selectively deliver therapies to cancer cells, maximizing treatment efficacy while minimizing side effects (Liu et al. 2020).
- Personalized Janus nanoparticles have the potential to revolutionize cancer treatment by offering customized therapies based on the unique characteristics of each patient's tumor.

These breakthroughs in Janus nanoparticle research highlight their potential to transform cancer modalities by improving therapeutic efficacy, reducing side effects, overcoming drug resistance, enhancing imaging and diagnosis, and enabling personalized medicine approaches. Further research and clinical studies are needed to validate these findings and translate them into clinical practice. However, the results obtained thus far are promising and provide a solid foundation for the future development of Janus nanoparticles as effective cancer therapies.

These future prospects highlight the potential of Janus nanoparticles to revolutionize cancer therapy by offering targeted, personalized, and multifunctional approaches. Continued research and development in the field of Janus nanoparticles hold great promise for improving outcomes in cancer treatment.

14.7.6 Summary of the Design and Synthesis of Janus Nanoparticles

- Janus nanoparticles are designed and synthesized with two distinct sides or regions, enabling unique functionalities and applications.
- Various design methods, including surface functionalization, core-shell structures, and Janus nanoparticle assembly, are employed to optimize their properties for specific applications (Li et al. 2020; Zhang et al. 2019b; Wang et al. 2021g; Liu et al. 2020).

Here is a timeline outlining the key milestones and developments in the use of Janus nanoparticles in cancer research and therapy:

1. **2004: Discovery of Janus Particles:**
   - Janus particles, named after the two-faced Roman god Janus, were first introduced by researchers to describe particles with two distinct sides or regions.
2. **2005–2010: Early Research and Characterization:**
   - Scientists began exploring the synthesis and characterization of Janus nanoparticles, including their unique properties and potential applications.
   - Initial studies focused on understanding the self-assembly, surface modification, and functionalization of Janus nanoparticles.
3. **2011–2015: Targeted Drug Delivery and Imaging:**
   - Researchers started investigating the potential of Janus nanoparticles in targeted drug delivery to cancer cells.

   - The design and functionalization of Janus nanoparticles were optimized for specific targeting ligands, imaging agents, and controlled drug release.
   - Studies demonstrated the ability of Janus nanoparticles to enhance drug efficacy and minimize off-target effects.
4. **2016–2018: Combination Therapy and Immunotherapy:**
   - Advances in Janus nanoparticles enabled the combination of multiple therapeutic agents within a single nanoparticle.
   - Researchers explored the integration of chemotherapy drugs, immunotherapies, and gene therapies on different sides of Janus nanoparticles.
   - Combination therapies using Janus nanoparticles demonstrated synergistic effects and improved treatment outcomes in preclinical models.
5. **2019–2020: Personalized Medicine and Clinical Translation:**
   - Efforts were made to tailor Janus nanoparticles for personalized medicine approaches in cancer therapy.
   - Researchers focused on incorporating tumor-specific targeting ligands and biomarkers on Janus nanoparticles.
   - Preclinical studies showed the potential of personalized Janus nanoparticles in improving treatment efficacy and patient outcomes.
   - Clinical translation of Janus nanoparticles in cancer therapy began, with early-stage trials evaluating their safety and efficacy.
6. **Present and Future:**
   - Ongoing research is focused on further optimizing the design, synthesis, and functionalization of Janus nanoparticles for cancer therapy.
   - Advances in nanotechnology, material science, and imaging techniques continue to contribute to the development of Janus nanoparticles.
   - Future prospects include the integration of artificial intelligence and nanomedicine for precise tumor targeting and therapy guidance.

14.7.7 Potential Impact and Future Prospects of Janus Nanoparticles in Various Fields

- Janus nanoparticles have shown great potential in cancer therapeutics, diagnostics, gene therapy, photothermal therapy, and biosensing (Zhang et al. 2021b; Wang et al. 2021c; Wu et al. 2020; Zhang et al. 2019a; Wang et al. 2019a).
- They offer targeted drug delivery, overcoming drug resistance, combination therapy, molecular imaging, theranostic platforms, and personalized medicine approaches.
- Although challenges exist, ongoing research and development efforts hold promise for addressing these challenges and advancing the field of Janus nanoparticles.
- The future prospects of Janus nanoparticles include improved stability and biocompatibility, scalable production methods, rigorous preclinical and clinical evaluation, and exploration of emerging trends and directions.

14.8 Conclusion

Janus nanoparticles (JNPs) have been applied for cancer theranostics, nanomedicine, biosensing, and catalysis owing to their numerous distinctive benefits. This article encompasses state-of-the-art developments in JNPs for cancer theranostics and biomedical applications, including their classification, fabrication strategies, contrast over traditional nanoparticle/core-shell counterpart performances. First of all, various medications can be released independently on account of JNPs' asymmetric modification. Second, the signal interference and attenuation that often occur in core-shell or alloy nanoparticles are significantly reduced by the spatial separation of imaging materials. Thirdly, the composition's asymmetry contributes to the photothermal or potential gradient that can propel JNP motion for dynamic drug administration. Further, the challenges and prospects of JNPs in cancer theranostics are also covered.

Moreover, this review focuses on how the Janus nanoparticles have demonstrated significant potential in various fields, particularly in cancer therapy and diagnostics. Overcoming challenges related to stability, biocompatibility, scalability, and clinical evaluation will be crucial for their successful translation into clinical practice. With continued research and development, Janus nanoparticles have the potential to make a substantial impact on personalized medicine and improve patient outcomes in the future. Through this work, a comprehensive understanding of Janus nanoparticles for cancer theranostics will be established. The research will contribute to the advancement of personalized medicine approaches, improved targeted drug delivery, and enhanced molecular imaging in cancer therapy. The findings will shed light on the challenges and future directions of Janus nanoparticles, paving the way for their successful translation into clinical practice and ultimately improving patient outcomes in cancer treatment.

To conclude, Janus nanoparticles hold great promise in the field of cancer theranostics. Their dual-sided functionality, combining therapeutic and diagnostic capabilities, makes them highly versatile tools for cancer diagnosis, treatment, and monitoring. By selectively targeting cancer cells, delivering therapeutic agents, and providing real-time imaging feedback, Janus nanoparticles have the potential to revolutionize cancer care, improving patient outcomes and guiding personalized treatment strategies. However, further research and development, as well as comprehensive preclinical and clinical studies, are necessary to fully explore the potential of Janus nanoparticles in cancer theranostics and ensure their safe and effective translation into clinical practice.

References

Chen M et al (2020) Janus particles for cancer therapy: current advances and perspectives. Adv Healthc Mater 9(21):1901696

Huang X, El-Sayed IH, Qian W, El-Sayed MA (2006) Cancer cell imaging and photothermal therapy in the near-infrared region by using gold nanorods. J Am Chem Soc 128(6):2115–2120

Huang X, Zhang Y, Liang XJ (2018) Synergistic cancer therapy by nanomedicine-based combination drug delivery. Trends Biotechnol 33(9):480–486
Lee S, Cha EJ, Park K, Lee SY, Hong JK, Sun IC et al (2007) A near-infrared-fluorescence-quenched gold-nanoparticle imaging probe for in vivo drug screening and protease activity determination. Angew Chem Int Ed 46(28):5397–5401
Li X, Li J (2021) Janus nanoparticles for cancer theranostics. Nanomedicine 16(14):1163–1166
Li J et al (2018) Magnetic Janus nanoparticles for cancer therapy. Wiley Interdiscip Rev Nanomed Nanobiotechnol 10(6):e1519
Li J et al (2019) Silica-based Janus nanoparticles for biomedical applications: recent advances and perspectives. Adv Healthc Mater 8(11):1801356
Li J et al (2020) Janus nanoparticles for breast cancer imaging and therapy. Nano Today 35:100925
Li X, Xie C, Lu L, Wang X (2021a) Janus nanoparticles for enhanced cancer immunotherapy. Nano Today 39:101227
Li X, Zhang Y, Li J (2021b) Janus nanoparticles for combination cancer immunotherapy. Front Chem 9:668866
Li X, Chen Y, Li J (2021c) Janus nanoparticles for cancer immunotherapy. Adv Ther 4(3):2000223
Liu Y et al (2019) Janus liposomes for enhanced drug delivery: from design principles to clinical translation. J Control Release 311-312:233–249
Liu Y et al (2020) Janus nanoparticles for colorectal cancer imaging and therapy. Curr Med Chem 27(18):3029–3045
Liu Y, Chen Y, Li J, Yang X (2021a) Janus nanoparticles for targeted drug delivery in cancer therapy. J Mater Chem B 9(34):6903–6920
Liu Y, Chen Y, Li J (2021b) Janus nanoparticles for cancer imaging. Adv Healthc Mater 10(18):2100930
Liu Y, Zhang Y, Li J (2021c) Janus nanoparticles for cancer diagnosis. Front Chem 9:695883
Liu Y, Wang J, Li J (2021d) Janus nanoparticles for cancer radiotherapy. Front Chem 9:722563
Liu Y, Chen Y, Li J, Yang X (2021e) Janus nanoparticles for combination cancer therapy. J Mater Chem B 9(4):972–996
Wang Y, Li J (2021a) Janus nanoparticles for cancer therapy: from design to applications. Front Bioeng Biotechnol 9:650315
Wang Z, Li J (2021b) Janus nanoparticles for cancer combination therapy. Adv Healthc Mater 10(20):2100874
Wang Y, Li J (2021c) Janus nanoparticles for cancer photothermal therapy. Front Bioeng Biotechnol 9:687777
Wang M, Liang X, Cheng Y, Wang H (2018) Janus nanoplatforms for synergistic tumor therapy. Curr Med Chem 25(26):3126–3150
Wang M, Liang X, Cheng Y, Wang H (2019a) Janus nanoplatforms for synergistic tumor therapy. Curr Med Chem 26(38):6840–6857
Wang Z, Su Z, Chen M (2019b) Janus nanoparticles for cancer therapy: a promising approach. Curr Med Chem 26(38):6840–6857
Wang J et al (2020) Janus nanoparticles for pancreatic cancer imaging and therapy. Adv Healthc Mater 9(24):2000882
Wang J, Zhang Y, Li J, Huang X (2021a) Janus nanoparticles for synergistic cancer therapy. Adv Sci 8(15):2100655
Wang Y, Zhang Y, Li J (2021b) Janus nanoparticles for photothermal cancer therapy. Nano 11(6):1467
Wang Z, Su Z, Chen M (2021c) Janus nanoparticles for combination cancer therapy: principles, strategies, and challenges. J Mater Chem B 9(25):5167–5190
Wang J, Wang Y, Li J (2021d) Janus nanoparticles for cancer theranostics. Front Bioeng Biotechnol 9:672614
Wang Z, Su Z, Chen M (2021e) Janus nanoparticles for cancer photodynamic therapy. J Mater Chem B 9(16):3325–3344
Wang J, Zhang Y, Li J (2021f) Janus nanoparticles for cancer nanomedicine. Front Chem 9:675835

Wang Z et al (2021g) Janus nanoparticles for prostate cancer therapy: advances and perspectives. J Control Release 336:510–529

Wu Y, Zhang X, Zhang J, Liu Y (2020) Janus nanoparticles for enhanced cancer therapy: principles, progress, and perspectives. ACS Appl Mater Interfaces 12(41):45581–45599

Zhang S, Li J (2021a) Janus nanoparticles for cancer drug delivery. Adv Drug Deliv Rev 176:113871

Zhang Y, Li J (2021b) Janus nanoparticles for cancer imaging and therapy. Adv Funct Mater 31(5):2006089

Zhang Y, Liang XJ (2019) Janus nanoparticles for targeted therapy of cancer. Curr Opin Chem Eng 24:22–31

Zhang Y, Wang F, Liang XJ (2018a) Advances in drug delivery and biomedical applications of carbon nanotubes. Curr Med Chem 25(30):3598–3618

Zhang Y et al (2018b) Gold-based Janus nanoparticles for biomedical applications. Bioconjug Chem 29(10):3235–3251

Zhang Y, Li Y, Liang XJ (2019a) Multifunctional nanoparticles for targeted imaging and therapy in cancer. Expert Opin Drug Deliv 16(5):477–492

Zhang Y et al (2019b) Janus nanoparticles for lung cancer imaging and therapy. Adv Drug Deliv Rev 138:16–31

Zhang Y, Li J, Huang X (2021a) Janus nanoparticles for combination cancer therapy. Front Bioeng Biotechnol 9:701628

Zhang S, Li J, Lykotrafitis G, Li J (2021b) Janus nanoparticles for cancer therapy: recent progress and future perspectives. J Control Release 338:43–62

Zhang S, Li J, Li J (2021c) Janus nanoparticles for cancer therapeutics: challenges and opportunities. Front Bioeng Biotechnol 9:643343

Advanced Uses of Quantum Dots in Medical Imaging

15

Asghar Ali, Saman Saim Rahman, and Mohan Kamthan

Abstract

Quantum dots (QDs) have emerged as a promising nanotechnology-based tool for medical imaging due to their unique properties, such as size-tunable fluorescence, high brightness, photostability, and biocompatibility. The advanced uses of QDs in medical imaging include their application in fluorescence imaging, magnetic resonance imaging (MRI), computed tomography (CT), and positron emission tomography (PET). QDs can be functionalized with specific ligands, antibodies, or peptides to selectively target biomolecules or cells, enabling specific imaging of disease states or molecular events in living organisms. QDs also have the potential to act as drug carriers or contrast agents for imaging-guided therapy. However, the clinical translation of QDs requires further investigation and optimization regarding toxicity, biodegradability, and clearance. This chapter provides an overview of the recent advancements in QD-based medical imaging and discusses the challenges and opportunities in their translation from bench to bedside.

Keywords

Quantum dots · Medical imaging · Fluorescence imaging · Magnetic resonance imaging · Computed tomography · Targeted imaging · Drug delivery · Personalized medicine

A. Ali · S. S. Rahman · M. Kamthan (✉)
Clinical Biochemistry Laboratory, Department of Biochemistry, School of Chemical and Life Sciences, Jamia Hamdard, New Delhi, India
e-mail: asgharali@jamiahamdard.ac.in; mohan.kamthan@jamiahamdard.ac.in

M. Mohsin, N. Soleja (eds.), *Nanoscale Sensors and their Applications in Biomedical Imaging*, https://doi.org/10.1007/978-981-97-3144-2_15

Abbreviations

CdSe	Cadmium selenide
CT	Computed tomography
DTI	Diffusion tensor imaging
FACTS	Fluorescence-activated cell sorting
FDA	Food and Drug Administration
FDG	Fluorodeoxyglucose
FP	Fluorescent proteins
FRET	Forster resonance energy transfer
IVUS	Intravascular ultrasound
MRI	Magnetic resonance imaging
nm	Nanometer
OCT	Optical coherence tomography
PbS	Lead sulfide
PET	Positron-emitted tomography
PL	Photoluminescence
QDs	Quantum dots
SPECT	Single-photon emission-computed tomography

15.1 Introduction

Quantum dots, also known as "artificial atoms," have discrete energy levels, varying according to size. Colloidal quantum dots are now commonly used in biological investigations. These dots can be used to pattern inorganic materials on the nanoscale, creating new sensing methods. The intersection between nanotechnology and biology is significant due to the complexity of biological components at the nanoscale (Alivisatos 2004; Ashoori 1996; Clapp 2008). QDs refer to semiconductor nanoparticles that possess physical dimensions that are either close to or less than the Bohr radius. The spatial confinement of intrinsic electron and hole carries leads to increased band-gap energy and to a splitting of the continuous energy bands in discrete energy levels. This property of QDs makes it an intermediate between the bulk materials and the molecules (Wegner and Hildebrandt 2015).

One of the reasons QDs are used for bioimaging purposes is that fluorescent proteins (FPs) and fluorescent dyes have some limitations such as small effective line shifts, narrow excitation spectra, susceptibility to photobleaching, and wide fluorescence bands (Biju et al. 2010; Wegner and Hildebrandt 2015). A comparative study between QDs and organic dyes shows that QDs exhibit superior optical properties but also some limitations such as nano-colloidal behavior, blinking, and controversial toxicity. Despite QD's photophysical advantages it will not replace organic dyes and fluorescence proteins for bioimaging applications however, can be used along with them for different purposes (Resch-Genger et al. 2008). Unlike QDs, most organic dyes are small and easy to conjugate into various biomolecules without altering the biological function of the dye's bioconjugate. In addition, many dyes can bind to larger biomolecules (e.g., antibodies), increasing the overall brightness of the bio-conjugated PL. The commercial availability of a wide range (biological and spectral) of bio-conjugated dyes is another important advantage of organic dyes. For bioimaging applications, it is important to evaluate the advantages of QD against their disadvantages and to choose the right fluorophore (QD or not) for each specific biosensor (Wegner and Hildebrandt 2015).

Quantum dots possess remarkable optical and electronic properties despite their small size. These properties, such as bright and stable fluorescence emission and high photostability, have paved the way for their application in diverse medical imaging techniques like fluorescence imaging, magnetic resonance imaging (MRI), and computed tomography (CT). This chapter aims to explore the advanced applications of QDs in medical imaging and their potential implications for personalized medicine (Matea et al. 2017).

15.2 Potential Applications of Quantum Dots

QDs have shown significant potential for various applications, including medical imaging, drug delivery, solar cells, and electronics. In the field of medical imaging, QDs have already been developed as contrast agents for fluorescence imaging, magnetic resonance imaging (MRI), and computed tomography (CT). Their ability to emit light at different wavelengths based on their size and their exceptional stability makes them promising fluorescent probes for in vivo imaging of cells and tissues. Quantum dots can be functionalized with metal chelators, enabling their use as contrast agents in MRI. This modification enhances their sensitivity and contrast, making them valuable for detecting tumors and other abnormalities (Tang et al. 2019) Furthermore, QD-based imaging and drug delivery systems have the potential to revolutionize personalized medicine by enabling more targeted and effective treatment of diseases. Additionally, quantum dots have been used to target specific biomolecules like proteins and nucleic acids, allowing researchers to observe and understand cellular and molecular interactions. Due to their unique electronic properties, QDs have been studied as a potential platform for solar cells and energy storage devices. Additionally, by coating quantum dots with high atomic numbers materials like gold or iodine, they can be utilized in CT imaging to achieve improved contrast and spatial resolution (Clapp 2008) (Table 15.1).

15.2.1 Fluorescence Imaging (FI)

Fluorescence imaging is a non-invasive imaging technique that utilizes fluorescent molecules to visualize biological processes in vivo. QDs have unique optical properties that make them ideal for use in fluorescence imaging. They emit bright and stable fluorescence over a broad range of wavelengths, enabling multiplexed imaging of multiple targets within the same sample. Additionally, QDs can be conjugated

Table 15.1 General information on each imaging technique

Imaging techniques	Principle	Advantages	Disadvantages	Applications	References
Fluorescence imaging (FI)	Detection of fluorescent markers	1. High sensitivity and specificity 2. Real-time imaging capabilities 3. Non-ionizing radiation	1. Limited penetration 2. Tissue autofluorescence interference 3. Limited spatial resolution	1. Bioimaging and cell tracking 2. Cancer imaging and diagnosis 3. Drug delivery and therapeutics	Helmchen and Denk (2005), Ntziachristos and Weissleder (2001), Weissleder (2001)
Magnetic resonance imaging (MRI)	Interaction of hydrogen nuclei with magnetic fields	1. Suitable for cellular and molecular level imaging 2. Excellent soft tissue contrast 3. Multiplanar imaging capability 4. Non-ionizing radiation	1. Challenges in quantification 2. Expensive and complex 3. Relatively long scan times 4. Claustrophobic for some patients 5. Metallic implants can cause artifacts	1. Contrast agents 2. Multimodal imaging 3. Theragnostic functional imaging	Bydder and Young (1985), Torkzad et al. (2010)
Computed tomography (CT)	X-ray absorption	1. Functional and anatomical imaging 2. Fast imaging with high spatial resolution 3. Suitable for bone and lung imaging Valuable in emergency settings	1. Limited availability in some regions 2. Ionizing radiation exposure 3. Lower soft tissue contrast compared to MRI 4. Used as contrast agent for some applications 5. Limited ability to distinguish certain soft tissues	1. Contrast agents 2. Multispectral agents 3. Diagnostics and therapeutic functions 4. Image-guided surgery	Alric et al. (2008), Fang et al. (2012), Wegner and Hildebrandt (2015)

Positron emission tomography (PET)	Detection of gamma rays emitted by radiotracers	1. Functional and molecular imaging 2. 3D image reconstruction capability 3. Functional and molecular imaging 4. Early disease detection and treatment monitoring 5. Valuable in oncology and neurology 6. Quantitative measurements of metabolic processes 7. Whole-body imaging	1. Artifacts from patient motion and metal objects 2. Ionizing radiation exposure 3. Limited spatial resolution 4. Expensive and limited availability 5. Relatively long acquisition time 6. Dependency on radiotracers and cyclotron facilities	Cancer imaging Neuroimaging Cardiovascular imaging Drug delivery and theranostics Image-guided surgery	Jin et al. (2011), Kim et al. (2011), Kumar et al. (2023), Liu and Tang (2020)

with specific targeting ligands to selectively image and deliver therapeutic agents to diseased cells (Li and Zhu 2013).

15.2.1.1 Principles of Fluorescence Imaging with Quantum Dots

Fluorescence imaging with quantum dots (QDs) is a powerful technique used in various fields, including biological and biomedical research. Quantum dots exhibit unique optical properties, such as high fluorescence brightness, tunable emission spectra, and exceptional photostability. This makes them ideal for fluorescence imaging applications.

Fluorescence imaging with quantum dots involves key principles such as quantum confinement, size-dependent emission, broad absorption, narrow emission, brightness, and photostability. Quantum dots are nanoscale semiconductor crystals that exhibit size-dependent electronic and optical properties, allowing them to emit and absorb light at specific wavelengths. Their tunable emission property enables simultaneous detection of different QDs, while their broad absorption spectrum allows for excitation with a single light source. Quantum dots possess high brightness and photostability, making them suitable for long-term imaging applications (Michalet et al. 2005; Chan and Nie 1998).

15.2.1.2 Advantages of Fluorescence Imaging with Quantum Dots

Quantum dots (QDs) exhibit higher fluorescence brightness and photostability than traditional organic dyes. Their unique semiconductor properties allow for efficient light absorption and emission, producing bright and long-lasting fluorescence signals. This brightness is particularly advantageous for detecting low-abundance targets or visualizing fine structures in biological samples (Alivisatos et al. 2005). QDs have a broad absorption spectrum and narrow emission spectra, allowing for multiplexing capabilities in fluorescence imaging. By varying the size and composition of QDs, their emission wavelengths can be precisely tuned, enabling simultaneous visualization of multiple targets in a single sample. This feature is valuable for multiplexed labeling and colocalization studies (Michalet et al. 2005).

Quantum dots exhibit minimal background fluorescence, resulting in a high signal-to-noise ratio. This property is particularly advantageous for imaging applications requiring high sensitivity and precise quantification. It allows for detecting weak signals and improves the accuracy of quantitative analyses (Zhang and Wang 2012). Quantum dots exhibit excellent photostability, meaning they can emit light for extended periods without significant loss of fluorescence signal. This stability allows for long-term imaging experiments without photobleaching, making them suitable for dynamic live-cell imaging and tracking of biological processes (Pathak et al. 2007).

15.2.1.3 Applications of Fluorescence Imaging with Quantum Dots

Fluorescence imaging with quantum dots (QDs) has found various applications in a wide range of fields, including biomedical research, bioimaging, diagnostics, and sensing. Some of the applications of fluorescence imaging with quantum dots are as follows:

(a) *Bioimaging and Cell Tracking*: Quantum dots have been extensively used for cellular imaging and tracking due to their superior photophysical properties. They offer excellent brightness, resistance to photobleaching, and narrow emission spectra, allowing multiplexed imaging. QDs have been employed to track single cells, study cellular processes, and visualize subcellular structures (Chan and Nie 1998).

(b) *Cancer Imaging and Diagnosis*: Quantum dots have shown great promise in cancer imaging and diagnosis. They can be functionalized with tumor-targeting ligands to specifically label cancer cells or tissues, enabling early detection, precise imaging, and accurate diagnosis of cancer. QDs have been utilized in various imaging modalities, including fluorescence microscopy, optical coherence tomography (OCT), and in vivo imaging (Smith et al. 2008).

(c) *Drug Delivery and Therapeutics*: Quantum dots have been explored as carriers for drug delivery and therapeutics. Their large surface area allows for the attachment of drug molecules, while their fluorescent properties enable real-time drug release and distribution monitoring. QDs have been used for targeted drug delivery, photothermal therapy, and photodynamic therapy (Algar et al. 2011).

(d) *Sensing and Biosensing*: Quantum dots have been employed as sensitive probes for various sensing applications. Their fluorescence can be modulated by changes in the local environment or through interactions with analytes. QDs have been used in biosensing platforms for the detection of proteins, DNA, ions, and small molecules. They offer high sensitivity, selectivity, and multiplexing capabilities (Ortega-Arizmendi et al. 2013).

(e) *Environmental Monitoring*: Quantum dots have been employed in environmental monitoring applications. They can be used as fluorescent probes for detecting and monitoring pollutants, heavy metals, and contaminants in water, soil, and air. QDs offer high sensitivity, rapid detection, and the ability to differentiate multiple analytes simultaneously (Walling et al. 2009).

15.2.2 Magnetic Resonance Imaging (MRI)

Magnetic resonance imaging (MRI) is a widely used imaging modality that provides high-resolution human body images. QDs have been used in MRI as contrast agents due to their magnetic properties. Researchers have developed QD-based contrast agents that provide higher contrast and resolution than traditional contrast agents. Furthermore, QDs can be conjugated with specific targeting ligands to image and deliver therapeutic agents to diseased cells selectively (Ding et al. 2014; Matea et al. 2017).

15.2.2.1 Principles of Magnetic Resonance Imaging with Quantum Dots

MRI can be improved by using quantum dots as contrast agents. When functionalized with targeting molecules, quantum dots act as sensitive markers that can enhance MRI imaging capabilities and provide detailed imaging of specific areas.

This approach, known as QD-MRI, combines the strengths of traditional MRI with the unique properties of quantum dots for enhanced sensitivity and specificity (Ding et al. 2014).

15.2.2.2 Advantages of Magnetic Resonance Imaging with Quantum Dots

Magnetic resonance imaging (MRI) is a powerful medical imaging technique that provides detailed anatomical and functional information about the human body. By combining MRI with quantum dots, several advantages and limitations arise. Quantum dots can be engineered to have high magnetic moments, resulting in enhanced contrast in MRI images. This improved contrast allows for better visualization and tissue differentiation, making identifying abnormalities or lesions easier. Quantum dots can be functionalized to emit light at different wavelengths, making them suitable for multimodal imaging. Combining MRI with other imaging techniques like fluorescence imaging or positron emission tomography (PET) makes it possible to simultaneously obtain complementary information about the tissue's structure, function, and metabolism (Matea et al. 2017).

Quantum dots can be modified to attach specific targeting molecules such as antibodies or peptides. This enables targeted imaging, where quantum dots accumulate at specific sites, such as tumor cells, providing higher sensitivity and specificity in detecting pathological conditions. Quantum dots have excellent stability and can withstand harsh biological conditions, allowing long-term imaging studies. This stability ensures that the imaging agent remains intact and provides accurate information over extended periods (Katwal 2023; Koole et al. 2009).

15.2.2.3 Applications of Magnetic Resonance Imaging with Quantum Dots

Quantum dots can enhance MRI images by exhibiting strong magnetic properties. They can be used in contrast agents to improve sensitivity and specificity for detecting diseases. Quantum dots can also emit strong fluorescence signals for fluorescence microscopy. Combining these properties can lead to the development of multimodal imaging techniques for precise localization in surgical interventions (Revia and Zhang 2016).

They can be used as theragnostic agents by incorporating therapeutic agents onto them. MRI can guide their targeted drug delivery and monitor their distribution. They can also be engineered to respond to physiological or biochemical changes, creating functional imaging techniques for medical use. Magnetic resonance imaging with quantum dots is promising for medical applications. More research is needed to ensure biocompatibility and optimize magnetic properties. The advancements in this field have great potential for medical imaging (Gour et al. 2021; Katwal 2023).

15.2.3 Computed Tomography (CT)

Computed tomography (CT) is an imaging technique that uses X-rays to create detailed images of the body. QDs have been used in CT as contrast agents due to their high X-ray attenuation coefficient. QDs provide higher contrast and resolution than traditional contrast agents, making them ideal for imaging and monitoring biological processes in vivo (Cierniak 2011).

15.2.3.1 Principles of Computed Tomography with Quantum Dots

Computed tomography (CT) is a medical imaging technique that uses X-ray technology to generate cross-sectional images of the body. Traditional CT scanners use iodine-based contrast agents to enhance the visibility of certain tissues and structures. However, there has been growing interest in using quantum dots as contrast agents in CT imaging due to their unique properties. Quantum dots (QDs) are nanoscale semiconductor particles that exhibit size-dependent fluorescence and strong X-ray attenuation properties. When excited by X-rays, QDs emit fluorescent light that can be detected and used to generate high-resolution images (Cierniak 2011; Molkenova et al. 2022).

15.2.3.2 Advantages of Computed Tomography with Quantum Dots

Computed tomography (CT) is a widely used imaging technique that provides detailed cross-sectional images of the human body. Conventional CT scans typically use X-rays to create these images, but there is ongoing research and development in using quantum dots (QDs) as a contrast agent in CT imaging. Quantum dots can provide enhanced contrast compared to traditional iodine-based contrast agents used in CT imaging. QDs have strong X-ray attenuation properties, allowing for better differentiation between various tissues and structures in the body. This improved contrast can aid in detecting and characterizing lesions, tumors, and other abnormalities (Alric et al. 2008; Fang et al. 2012). Quantum dots can be engineered to emit light across a range of wavelengths, enabling multimodal imaging. They can be used simultaneously for CT imaging and fluorescence imaging, allowing for the integration of anatomical and functional information. This combination of imaging modalities can enhance diagnostic accuracy and improve the understanding of disease processes (Wegner and Hildebrandt 2015).

Quantum dots can be functionalized with specific ligands or antibodies to bind to molecules or cell surface markers selectively. This targeted imaging approach allows for the visualization of specific molecular targets within the body, such as cancer cells or inflammatory markers. It has the potential to improve early detection, characterization, and monitoring of diseases (Fang et al. 2012). Quantum dots have prolonged fluorescence lifetimes compared to organic dyes, enabling longer imaging windows. This property is particularly advantageous for CT imaging, as it allows for extended scanning times without a significant loss of signal intensity. It can also facilitate dynamic imaging studies, capturing temporal changes within the body (Wegner and Hildebrandt 2015).

15.2.3.3 Applications of Computed Tomography with Quantum Dots

Computed tomography (CT) is a medical imaging technique that uses X-rays to produce cross-sectional images of the body. Quantum dots can be engineered to emit strong and specific fluorescence signals. Attaching these quantum dots to targeted biomolecules can act as contrast agents in CT imaging. The quantum dots' fluorescence properties can enhance the visibility of specific tissues or organs, improving the detection of abnormalities or diseases. They can emit light at different wavelengths based on their size and composition. Using quantum dots with different emission spectra makes it possible to perform multispectral CT imaging. This technique can provide additional information about tissue composition, blood flow, or functional parameters, leading to more comprehensive diagnostic capabilities (Gour et al. 2021).

Quantum dots can serve dual purposes by combining diagnostic and therapeutic functions. Loading quantum dots with therapeutic agents can be used for targeted drug delivery. Simultaneously, CT imaging can enable real-time drug distribution and efficacy monitoring through the quantum dots' fluorescent properties. Quantum dots can be utilized to improve surgical procedures by providing real-time imaging guidance. By labeling specific anatomical structures or tumor margins with quantum dots, surgeons can visualize the precise location of these structures during procedures. This can enhance the accuracy and safety of surgical interventions (Pericleous et al. 2012; Revia and Zhang 2016).

15.2.4 Positron Emission Tomography (PET)

Positron emission tomography (PET) is a medical imaging technique that uses radioactive tracers to visualize and measure various physiological processes in the body. It is commonly used for diagnosing and monitoring diseases such as cancer, cardiovascular conditions, and neurological disorders. Quantum dots possess unique optical and electronic properties. While quantum dots have been extensively investigated for their applications in imaging and sensing, they have not yet been widely used in PET imaging (Chen et al. 2008). Gamma rays emitted by radioactive tracers like FDG can be detected using PET scanners. Quantum dots, which are nanoparticles, have the potential to be utilized in PET imaging by binding them to appropriate positron-emitting radionuclides. However, efficient radionuclide attachment requires suitable surface chemistry, and the quantum dots must emit gamma rays upon decay of the attached radionuclide (Liu et al. 2010).

15.2.4.1 Principles of Positron Emission Tomography with Quantum Dots

Quantum dots can be labeled with radiotracers for PET imaging. Various techniques can be used to radiolabel quantum dots with positron-emitting isotopes. The radiolabelled quantum dots are administered, and gamma rays emitted are detected to generate a 3D image. PET images can be analyzed to assess radiotracer-labeled quantum dots' concentration and spatial distribution in specific tissues or organs (Cai and Chen 2007; Kinahan and Fletcher 2010; Molaei 2020; Phelps 2000).

15.2.4.2 Advantages of Positron Emission Tomography with Quantum Dots

Quantum dots have high quantum efficiency and brightness, allowing for enhanced detection sensitivity. This means that even small amounts of the tracer can be detected, leading to improved imaging accuracy. Quantum dots can emit light at different wavelengths simultaneously, allowing for multiplexing. This means that multiple tracers can be used in a single imaging session, providing more comprehensive and detailed information about different biological processes (Tu et al. 2011). Quantum dots are highly stable and can emit light for extended periods. This stability ensures that the imaging signal remains strong throughout the imaging process, allowing for longer imaging sessions and better image quality. Quantum dots have a large surface area, which can be easily modified to attach various biomolecules, such as antibodies or peptides. This enables targeted imaging of specific molecules or receptors in the body, leading to improved specificity and accuracy of PET imaging (Tan et al. 2011).

15.2.4.3 Applications of Positron Emission Tomography with Quantum Dots

Positron emission tomography (PET) is a medical imaging technique that uses radioactive tracers to visualize and measure metabolic processes in the body. PET with quantum dots has shown promise in cancer imaging. Researchers can create targeted imaging probes by conjugating quantum dots with tumor-targeting ligands or antibodies. These probes can specifically bind to cancer cells and provide high-resolution images, enabling accurate tumor detection and characterization. Quantum dots' tunable emission properties allow for multiplexed imaging, where different types of cancer cells can be simultaneously detected using different colored quantum dots (Kumar et al. 2023). PET with quantum dots can be used to study brain function and neurodegenerative disorders. Quantum dots can be functionalized to target specific biomarkers associated with neurodegenerative diseases like Alzheimer's or Parkinson's. By tracking the accumulation of these quantum dot probes in the brain, researchers can gain insights into disease progression and develop new therapeutic strategies (Jin et al. 2011). Quantum dots can be utilized in PET to study cardiovascular diseases. By labeling quantum dots with specific targeting molecules that bind to markers of atherosclerosis or thrombosis, researchers can visualize the distribution and progression of these conditions. This can aid in early diagnosis, risk assessment, and monitoring of therapeutic interventions (Kim et al. 2011).

Quantum dots can serve as multifunctional platforms for both imaging and targeted drug delivery. By attaching therapeutic agents to quantum dots and monitoring their distribution using PET, researchers can assess the efficacy of drug delivery systems in real time. Additionally, quantum dots can be used to monitor and evaluate the response to treatment, enabling personalized medicine approaches (Liu et al.

2013). Quantum dots can be used as fluorescent markers during surgical procedures. By incorporating them into PET tracers, surgeons can use PET imaging to locate and visualize tumor margins or sentinel lymph nodes precisely. This can aid in surgical planning and improve the accuracy of tumor resection, leading to better outcomes for patients (Jin et al. 2011).

15.2.5 Functionalized QDs with Specific Ligands, Antibodies, or Peptides

Functionalized quantum dots (QDs) refer to QDs that have been modified or coated with specific ligands, antibodies, or peptides. This modification allows for their selective binding to specific targets, such as cells, proteins, or biomolecules, opening a wide range of applications in various fields, including biotechnology, medicine, and materials science (Karakoti et al. 2015).

(a) *Ligand-Functionalized QDs:* Ligands, such as small organic molecules or polymers, can be attached to the surface of QDs to provide stability and facilitate their interaction with target molecules. For instance, a ligand with a high affinity for a specific receptor can be used to functionalize QDs, enabling their targeted delivery to cells expressing that receptor (Karakoti et al. 2015).
(b) *Antibody-Functionalized QDs:* Antibodies, which are highly specific for their target antigens, can be conjugated to QDs to create antibody-QD conjugates. These conjugates can be used for targeted imaging, diagnostics, and therapy by choosing antibodies that recognize specific cell surface markers or proteins. They can also be employed in techniques such as fluorescence-activated cell sorting (FACS) for cell separation or in immunoassays for detecting specific analytes (Pathak et al. 2007).
(c) *Peptide-Functionalized QDs:* Peptides are short chains of amino acids that can be designed to bind to specific molecules or receptors. QDs can be functionalized with peptides that have an affinity for certain targets, enabling their precise recognition and localization. Peptide-functionalized QDs have been utilized in diverse applications, such as cell imaging, drug delivery, and biosensing (Amber Nagy et al. 2014; Delehanty et al. 2011).

15.2.5.1 Principles

Functionalized quantum dots (QDs) operate on the principle of altering the surface of these nanoscale semiconductor particles using diverse functional groups or ligands. This modification serves to enhance stability, solubility, and the ability to interface with various biological, chemical, or electronic systems. The procedure entails affixing molecules or polymers to the QD surface, resulting in specific attributes that facilitate interactions with targets. This principle underpins a wide array of applications in biology, sensing, optoelectronics, and photonics, allowing customization of QD behavior for specific uses (Molaei 2020; Yang et al. 2020).

15.2.5.2 Advantages

Functionalized quantum dots (QDs) have gained significant attention in various fields due to their unique properties and potential applications. Functionalized QDs can exhibit size-dependent emission wavelengths, covering a wide range of the electromagnetic spectrum. By controlling the size and composition of the QDs, their optical properties can be tuned, enabling applications such as multicolor imaging, optoelectronics, and display technologies (Pinaud et al. 2006). QDs possess excellent photostability compared to conventional organic dyes and fluorescent proteins. They exhibit minimal photobleaching and blinking, allowing for long-term imaging and sensing applications (Tiwari et al. 2011).

QDs have a broad absorption spectrum, which means they can absorb photons across a wide range of wavelengths. This feature makes them suitable for various applications, including solar cells, photocatalysis, and light-emitting devices (Alivisatos 1996). Functionalized QDs can efficiently transfer energy to nearby molecules or other QDs through a process called Förster resonance energy transfer (FRET). This property enables the development of QD-based sensors, bioimaging agents, and light-harvesting systems (Tiwari et al. 2011).

QDs can be functionalized with biocompatible surface coatings, such as polymers or biomolecules, allowing them to be easily integrated into biological systems. This feature facilitates applications in bioimaging, biosensing, and drug delivery (Medintz et al. 2005; Tiwari et al. 2011).

15.2.5.3 Applications

Functionalized quantum dots (QDs) have several applications in medical imaging due to their unique optical properties and biocompatibility. Here are some of the key applications of functionalized QDs in medical imaging:

(a) *Fluorescence Imaging:* QDs can be functionalized with specific biomolecules such as antibodies or peptides to target specific tissues or cells. By conjugating QDs to these targeting molecules, they can be used as fluorescent probes for imaging specific disease markers or cellular processes. QDs offer high brightness, photostability, and a broad range of emission wavelengths, allowing for multiplexed imaging of multiple targets simultaneously (Michalet et al. 2005).

(b) In Vivo *Imaging*: Functionalized QDs can be used for in vivo imaging to visualize tissues or organs deep within the body. QDs can be surface-modified to enhance their stability, biocompatibility, and targeting capabilities. They can be administered systemically and targeted to specific organs or tumor sites, allowing for non-invasive imaging and tracking of disease progression in real time (Gao et al. 2004).

(c) *Sentinel Lymph Node Mapping:* Sentinel lymph node mapping is a technique used in cancer surgery to identify the first lymph node(s) that cancer cells are most likely to spread. Functionalizing QDs with appropriate targeting molecules can be injected near the tumor site and tracked in real-time using fluorescence imaging. This helps surgeons precisely identify and remove only the affected lymph nodes, reducing the invasiveness of the procedure (Wu et al. 2003).

(d) *Drug Delivery and Tracking:* Functionalized QDs can be used as drug delivery system carriers. Attaching therapeutic drugs to the surface of QDs can be delivered to specific cells or tissues. The fluorescence of QDs can be used to track the distribution and release of the drug within the body, enabling real-time monitoring of the therapeutic efficacy (Wu et al. 2003).

(e) *Image-Guided Surgery:* QDs can serve as imaging agents during surgical procedures, providing real-time guidance to surgeons. By incorporating QDs into surgical instruments or as fluorescent dyes directly applied to tissues, surgeons can visualize the precise location of tumors, blood vessels, or other critical structures during the operation, improving surgical accuracy and reducing the risk of complications (Frangioni 2003).

15.2.6 Targeted Imaging and Drug Delivery

One of the major advantages of QDs is their ability to be conjugated with specific targeting ligands to image and deliver therapeutic agents to diseased cells selectively. Researchers have developed QD-based imaging and drug delivery systems that can target specific cells or tissues. This has the potential to revolutionize personalized medicine by allowing for more targeted and effective treatment of diseases (Wang et al. 2014; Yang et al. 2014).

15.2.6.1 Principles of Targeted Imaging Drug Delivery

Targeted imaging and drug delivery refer to the precise and selective transport of imaging agents or therapeutic drugs to specific cells, tissues, or organs in the body. The main aim is to enhance the effectiveness of imaging or drug treatments while minimizing unwanted side effects, achieved by improving specificity and localized action. In the context of targeted imaging, the principle involves the use of imaging agents that gather selectively in the desired target cells or tissues. This facilitates better detection and visualization of specific abnormalities. These imaging agents could encompass contrast agents suitable for various techniques like MRI, CT, PET, and fluorescent dyes tailored for optical imaging (Qi and Gao 2008).

15.2.6.2 Advantages of Targeted Imaging Drug Delivery

Targeted imaging drug delivery involves administering imaging agents or contrast agents designed to accumulate at specific sites like tumors or organs of interest. This approach offers advantages including enhanced specificity, which localizes imaging agents accurately for precise disease diagnosis (Bagalkot et al. 2007). This reduces background noise, improves signal-to-noise ratio, and enhances imaging quality and detection sensitivity. The technique also increases sensitivity by concentrating contrast agents at the site of interest, boosting signal intensity for better disease detection. Targeted delivery lowers systemic toxicity by minimizing exposure of healthy tissues to imaging agents, reducing potential side effects. This approach enables the use of lower agent doses, beneficial for patients with compromised organ function. Furthermore, targeted imaging drug delivery supports personalized medicine by tailoring approaches to individual patients, considering their specific disease characteristics. This customization can lead to more accurate diagnoses and better treatment outcomes (Bilan et al. 2016).

15.2.6.3 Applications of Targeted Imaging Drug Delivery

Targeted imaging and drug delivery are important techniques in the field of medicine and have a wide range of applications. Here are some notable applications of target imaging and drug delivery:

(a) *Cancer Diagnosis and Treatment:* Targeted imaging techniques, such as positron emission tomography (PET) and magnetic resonance imaging (MRI), can be used to identify and visualize specific cancerous cells or tumors in the body. These imaging techniques help in early detection, accurate staging, and monitoring of cancer progression. Additionally, targeted drug delivery systems can deliver chemotherapy drugs directly to the tumor site, minimizing damage to healthy tissues and improving treatment effectiveness (Bagalkot et al. 2007).

(b) *Cardiovascular Disease:* Targeted imaging and drug delivery play a crucial role in diagnosing and treating cardiovascular diseases. Imaging techniques like computed tomography (CT) angiography, and intravascular ultrasound (IVUS) can provide detailed images of blood vessels and help identify blockages or abnormalities. Targeted drug delivery systems can be used to deliver drugs directly to the affected areas, such as stents coated with anti-inflammatory or anti-restenotic drugs for treating arterial blockages (Sabir et al. 2021).

(c) *Neurological Disorders:* Targeted imaging and drug delivery techniques are used in the diagnosis and treatment of various neurological disorders, including Alzheimer's disease, Parkinson's disease, and brain tumors (Ashrafizadeh et al. 2020). Imaging methods like functional MRI (fMRI) and diffusion tensor imaging (DTI) help in visualizing brain structures and identifying abnormalities. Targeted drug delivery can deliver drugs across the blood–brain barrier and specifically target affected brain regions, improving treatment outcomes (Calabrese et al. 2021).

(d) *Infectious Diseases:* Targeted imaging and drug delivery can be applied to infectious diseases such as bacterial infections and viral infections. Imaging techniques like single-photon emission computed tomography (SPECT) and molecular imaging can be used to track the location and progression of infections. Targeted drug delivery systems can deliver antimicrobial agents directly to the infection site, enhancing therapeutic efficacy and reducing systemic side effects (Yukawa et al. 2023).

(e) *Regenerative Medicine:* Targeted drug delivery plays a crucial role in regenerative medicine, which focuses on repairing or replacing damaged tissues or organs. Controlled-release systems can deliver growth factors, stem cells, or other therapeutic agents directly to the desired site of tissue regeneration, promoting healing and tissue repair (Onoshima et al. 2015; Yukawa et al. 2023).

15.3 Challenges

Although there have been promising results, there are still challenges that need to be addressed before quantum dots (QDs) can be widely used in various applications. One of the main challenges is the toxicity of QDs. Studies have shown that QDs can cause toxicity and inflammation in vivo, which may limit their use in clinical settings. Additionally, the stability and biocompatibility of QDs need to be improved to ensure their safety for clinical use. Moreover, the scalability and cost-effectiveness of QDs need to be considered for large-scale applications. The challenges related to QDs are complex and encompass various aspects such as pharmaceutical, industrial, technical, and biological, which have contributed to the scarcity of clinical trials. These challenges are depicted in Fig. 15.1 and elaborated upon in the following discussion (Abdellatif et al. 2022; Liang et al. 2022; Reshma and Mohanan 2019; Tsoi et al. 2013).

(a) *Pharmaceutical Issues:* Quantum dots present challenges in pharmaceuticals due to their small size, surface area, and metallic properties. They are prone to issues like aggregation and degradation, which affect their optical properties. Significant effort has been put into improving their physical and chemical stability (Abdellatif et al. 2022). Zhang et al. (2021) found a way to stabilize PbS QDs by passivating surface S atoms with lead mono-carboxylate using trioctylphosphine. This improved their quantum properties and shielded them from oxygen regardless of size or original surface ligands (Zhang et al. 2021).

(b) *Industrial Issues:* Large-scale production of quantum dots (QDs) faces challenges such as complex surface modifications, inconsistent physicochemical properties, trace impurities, and environmental risks from toxic heavy metals and solvents. To mitigate these concerns, cadmium-free alternatives like carbon, graphene, and silicon QDs are being developed. These alternatives offer

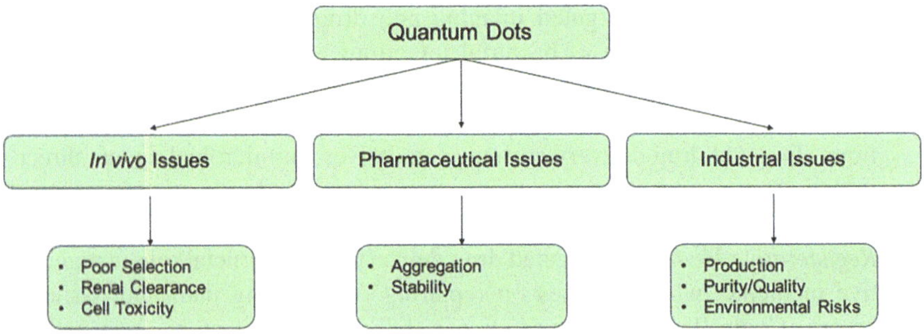

Fig. 15.1 Several common obstacles hinder the clinical application of quantum dots (QDs) (Abdellatif et al. 2022)

comparable optical performance while minimizing the environmental hazards associated with heavy metals (Ranjbar-Navazi et al. 2019).

(c) In Vivo *Issues:* The use of quantum dots faces challenges due to their lack of organ specificity and potential toxicity, especially with heavy metals like lead and cadmium, which can accumulate in the body (Abdellatif et al. 2022).

15.3.1 Toxicity of the Quantum Dots

QDs have gained significant interest in medical imaging due to their bright and tunable fluorescence properties, high photostability, and long fluorescence lifetime. However, the potential toxicity of quantum dots is an important consideration for their use in biomedical applications. Several studies have investigated the toxicity of quantum dots, particularly in the context of medical imaging. Various factors influence the toxicity of quantum dots (QDs) such as size, charge, concentration, outer coating bioactivity, and stability. The physicochemical characteristics of QDs determine their toxicity, specifically the functional coating and core stability. These factors should be considered when assessing the risk of QD toxicity in real-world exposure scenarios (Hardman 2006).

Some of the toxicity of quantum dots in medical imaging is discussed below:

(a) Quantum dots made of CdSe possess special optical properties that make them valuable, but they also raise concerns about health and the environment due to the presence of toxic cadmium. Exposure to CdSe quantum dots containing cadmium can lead to inflammation in the lungs (Gidwani et al. 2021). Researchers are looking for non-cadmium-based materials for quantum dots to minimize the health and environmental risks associated with cadmium-based quantum dots (Liu and Tang 2020).
(b) CdSe quantum dots can release toxic cadmium ions that disrupt cellular processes and cause cytotoxicity, particularly in liver cells (Gidwani et al. 2021).
(c) Studies on Ag-based QDs in vivo toxicity show generally mild effects, with low acute toxicity, limited organ accumulation, and systemic distribution in animal models. However, toxicity depends on factors like size, shape, coating, concentration, and exposure duration. Mechanisms include oxidative stress, mitochondrial disruption, DNA damage, and activation of apoptotic pathways (Liu and Tang 2020).

15.3.2 Biocompatibility

The process of designing quantum dots involves three main steps: synthesizing the quantum dot itself, modifying its surface to ensure biological compatibility, and subsequently modifying the surface to enable targeted delivery. There are three main methods for targeting biocompatible quantum dots: using antibodies, peptides, or small molecules.

(a) *Antibody Targeting:* Antibody-quantum dot conjugates are widely used in various fields. They can be used by combining biotinylated antibodies with streptavidin-coated quantum dots or by directly linking antibodies to quantum dots. These conjugates have shown great versatility in different applications. For example, researchers have used antibodies that are specific to F and G proteins, along with quantum dots of different sizes, to detect and monitor respiratory syncytial viral infection in vitro (Rosenthal et al. 2011).
(b) *Peptide Targeting:* When it comes to labeling, peptide-quantum dot nanoconjugates are a smart choice because the peptide epitopes of proteins or known peptides can help with ligand–target interactions. The strength of these interactions can vary, much like antibody recognition. But by using peptides in quantum dot probe design, synthesis costs can be reduced and the size of the nanoconjugate can be smaller. It is possible to improve the binding capacity of the resulting complexes to the target through multivalent interactions by covalently conjugating multiple peptides to a single quantum dot. Peptide-functionalized quantum dots have been used successfully to target a variety of cellular proteins, including growth factor receptors, G protein-coupled receptors, integrins, and ion channels (Orndorff and Rosenthal 2009).
(c) *Small Molecule Targeting:* Ligand-conjugated quantum dots can track protein dynamics and aid in drug discovery through displacement assays. They can also serve as probes for studying allosteric modulation but require advanced organic chemistry skills for optimal customization (Rosenthal et al. 2002, 2011).
(d) *High-Affinity Fusion Tag Targeting:* Several novel approaches utilizing high-affinity fusion tags have been recently developed to enhance the options available for precise and selective labeling of cellular targets using quantum dots. These approaches involve the expression of a cellular target of interest with a genetically engineered fusion tag, typically a short peptide sequence containing a recognition epitope for a complementary binding partner (Rosenthal et al. 2011).
(e) *Nonspecific Binding:* The surface functional group on quantum dots can impact their binding behavior with cells, which may lead to nonspecific binding. This can happen because of electrostatic interactions between surface groups and the cell surface or the adsorption of serum protein. To guarantee precise outcomes, researchers should evaluate the potential for nonspecific interactions and choose a surface functionalization technique that reduces nonspecific binding. Nonspecific binding of quantum dots to cells presents a problem in many biological applications, and it is essential to conduct thorough control experiments to address this issue. One must exercise caution and skepticism when interpreting results without robust control experiments (Rosenthal et al. 2011; Warnement et al. 2008).

15.3.3 Cost-Effectiveness

Quantum dots have shown great potential in medical imaging applications, but their cost-effectiveness remains critical for widespread adoption in the healthcare industry. When exploring the cost-effectiveness of quantum dots in medical imaging, it is important to keep in mind the following factors:

(a) *Precision and Sensitivity:* Quantum dots offer several advantages over traditional imaging agents, such as organic dyes or nanoparticles. They can be precisely engineered to emit specific wavelengths of light, which allows for improved sensitivity and specificity in imaging. This ability to tune quantum dots' properties can lead to more accurate and detailed imaging results, potentially reducing the need for additional imaging procedures (Zhang et al. 2023).
(b) *Longevity:* Quantum dots have a longer shelf life and better photostability compared to conventional imaging agents. This increased stability can lead to reduced waste and lower replacement costs, potentially contributing to cost-effectiveness over time (Jin et al. 2011).
(c) *Multiplexing Capability:* Quantum dots can be designed to emit different colors of light simultaneously, enabling multiplexed imaging. This means that multiple targets or biomarkers can be visualized in a single imaging session, potentially reducing the overall imaging cost and time required for diagnosis and treatment planning (Pandey and Bodas 2020).
(d) *Manufacturing and Production Costs:* The cost of manufacturing quantum dots has been a limiting factor in their widespread adoption. However, ongoing research and technological advancements are focused on developing more cost-effective synthesis methods and scalable production processes. As these manufacturing improvements continue, the cost of quantum dots for medical imaging applications may decrease (Ezati et al. 2022).

15.3.4 Safety Concerns

Quantum dots, which are tiny semiconductor particles at the nanoscale, have garnered considerable interest in the field of medical imaging because of their exceptional optical characteristics. While they offer promising advantages in terms of brightness, stability, and tunable emission wavelengths, it is important to consider the potential safety concerns associated with their use in medical applications (Reshma and Mohanan 2019).

(a) *Toxicity:* Some quantum dots contain heavy metals, such as cadmium or lead, which can be toxic to living organisms. The potential release of these heavy metals from quantum dots raises concerns about their biocompatibility and long-term safety. Extensive research is being conducted to develop non-toxic or less toxic quantum dot materials to address this concern (Tsoi et al. 2013; Wu and Tang 2014).

(b) *Biodegradability:* Quantum dots are typically not biodegradable, meaning they persist in the body for an extended period. This raises concerns about their potential accumulation in tissues and potential long-term adverse effects. Biodegradable quantum dots are being investigated as a potential solution to mitigate this concern (Reshma and Mohanan 2019).
(c) *Clearance:* Efficient clearance from the body is important to minimize potential toxicity. If quantum dots cannot be efficiently eliminated from the body, they may accumulate in organs or tissues, potentially causing harm. Research is ongoing to develop strategies to enhance the clearance of quantum dots from the body (Salykina et al. 2010; Soo Choi et al. 2007).
(d) *Immunogenicity:* Quantum dots can potentially trigger an immune response in the body, leading to inflammation or other adverse reactions. This immunogenic response may limit their long-term use or require additional modifications to minimize immune reactions (Liang et al. 2022).
(e) *Stability:* Quantum dots may undergo degradation over time, leading to the release of toxic components or loss of their imaging properties. Ensuring the stability of quantum dots under physiological conditions is crucial for their safe and reliable use (Zhu et al. 2011).
(f) *Regulatory Approval:* Quantum dots in medical imaging require regulatory approval to ensure their safety and efficacy. This process involves rigorous testing and evaluation to assess their potential risks and benefits (Gidwani et al. 2021).

Research is being conducted to address safety concerns related to the use of quantum dots in medical imaging. Scientists are working on developing safer materials, improving compatibility with biological systems, enhancing clearance mechanisms, and exploring alternative imaging agents. It is crucial to understand potential risks thoroughly and to continue researching to ensure the safe use of quantum dots in medical imaging, like any new technology (Gidwani et al. 2021; Reshma and Mohanan 2019).

15.3.5 Biodegradability and Clearance

Quantum dots have a potential environmental impact due to their toxicity. To address this concern, researchers are developing alternative quantum dots that are more environmentally friendly and biodegradable. Strategies include using non-toxic materials, encapsulating with biocompatible coatings, and developing easily metabolized quantum dots. The biodegradability of quantum dots varies and requires careful assessment before widespread use. The clearance of quantum dots in medical imaging depends on their size, surface properties, and route of administration (Hardman 2006).

Here are some key points regarding the clearance of quantum dots in medical imaging:

(a) *Renal Clearance:* To ensure efficient and rapid clearance through the kidneys, small quantum dots, usually less than 5–6 nm, require surface modification for enhanced solubility and stability. Once administered, they are promptly filtered and excreted through urine, effectively eliminating them from the body (Soo Choi et al. 2007).
(b) *Hepatic Clearance:* Larger quantum dots or those with different surface modifications may be cleared through the liver. In this case, quantum dots are taken up by liver cells (hepatocytes) and metabolized or excreted into the bile. From there, they can be eliminated through feces (Reshma and Mohanan 2019).
(c) *Lymphatic Clearance:* Some quantum dots can be cleared through the lymphatic system, particularly if they are administered subcutaneously or in tissues close to lymph nodes. Quantum dots taken up by lymphatic vessels can be transported to lymph nodes and eventually cleared from the body (Reshma and Mohanan 2019; Salykina et al. 2010).
(d) *Metabolism and Excretion:* Quantum dots can undergo bodily metabolic transformations, depending on their composition. The by-products or metabolites resulting from the breakdown of quantum dots are typically cleared through normal metabolic pathways (Hardman 2006).
(e) *Long-Term Retention:* It is crucial to note that the long-term retention of quantum dots in the body is a concern due to potential toxicity. While some quantum dots may be rapidly cleared, others can accumulate in certain organs or tissues, leading to potentially adverse effects. Therefore, extensive research is being conducted to optimize the design and surface modifications of quantum dots to minimize long-term retention and enhance their biocompatibility (Tsoi et al. 2013).

15.4 Research Directions

Several research directions must be explored to overcome the challenges and progress in the quantum dots (QDs) field. One of the most significant research areas is optimizing QDs for specific applications such as drug delivery, medical imaging, and energy devices. This involves enhancing the stability, biocompatibility, and toxicity of QDs. Moreover, creating new QD-based materials with better properties is necessary (Koole et al. 2009; Molaei 2020; Pardo et al. 2018). The development of affordable and scalable synthesis methods for quantum dots (QDs) is an important area of research. QDs can revolutionize medical imaging by improving imaging sensitivity, allowing for more diverse imaging capabilities, and enhancing contrast. Some of the research directions in the field of medical imaging are discussed as below:

(a) *Contrast-Enhanced Imaging:* QDs can act as contrast agents in different imaging methods like fluorescence imaging, OCT, and photoacoustic imaging. Scientists are studying QDs with improved surface coatings to enhance their stability, targeting abilities, and biocompatibility for specific imaging uses (Dirheimer et al. 2022).

(b) *Multimodal Imaging:* QDs emit light at multiple wavelengths, enabling simultaneous imaging with different modalities. This allows for complementary information to be obtained from a single session, such as fluorescence and MRI (Koole et al. 2009).

(c) *Targeted Imaging and Drug Delivery:* QDs can be modified with targeting ligands to bind to diseased cells or tissues for targeted imaging or drug delivery. They are being researched for use in diseases such as cancer, cardiovascular disorders, and neurological conditions (Pardo et al. 2018).

(d) *Real-Time Intraoperative Imaging:* QDs are used in real-time imaging during surgery. They help with tumor localization, lymph node mapping, and image-guided surgical interventions (Ohnishi et al. 2005).

(e) *Diagnostic and Therapeutic:* QDs can have both diagnostic and therapeutic functions, making them useful for diagnostic and therapeutic applications. They can be used to monitor drug delivery and efficacy and deliver therapeutic payloads to the target site. This approach has the potential to improve personalized medicine and treatment outcomes (Wang et al. 2012).

(f) *Biosensing and Molecular Imaging:* Scientists are using quantum dots as fluorescent probes to detect biological analytes and develop biosensors and molecular imaging platforms for disease detection and investigating molecular interactions in living systems (Chen et al. 2012).

15.5 Conclusion

In conclusion, the advanced applications of quantum dots in medical imaging can potentially revolutionize the field of personalized medicine. The unique optical and electronic properties of QDs have enabled the development of novel imaging modalities that provide higher contrast, resolution, and multiplexing capabilities than traditional imaging agents. Furthermore, the ability to conjugate QDs with specific targeting ligands has facilitated targeted imaging and drug delivery to diseased cells. Despite the promising results, the safety of QDs for clinical use is still a concern. Further research is needed to optimize their performance and ensure their safety for clinical use. Nonetheless, developing QD-based imaging and drug delivery systems is a significant step toward more personalized and effective treatment of diseases. As such, the advanced applications of quantum dots in medical imaging hold great promise for the future of healthcare.

Acknowledgments Mohan Kamthan gratefully acknowledges the Core Research Grant from the Science and Engineering Research Board (SERB), Govt. of India. Asghar Ali acknowledges the Department of Health (DHR) (File no: R.12014/61/2011-HR) for the Young Scientist Award.

References

Abdellatif AA, Younis MA, Alsharidah M, Al Rugaie O, Tawfeek HM (2022) Biomedical applications of quantum dots: overview, challenges, and clinical potential. Int J Nanomedicine 17:1951–1970. https://doi.org/10.2147/IJN.S357980

Algar WR, Prasuhn DE, Stewart MH, Jennings TL, Blanco-Canosa JB, Dawson PE et al (2011) The controlled display of biomolecules on nanoparticles: a challenge suited to bioorthogonal chemistry. Bioconjug Chem 22:825–858. https://doi.org/10.1021/bc200065z

Alivisatos AP (1996) Semiconductor clusters, nanocrystals, and quantum dots. Science (1979) 271:933–937. https://doi.org/10.1126/science.271.5251.933

Alivisatos P (2004) The use of nanocrystals in biological detection. Nat Biotechnol 22:47–52. https://doi.org/10.1038/nbt927

Alivisatos AP, Gu W, Larabell C (2005) Quantum dots as cellular probes. Annu Rev Biomed Eng 7:55–76. https://doi.org/10.1146/annurev.bioeng.7.060804.100432

Alric C, Taleb J, Le Duc G, Mandon C, Billotey C, Le Meur-Herland A et al (2008) Gadolinium chelate coated gold nanoparticles as contrast agents for both X-ray computed tomography and magnetic resonance imaging. J Am Chem Soc 130:5908–5915. https://doi.org/10.1021/ja078176p

Ashoori RC (1996) Electrons in artificial atoms. Nature 379:413–419. https://doi.org/10.1038/379413a0

Ashrafizadeh M, Mohammadinejad R, Kailasa SK, Ahmadi Z, Afshar EG, Pardakhty A (2020) Carbon dots as versatile nanoarchitectures for the treatment of neurological disorders and their theranostic applications: a review. Adv Colloid Interf Sci 278:102123. https://doi.org/10.1016/j.cis.2020.102123

Bagalkot V, Zhang L, Levy-Nissenbaum E, Jon S, Kantoff PW, Langer R et al (2007) Quantum dot–aptamer conjugates for synchronous cancer imaging, therapy, and sensing of drug delivery based on bi-fluorescence resonance energy transfer. Nano Lett 7:3065–3070. https://doi.org/10.1021/nl071546n

Biju V, Itoh T, Ishikawa M (2010) Delivering quantum dots to cells: bioconjugated quantum dots for targeted and nonspecific extracellular and intracellular imaging. Chem Soc Rev 39:3031. https://doi.org/10.1039/b926512k

Bilan R, Nabiev I, Sukhanova A (2016) Quantum dot-based nanotools for bioimaging, diagnostics, and drug delivery. Chembiochem 17:2103–2114. https://doi.org/10.1002/cbic.201600357

Bydder GM, Young IR (1985) MR imaging: clinical use of the inversion recovery sequence. J Comput Assist Tomogr 9:659–675

Cai W, Chen X (2007) Nanoplatforms for targeted molecular imaging in living subjects. Small 3:1840–1854. https://doi.org/10.1002/smll.200700351

Calabrese G, De Luca G, Nocito G, Rizzo MG, Lombardo SP, Chisari G et al (2021) Carbon dots: an innovative tool for drug delivery in brain tumors. Int J Mol Sci 22:11783. https://doi.org/10.3390/ijms222111783

Chan WCW, Nie S (1998) Quantum dot bioconjugates for ultrasensitive nonisotopic detection. Science (1979) 281:2016–2018. https://doi.org/10.1126/science.281.5385.2016

Chen K, Li Z-B, Wang H, Cai W, Chen X (2008) Dual-modality optical and positron emission tomography imaging of vascular endothelial growth factor receptor on tumor vasculature using quantum dots. Eur J Nucl Med Mol Imaging 35:2235–2244. https://doi.org/10.1007/s00259-008-0860-8

Chen N-T, Cheng S-H, Liu C-P, Souris J, Chen C-T, Mou C-Y et al (2012) Recent advances in nanoparticle-based Förster resonance energy transfer for biosensing, molecular imaging and drug release profiling. Int J Mol Sci 13:16598–16623. https://doi.org/10.3390/ijms131216598

Cierniak R (2011) X-ray computed tomography in biomedical engineering. Springer London, London. https://doi.org/10.1007/978-0-85729-027-4

Clapp A (2008) Potential clinical applications of quantum dots. Int J Nanomedicine:151. https://doi.org/10.2147/IJN.S614

Delehanty JB, Bradburne CE, Susumu K, Boeneman K, Mei BC, Farrell D et al (2011) Spatiotemporal multicolor labeling of individual cells using peptide-functionalized quantum dots and mixed delivery techniques. J Am Chem Soc 133:10482–10489. https://doi.org/10.1021/ja200555z

Ding K, Jing L, Liu C, Hou Y, Gao M (2014) Magnetically engineered cd-free quantum dots as dual-modality probes for fluorescence/magnetic resonance imaging of tumors. Biomaterials 35:1608–1617. https://doi.org/10.1016/j.biomaterials.2013.10.078

Dirheimer L, Pons T, Marchal F, Bezdetnaya L (2022) Quantum dots mediated imaging and phototherapy in cancer spheroid models: state of the art and perspectives. Pharmaceutics 14:2136. https://doi.org/10.3390/pharmaceutics14102136

Ezati P, Priyadarshi R, Rhim J-W (2022) Prospects of sustainable and renewable source-based carbon quantum dots for food packaging applications. Sustain Mater Technol 33:e00494. https://doi.org/10.1016/j.susmat.2022.e00494

Fang M, Peng C-W, Pang D-W, Li Y (2012) Quantum dots for cancer research: current status, remaining issues, and future perspectives. Cancer Biol Med 9:151–163. https://doi.org/10.7497/j.issn.2095-3941.2012.03.001

Frangioni J (2003) In vivo near-infrared fluorescence imaging. Curr Opin Chem Biol 7:626–634. https://doi.org/10.1016/j.cbpa.2003.08.007

Gao X, Cui Y, Levenson RM, Chung LWK, Nie S (2004) In vivo cancer targeting and imaging with semiconductor quantum dots. Nat Biotechnol 22:969–976. https://doi.org/10.1038/nbt994

Gidwani B, Sahu V, Shukla SS, Pandey R, Joshi V, Jain VK et al (2021) Quantum dots: Prospectives, toxicity, advances and applications. J Drug Deliv Sci Technol 61:102308. https://doi.org/10.1016/j.jddst.2020.102308

Gour A, Ramteke S, Jain NK (2021) Pharmaceutical applications of quantum dots. AAPS PharmSciTech 22:233. https://doi.org/10.1208/s12249-021-02103-w

Hardman R (2006) A toxicologic review of quantum dots: toxicity depends on physicochemical and environmental factors. Environ Health Perspect 114:165–172. https://doi.org/10.1289/ehp.8284

Helmchen F, Denk W (2005) Deep tissue two-photon microscopy. Nat Methods 2:932–940. https://doi.org/10.1038/nmeth818

Jin S, Hu Y, Gu Z, Liu L, Wu H-C (2011) Application of quantum dots in biological imaging. J Nanomater 2011:1–13. https://doi.org/10.1155/2011/834139

Karakoti AS, Shukla R, Shanker R, Singh S (2015) Surface functionalization of quantum dots for biological applications. Adv Colloid Interf Sci 215:28–45. https://doi.org/10.1016/j.cis.2014.11.004

Katwal R (2023) A review: properties and diverse applications of smart magnetic quantum dots. Nano-Struct Nano-Objects 35:101001. https://doi.org/10.1016/j.nanoso.2023.101001

Kim BH, Lee N, Kim H, An K, Il PY, Choi Y et al (2011) Large-scale synthesis of uniform and extremely small-sized iron oxide nanoparticles for high-resolution T_1 magnetic resonance imaging contrast agents. J Am Chem Soc 133:12624–12631. https://doi.org/10.1021/ja203340u

Kinahan PE, Fletcher JW (2010) Positron emission tomography-computed tomography standardized uptake values in clinical practice and assessing response to therapy. Semin Ultrasound CT MRI 31:496–505. https://doi.org/10.1053/j.sult.2010.10.001

Koole R, Mulder WJM, van Schooneveld MM, Strijkers GJ, Meijerink A, Nicolay K (2009) Magnetic quantum dots for multimodal imaging. Wiley Interdiscip Rev Nanomed Nanobiotechnol 1:475–491. https://doi.org/10.1002/wnan.14

Kumar VB, Sher I, Rencus-Lazar S, Rotenstreich Y, Gazit E (2023) Functional carbon quantum dots for ocular imaging and therapeutic applications. Small 19:2205754. https://doi.org/10.1002/smll.202205754

Li J, Zhu J-J (2013) Quantum dots for fluorescent biosensing and bio-imaging applications. Analyst 138:2506. https://doi.org/10.1039/c3an36705c

Liang Y, Zhang T, Tang M (2022) Toxicity of quantum dots on target organs and immune system. J Appl Toxicol 42:17–40. https://doi.org/10.1002/jat.4180

Liu N, Tang M (2020) Toxicity of different types of quantum dots to mammalian cells in vitro: An update review. J Hazard Mater 399:122606. https://doi.org/10.1016/j.jhazmat.2020.122606
Liu H, Zhang X, Xing B, Han P, Gambhir SS, Cheng Z (2010) Radiation-luminescence-excited quantum dots for in vivo multiplexed optical imaging. Small 6:1087–1091. https://doi.org/10.1002/smll.200902408
Liu L, Miao Q, Liang G (2013) Quantum dots as multifunctional materials for tumor imaging and therapy. Materials 6:483–499. https://doi.org/10.3390/ma6020483
Matea C, Mocan T, Tabaran F, Pop T, Mosteanu O, Puia C et al (2017) Quantum dots in imaging, drug delivery and sensor applications. Int J Nanomedicine 12:5421–5431. https://doi.org/10.2147/IJN.S138624
Medintz IL, Uyeda HT, Goldman ER, Mattoussi H (2005) Quantum dot bioconjugates for imaging, labelling and sensing. Nat Mater 4:435–446. https://doi.org/10.1038/nmat1390
Michalet X, Pinaud FF, Bentolila LA, Tsay JM, Doose S, Li JJ et al (2005) Quantum dots for live cells, in vivo imaging, and diagnostics. Science (1979) 307:538–544. https://doi.org/10.1126/science.1104274
Molaei MJ (2020) Principles, mechanisms, and application of carbon quantum dots in sensors: a review. Anal Methods 12:1266–1287. https://doi.org/10.1039/C9AY02696G
Molkenova A, Kairova M, Zhussupbekova A, Zhussupbekov K, Duisenbayeva B, Shvets IV et al (2022) Carbon dots doped with barium as a novel contrast agent for efficient CT X-ray attenuation. Nano-Struct Nano-Objects 29:100839. https://doi.org/10.1016/j.nanoso.2022.100839
Nagy A, Gemmill KB, Delehanty JB, Medintz IL, Sapsford KE (2014) Peptide-functionalized quantum dot biosensors. IEEE J Sel Top Quantum Electron 20:115–126. https://doi.org/10.1109/JSTQE.2013.2284427
Ntziachristos V, Weissleder R (2001) Experimental three-dimensional fluorescence reconstruction of diffuse media by use of a normalized born approximation. Opt Lett 26:893–895. https://doi.org/10.1364/ol.26.000893
Ohnishi S, Lomnes SJ, Laurence RG, Gogbashian A, Mariani G, Frangioni JV (2005) Organic alternatives to quantum dots for intraoperative near-infrared fluorescent sentinel lymph node mapping. Mol Imaging 4:153535002005051. https://doi.org/10.1162/15353500200505127
Onoshima D, Yukawa H, Baba Y (2015) Multifunctional quantum dots-based cancer diagnostics and stem cell therapeutics for regenerative medicine. Adv Drug Deliv Rev 95:2–14. https://doi.org/10.1016/j.addr.2015.08.004
Orndorff RL, Rosenthal SJ (2009) Neurotoxin quantum dot conjugates detect endogenous targets expressed in live cancer cells. Nano Lett 9:2589–2599. https://doi.org/10.1021/nl900789e
Ortega-Arizmendi AI, Aldeco-Pérez E, Cuevas-Yañez E (2013) Alkyne-azide cycloaddition catalyzed by silver chloride and "abnormal" silver *N* -heterocyclic Carbene complex. Sci World J 2013:1–8. https://doi.org/10.1155/2013/186537
Pandey S, Bodas D (2020) High-quality quantum dots for multiplexed bioimaging: a critical review. Adv Colloid Interf Sci 278:102137. https://doi.org/10.1016/j.cis.2020.102137
Pardo J, Peng Z, Leblanc R (2018) Cancer targeting and drug delivery using carbon-based quantum dots and nanotubes. Molecules 23:378. https://doi.org/10.3390/molecules23020378
Pathak S, Davidson MC, Silva GA (2007) Characterization of the functional binding properties of antibody conjugated quantum dots. Nano Lett 7:1839–1845. https://doi.org/10.1021/nl062706i
Pericleous P, Gazouli M, Lyberopoulou A, Rizos S, Nikiteas N, Efstathopoulos EP (2012) Quantum dots hold promise for early cancer imaging and detection. Int J Cancer 131:519–528. https://doi.org/10.1002/ijc.27528
Phelps ME (2000) Positron emission tomography provides molecular imaging of biological processes. Proc Natl Acad Sci 97:9226–9233. https://doi.org/10.1073/pnas.97.16.9226
Pinaud F, Michalet X, Bentolila LA, Tsay JM, Doose S, Li JJ et al (2006) Advances in fluorescence imaging with quantum dot bio-probes. Biomaterials 27:1679–1687. https://doi.org/10.1016/j.biomaterials.2005.11.018
Qi L, Gao X (2008) Emerging application of quantum dots for drug delivery and therapy. Expert Opin Drug Deliv 5:263–267. https://doi.org/10.1517/17425247.5.3.263

Ranjbar-Navazi Z, Omidi Y, Eskandani M, Davaran S (2019) Cadmium-free quantum dot-based theranostics. TrAC Trends Anal Chem 118:386–400. https://doi.org/10.1016/j.trac.2019.05.041

Resch-Genger U, Grabolle M, Cavaliere-Jaricot S, Nitschke R, Nann T (2008) Quantum dots versus organic dyes as fluorescent labels. Nat Methods 5:763–775. https://doi.org/10.1038/nmeth.1248

Reshma VG, Mohanan PV (2019) Quantum dots: applications and safety consequences. J Lumin 205:287–298. https://doi.org/10.1016/j.jlumin.2018.09.015

Revia RA, Zhang M (2016) Magnetite nanoparticles for cancer diagnosis, treatment, and treatment monitoring: recent advances. Mater Today 19:157–168. https://doi.org/10.1016/j.mattod.2015.08.022

Rosenthal SJ, Tomlinson I, Adkins EM, Schroeter S, Adams S, Swafford L et al (2002) Targeting cell surface receptors with ligand-conjugated nanocrystals. J Am Chem Soc 124:4586–4594. https://doi.org/10.1021/ja003486s

Rosenthal SJ, Chang JC, Kovtun O, McBride JR, Tomlinson ID (2011) Biocompatible quantum dots for biological applications. Chem Biol 18:10–24. https://doi.org/10.1016/j.chembiol.2010.11.013

Sabir F, Barani M, Mukhtar M, Rahdar A, Cucchiarini M, Zafar MN et al (2021) Nanodiagnosis and Nanotreatment of cardiovascular diseases: an overview. Chemosensors 9:67. https://doi.org/10.3390/chemosensors9040067

Salykina YF, Zherdeva VV, Dezhurov SV, Wakstein MS, Shirmanova MV, Zagaynova EV et al (2010) In: Tuchin VV, Genina EA (eds) Biodistribution and clearance of quantum dots in small animals, p 799908. https://doi.org/10.1117/12.888959

Smith A, Duan H, Mohs A, Nie S (2008) Bioconjugated quantum dots for in vivo molecular and cellular imaging☆. Adv Drug Deliv Rev 60:1226–1240. https://doi.org/10.1016/j.addr.2008.03.015

Soo Choi H, Liu W, Misra P, Tanaka E, Zimmer JP, Itty Ipe B et al (2007) Renal clearance of quantum dots. Nat Biotechnol 25:1165–1170. https://doi.org/10.1038/nbt1340

Tan A, Yildirimer L, Rajadas J, De La Peña H, Pastorin G, Seifalian A (2011) Quantum dots and carbon nanotubes in oncology: a review on emerging theranostic applications in nanomedicine. Nanomedicine 6:1101–1114. https://doi.org/10.2217/nnm.11.64

Tang T, Wei Y, Yang Q, Yang Y, Sailor MJ, Pang H-B (2019) Rapid chelator-free radiolabeling of quantum dots for *in vivo* imaging. Nanoscale 11:22248–22254. https://doi.org/10.1039/C9NR08508D

Tiwari P, Vig K, Dennis V, Singh S (2011) Functionalized gold nanoparticles and their biomedical applications. Nano 1:31–63. https://doi.org/10.3390/nano1010031

Torkzad MR, Påhlman L, Glimelius B (2010) Magnetic resonance imaging (MRI) in rectal cancer: a comprehensive review. Insights Imaging 1:245–267. https://doi.org/10.1007/s13244-010-0037-4

Tsoi KM, Dai Q, Alman BA, Chan WCW (2013) Are quantum dots toxic? Exploring the discrepancy between cell culture and animal studies. Acc Chem Res 46:662–671. https://doi.org/10.1021/ar300040z

Tu C, Ma X, House A, Kauzlarich SM, Louie AY (2011) PET imaging and biodistribution of silicon quantum dots in mice. ACS Med Chem Lett 2:285–288. https://doi.org/10.1021/ml1002844

Walling M, Novak J, Shepard JRE (2009) Quantum dots for live cell and in vivo imaging. Int J Mol Sci 10:441–491. https://doi.org/10.3390/ijms10020441

Wang J, Han S, Ke D, Wang R (2012) Semiconductor quantum dots surface modification for potential cancer diagnostic and therapeutic applications. J Nanomater 2012:1–8. https://doi.org/10.1155/2012/129041

Wang X, Sun X, Lao J, He H, Cheng T, Wang M et al (2014) Multifunctional graphene quantum dots for simultaneous targeted cellular imaging and drug delivery. Colloids Surf B Biointerfaces 122:638–644. https://doi.org/10.1016/j.colsurfb.2014.07.043

Warnement MR, Tomlinson ID, Chang JC, Schreuder MA, Luckabaugh CM, Rosenthal SJ (2008) Controlling the reactivity of amphiphilic quantum dots in biological assays through hydrophobic assembly of custom PEG derivatives. Bioconjug Chem 19:1404–1413. https://doi.org/10.1021/bc800104n

Wegner KD, Hildebrandt N (2015) Quantum dots: bright and versatile in vitro and in vivo fluorescence imaging biosensors. Chem Soc Rev 44:4792–4834. https://doi.org/10.1039/C4CS00532E

Weissleder R (2001) A clearer vision for in vivo imaging. Nat Biotechnol 19:316–317. https://doi.org/10.1038/86684

Wu T, Tang M (2014) Toxicity of quantum dots on respiratory system. Inhal Toxicol 26:128–139. https://doi.org/10.3109/08958378.2013.871762

Wu X, Liu H, Liu J, Haley KN, Treadway JA, Larson JP et al (2003) Immunofluorescent labeling of cancer marker Her2 and other cellular targets with semiconductor quantum dots. Nat Biotechnol 21:41–46. https://doi.org/10.1038/nbt764

Yang J, Yao M-H, Wen L, Song J-T, Zhang M-Z, Zhao Y-D et al (2014) Multifunctional quantum dot–polypeptide hybrid nanogel for targeted imaging and drug delivery. Nanoscale 6:11282–11292. https://doi.org/10.1039/C4NR03058C

Yang Y, Mao G, Ji X, He Z (2020) DNA-templated quantum dots and their applications in biosensors, bioimaging, and therapy. J Mater Chem B 8:9–17. https://doi.org/10.1039/C9TB01870K

Yukawa H, Sato K, Baba Y (2023) Theranostics applications of quantum dots in regenerative medicine, cancer medicine, and infectious diseases. Adv Drug Deliv Rev:114863. https://doi.org/10.1016/j.addr.2023.114863

Zhang Y, Wang T-H (2012) Quantum dot enabled molecular sensing and diagnostics. Theranostics 2:631–654. https://doi.org/10.7150/thno.4308

Zhang X, Chen Y, Lian L, Zhang Z, Liu Y, Song L et al (2021) Stability enhancement of PbS quantum dots by site-selective surface passivation for near-infrared LED application. Nano Res 14:628–634. https://doi.org/10.1007/s12274-020-3081-5

Zhang R, Zhang L, Yu R, Wang C (2023) Rapid and sensitive detection of methyl parathion in rice based on carbon quantum dots nano-fluorescence probe and inner filter effect. Food Chem 413:135679. https://doi.org/10.1016/j.foodchem.2023.135679

Zhu Z-J, Yeh Y-C, Tang R, Yan B, Tamayo J, Vachet RW et al (2011) Stability of quantum dots in live cells. Nat Chem 3:963–968. https://doi.org/10.1038/nchem.1177

SPRINGER NATURE

GPSR Compliance

The European Union's (EU) General Product Safety Regulation (GPSR) is a set of rules that requires consumer products to be safe and our obligations to ensure this.

If you have any concerns about our products, you can contact us on ProductSafety@springernature.com

In case Publisher is established outside the EU, the EU authorized representative is:

Springer Nature Customer Service Center GmbH
Europaplatz 3
69115 Heidelberg, Germany

The manufacturer's authorised representative in the EU is Springer Nature Customer Service Centre GmbH, Europaplatz 3, 69115 Heidelberg, Germany. If you have any concerns regarding our products, please contact ProductSafety@springernature.com

Printed and bound by CPI Group (UK) Ltd, Croydon, CR0 4YY
30/09/2026
02225300-0012